GLIM for Ecologists

METHODS IN ECOLOGY

Series Editors

J.H.LAWTON FRS
Imperial College at Silwood Park
Ascot, UK

G.E.LIKENS
The New York Botanical Garden
New York, USA

METHODS IN ECOLOGY

GLIM for Ecologists

MICHAEL J. CRAWLEY
Department of Biology, Imperial College
Silwood Park, Ascot, Berkshire SL5 7PY, UK

Blackwell
Science

Editorial Offices:
Osney Mead, Oxford OX2 0EL
25 John Street, London WC1N 2BL
23 Ainslie Place, Edinburgh EH3 6AJ
238 Main Street, Cambridge
Massachusetts 02142, USA
54 University Street, Carlton
Victoria 3053, Australia

Other Editorial Offices:
Arnette Blackwell SA
1, rue de Lille
75007 Paris
France

Blackwell Wissenschafts-Verlag GmbH
Kurfürstendamm 57
10707 Berlin
Germany

Zehetnergasse 6
A-1140 Wien
Austria

First published 1993
Reprinted 1996

Set by Setrite Typesetters, Hong Kong
Printed and bound in Great Britain
at the University Press, Cambridge

DISTRIBUTORS

Marston Book Services Ltd
PO Box 87
Oxford OX2 0DT
(*Orders*: Tel: 01865 791155
Fax: 01865 791927
Telex: 837515)

USA
Blackwell Science, Inc.
238 Main Street
Cambridge, MA 02142
(*Orders*: Tel: 800 215-1000
617 876-7000
Fax: 617 492-5263)

Canada
Oxford University Press
70 Wynford Drive
Don Mills
Ontario M3C 1J9
(*Orders*: Tel: 416 441 2941)

Australia
Blackwell Science Pty Ltd
54 University Street
Carlton, Victoria 3053
(*Orders*: Tel: 03 9347 0300
Fax: 03 9349 3016)

A catalogue record for his title
is available from the British Library

ISBN 0-632-03156-5

Library of Congress
Cataloging-in-Publication Data

Crawley, Michael J.
GLIM for ecologists/
by Michael J. Crawley.
p. cm. –
(Methods in ecology)
Includes bibliographical references
and index.
ISBN 0-632-03156-5
1. Ecology – Statistical methods –
Data processing. 2. GLIM.
3. Linear models (Statistics) –
Data processing. I. Title.
II. Series.
QH541.15.S72C73 1993
574.5′015195 – dc20

Contents

The Methods in Ecology Series

The explosion of new technologies has created the need for a set of concise and authoritative books to guide researchers through the wide range of methods and approaches that are available to ecologists. The aim of this series is to help graduate students and established scientists choose and employ a methodology suited to a particular problem. Each volume is not simply a recipe book, but takes a critical look at different approaches to the solution of a problem, whether in the laboratory or in the field, and whether involving the collection or the analysis of data.

Rather than reiterate established methods, authors have been encouraged to feature new technologies, often borrowed from other disciplines, that ecologists can apply to their work. Innovative techniques, properly used, can offer particularly exciting opportunities for the advancement of ecology.

Each book guides the reader through the range of methods available, letting ecologists know what they could, and could not, hope to learn by using particular methods or approaches. The underlying principles are discussed, as well as the assumptions made in using the methodology, and the potential pitfalls that could occur – the type of information usually passed on by word of mouth or learned by experience. The books also provide a source of reference to further detailed information in the literature. There can be no substitute for working in the laboratory of a real expert on a subject, but we envisage this Methods in Ecology Series as being the 'next best thing'. We hope that, by consulting these books, ecologists will learn what technologies and techniques are available, what their main advantages and disadvantages are, when and where not to use a particular method, and how to interpret the results.

Much is now expected of the science of ecology, as humankind struggles with a growing environmental crisis. Good methodology alone never solved any problem, but bad or inappropriate methodology can only make matters worse. Ecologists now have a powerful and rapidly growing set of methods and tools with which to confront fundamental problems of a theoretical and applied nature. We hope that this series will be a major contribution towards making these techniques known to a much wider audience.

John H. Lawton
Gene E. Likens

Preface

Learning GLIM will change the way you do statistics. Most ecologists are familiar with regression, analysis of variance and the statistics of contingency tables, but the problem is that they tend to use these techniques whether or not the models embodied are correct, or the error structure of the data is appropriate. Elementary statistics courses for biologists tend to lead to the use of a stereotyped set of tests:

1 without critical attention to the underlying model involved;
2 without due regard to the precise distribution of sampling errors;
3 with little concern for the scale of measurement;
4 careless of dimensional homogeneity;
5 without considering the ideal transformation;
6 without any attempt at model simplification;
7 with too much emphasis on hypothesis testing and too little emphasis on parameter estimation.

GLIM changes all that. It forces the user to think about all these problems and, in so doing, fosters a generally more critical approach to statistical analysis.

The problem with GLIM is its quite prodigious unfriendliness. When you switch on, you are simply greeted by a question mark. To GLIM, this means 'OK. What now?', but to the beginner it is the nightmare of 'writer's block'; the horror of the blank sheet of paper.

This unfriendliness is a direct consequence of the two great strengths of GLIM; its power and its generality. GLIM can handle most of the analysis that ecologists are likely to carry out: regression, analysis of variance, log-linear models of counts, models using logits or probits for the analysis of proportions, models in which the variance increases with the mean, models of survival and so on. All these models are defined, fitted and interpreted in essentially the same way, so that once the initially unfamiliar and rather daunting output from GLIM has been mastered, all the different model structures are handled in exactly the same manner. GLIM is also useful in encouraging good statistical habits by:

1 allowing a thorough graphical and tabular inspection of data before statistical analysis is begun;
2 permitting the choice of an appropriate error structure, rather than forcing the implicit assumption of normality of errors;
3 encouraging a conscious decision about the ideal transformation for linearizing the relationship between the response variable and the

explanatory variables;
4 demanding a precise specification of the model to be fitted to the data;
5 allowing tests of hypotheses and the construction of confidence intervals, using the correct standard errors, especially when there are missing data, offsets or unequal weights;
6 encouraging model simplification, and the search for a minimal adequate model;
7 focusing on model criticism, and on a comparison of different models for the same data (e.g. determining whether the assumptions made about the error structure, transformation and model definition are appropriate to the question in hand);
8 allowing the identification of data values that are particularly influential in the parameterization of a given model, and encouraging a willingness to present a range of models and to discuss the implications of the influential points (e.g. repeating the analysis with and without the influential data points included);
9 highlighting the need for extra data collection and the performance of new experiments.

One of the objectives of statistical analysis is to distil a long and complicated set of data into a small number of meaningful descriptive statistics. Many of the modern computer statistics packages, however, do exactly the opposite of this. They generate literally pages of output from the most meagre sets of data. This copious output has several major shortcomings: (i) it is open to uncritical acceptance; (ii) it can lead to overinterpretation of data; and (iii) it encourages the bad habit of data-trawling (dredging through the output looking for significant results without any prior notion of a testable hypothesis).

GLIM, on the other hand, tells you nothing unless you explicitly ask for it. But this strength of GLIM is also a major stumbling block for beginners. Because you need to know what you are doing when you use GLIM, and GLIM demands that you tell it everything that you want to be done, the language is outstandingly unfriendly on first acquaintance. Again, because the output is minimal, its interpretation takes a lot of getting used to. Without investing a certain amount of time in learning to understand the output from GLIM the exercise will be futile. On the other hand, the investment, once made, will be amply rewarded. There is no point investing masses of effort in collecting data, and then not analysing them properly.

This book is intended for use as an introduction to the methods of generalized linear modelling using the GLIM statistical language. It is aimed at all ecologists, from students to research workers. I assume that the reader has a working knowledge of:
1 linear regression;

2 analysis of variance (ANOVA);
3 significance tests (t, F and χ^2);
4 running programs on a desktop computer;
5 directory structure and file management in DOS or Windows.

Nobody ever learned statistics by reading a book about it. The way to get the most out of this book is to work through the examples while sitting at your computer, and to check the calculations by hand, so that you can see exactly what the program is doing. All the data you need are in ASCII files on the disk provided with the book.

Special thanks are due to Sir David Cox and to Professor J. A. Nelder, both of whom gave freely of their expert advice. John Nelder also provided several of the macros for model checking which are used throughout the book. The failings of the book are no fault of theirs.

Particular recognition must go to Tony Ludlow's next-door neighbour, for introducing Tony to GLIM and, of course, to Tony for introducing GLIM to me. The comments (mostly constructive) of successive generations of Silwood students on the annual GLIM course have greatly improved the clarity of the presentation, and have helped me to understand which bits of GLIM are particularly daunting for beginners.

Learning GLIM will not be easy, but you will not regret making the effort.

M. J. Crawley
Ascot

Copies of GLIM4 macros are available by e-mail from m.crawley@ic.ac.uk

CHAPTER 1

Introduction

1.1 History

GLIM was first developed to provide a software tool for professional statisticians that would enable them to fit generalized linear models to data. The theoretical background to generalized linear modelling is discussed by McCullagh & Nelder (1989), but you do not have to be an expert statistician or a mathematical prodigy to use GLIM effectively. Since its early beginnings, the development of the GLIM software has been overseen by the GLIM Working Party of the Royal Statistical Society, with the aim of providing a statistical package that would allow users to develop a critical approach to model building and model checking. While GLIM can do many of the things that an ecologist would want of a statistical package (see below), there are several things that GLIM makes no attempt to do; these include multivariate statistics, time series analysis and the analysis of complex experiments (like split-plot and other nested or confounded experimental designs that have several different error terms). Larger, more comprehensive, statistical packages like Genstat are required for these techniques.

The GLIM language has evolved from the rather specialist version 1 in the early 1970s, through its first commercial version GLIM 3.12 released in the mid 1970s, via the much improved and widely used GLIM 3.77 released in 1985, to the present version GLIM 4, which appeared in 1993.

GLIM differs from most of the popular statistics packages in its generality and its flexibility. Many software packages constrain the user to doing a limited set of tests in a completely inflexible way. Most statistics packages have rigid data structures that do not allow the analysis of summary tables or other structures that are produced during a computing session. A general failing of many user-friendly statistics packages is their tendency to produce reams of output from even the most meagre of data sets, whether you want it or not.

A more fundamental objection is that some widely used statistical packages encourage the dangerous notion that hypothesis testing is the be all and end all of data analysis. This plays into the hands of those who would like to turn the statistical analysis of ecological data into an automatic, recipe-following procedure, devoid of any thought or critical input. This in turn nurtures the tendency, rife amongst beginners in the art of data analysis, of not looking at the data, and of plunging straight into hypothesis testing without any idea of what the data are actually saying. Their battle cry is: 'Yes, but is it significant?'

1.2 Overview of the GLIM package

The idea behind GLIM was to produce a package that provides a powerful set of tools for preliminary data analysis, like flexible graphics and good tabulating facilities, and to couple this with a general maximum likelihood estimator for fitting linear models to data. The fitting algorithm estimates parameter values and standard errors for a wide range of frequently used mathematical models including regression, analysis of variance, contingency tables, log-linear models for counted data, logistic models for binary responses, models for data where the variance increases with the mean, and a variety of models for analysing survival data. There is a comprehensive set of facilities for assessing the performance of the model and for checking the validity of the assumptions made about the error structure.

GLIM is made up of three components. First and foremost, it is a powerful tool for statistical modelling. It enables you to specify and fit statistical models to your data, assess the goodness of fit and display the estimates, standard errors and predicted values derived from the model. It provides you with the means to define and manipulate your data, but the way you go about the job of modelling is not predetermined, and the user is left with maximum control over the model-fitting process.

Second, GLIM can be used for data exploration, in tabulating and sorting data, in drawing scatterplots to look for trends in your data, or to check visually for the presence of outliers.

Third, it can be used as a sophisticated calculator to evaluate complex arithmetic expressions, or as a programming language to perform more extensive data manipulation. As a calculator, GLIM operates on either scalars (single numbers) or vectors (lists of numbers). These may be combined in general expressions, involving arithmetic, relational and transformational operators such as sums, greater-than tests, logarithms or probability integrals. The ability to combine frequently used sequences of commands into sub-programs known as macros makes GLIM a powerful programming language, ideally suited for tailoring one's specific statistical requirements.

GLIM enables you to specify a statistical model for your data, to find the best subset from a set of models, and to examine the implications of fitting such models. A small set of commands allows you to specify each component of the model independently, to obtain the maximum-likelihood estimates and goodness-of-fit statistics for the model, and to display and access the results of the fitted model. GLIM is especially useful in handling difficult or unusual data sets, because its flexibility enables it to cope with such problems as unequal replication, missing values, non-orthogonal designs and so on. Furthermore, the open-ended style of GLIM is particularly appropriate for following through original ideas and developing new concepts.

One of the great advantages of learning GLIM is that the simple concepts which underlie it provide a unified framework for learning about statistical ideas in general. By viewing particular models in a general context, GLIM highlights the fundamental similarities between statistical techniques and helps play down their superficial differences.

The GLIM system is described in detail in the GLIM Manual, obtainable from:

The GLIM Coordinator
Numerical Algorithms Group
256 Banbury Road
Oxford OX2 7DE
UK
Tel: (0865) 511245

1.3 How does it work?

GLIM operates by executing commands termed directives (these are shown in **bold** in the text) that the user enters at the keyboard (this may be anything from a notebook PC to the terminal of a mainframe computer). Each command consists of a directive name specifying the action required, followed by items indicating how the action is to be performed. Directive names begin with a $ symbol (or a back-slash, \, which has the advantage of being a lower-case character, and hence quicker to type). The directive ends at the next $ sign. For example, the **print** directive:

```
$print x $
```

causes all the values stored in the vector called x to be printed in a row (or rows if there are more than nine of them) across the screen.

Input of numbers to GLIM is highly flexible; data can be entered directly through the keyboard, they can be read from external files (in free or fixed format) or they can be attributed as part of a GLIM program. Random numbers and factor levels can be generated internally.

A comprehensive **tabulation** facility permits sets of vectors to be compressed to smaller sets and either stored or printed as tables; **sort** produces ordered vectors or the permutation required to order them. Values can be inspected in columns (**look**) or in tables (**tprint**) while **print** offers flexible control over the layout of collections of scalars, vectors and macros. Scatterplots (**plot**) and histograms (**hist**) are available, with automatic scaling or your own choice of scales for the axes, selection of subsets of the data to be plotted and control of plotting symbols.

The central feature of GLIM is its method of specifying and fitting generalized linear models. In fitting classical linear models GLIM allows you to specify quantitative and qualitative explanatory variates in a simple manner and allows compound terms, such as interactions and polynomials, to be easily constructed. The program also provides a framework embracing log-linear models for contingency tables, logit and probit models for the analysis of proportions and models with gamma errors. These can be specified by giving the probability distribution of the data (the **error** structure) and the function which links their expected values to the explanatory variables (the **link** function).

The maximum-likelihood estimates for the current model are obtained by the **fit** directive. This directive fits the current model and displays its goodness-of-fit statistic and degrees of freedom. Further output on the parameter values and their standard errors may be obtained through the **display estimates (disp e)** directive, and extra statistical information like the variance–covariance matrix and the information matrix can be accessed if required.

Sequences of models can be fitted by incrementally adding or removing explanatory variables or their interactions. It is easy to change other aspects of the model (e.g. to fix parameter values, respecify the error distribution or link function), or to modify the data (e.g. to edit or omit points or change their weights) or alter the model-fitting algorithm (e.g. the starting values or the number of iterations).

A useful feature of GLIM is that the current program state can be dumped in an external file using the **dump** directive, so that if you want to return to a modelling session several times you do not need to go through the entire procedure from scratch, but can simply take up where you left off by means of the **restore** directive.

Another useful feature of GLIM is that it keeps a full record of each session in a **transcript** file (called glim.log on most installations). This enables you not only to save paper (on output you do not want), but provides a computer-readable form of your output that can be incorporated directly into reports and summary tables in your word-processor. This means that you do not need to retype any of the statistical output; you simply extract the bits you want from the ASCII file containing the glim.log. You can control the material that is recorded in glim.log by means of the **transcript** directive. With big data sets, for example, it is often a good idea not to write the raw data into glim.log because it takes up so much space.

Sequences of GLIM directives can be stored in macros, and macros can then be executed by a single directive (**use**). Macros can invoke other macros (including themselves) and there are GLIM directives to execute macros repeatedly (**while**), to switch between macros (**switch**), or to jump back between them (**exit**). A library of predefined macros is supplied with the program, and is read into GLIM through the primary library channel. The library macros can be supplemented with frequently used macros that you have written for yourself, or, like the macros that come with this book, have been written by fellow ecologists for particular purposes.

1.4 Using GLIM

Once you have switched on your PC or logged on to your terminal and selected the appropriate directory, GLIM is invoked simply by typing:

```
glim
```

followed by the Return key. If you wish, you may type:

```
glim <t> <l>
```

where <t> is an optional name for the transcript file (glim.log is usually the default name) and <l> is an optional name for the file containing the macro library to be used during the run (the default path and filename are \glim\maclib.glm).

If you are in Windows then simply click on the GLIM icon or, if there is no icon, click on File Run then type c:\glim\glim.bat.

If GLIM is working properly, you will get a banner headline, which looks like this for Version 3.77:

```
GLIM 3.77 update 1 (copyright)1985 Royal Statistical Society, London
```

followed by a question mark. In following the worked examples, it is important that your computer is in exactly the same state as mine was when I produced the listings. To this end, it is vital that you stop your machine at the end of each exercise. Wherever you see the bullet (•) on the right-hand side of the page, type **$stop** and save glim.log under a new name if you want to keep a transcript of the session. Restart when you want to begin the next exercise. Where you see an open bullet (○) on the right-hand side of the page, you should not type the following GLIM directives (they are for illustration only, and do not stand alone). If you want to save your working GLIM session so that you can return to it later, then use the **$dump** directive. This will prompt you for the name of a file in which you want to save the current session. In order to begin where you left off, start up GLIM then type **$restore**. This will prompt you for the name of the file where the session is stored, and you can take up where you left off. At this point, you are ready to go. Now read on.

CHAPTER 2

An ecological example

We begin with a worked example. The idea is to use a typical data set to show how simple and how powerful the GLIM language is, and to demonstrate the advantages of investing the energy necessary to learn how to use it. However, there is no getting away from a major stumbling block: GLIM is not a user-friendly package. When it is ready to work with you, for example, it just says ? (which means 'what do you want to do now?'). New versions of GLIM will certainly be more friendly in years to come, but at present it does take a considerable commitment to become acquainted with it, and several months of experience to learn to love it. It is the object of this chapter to convince you that the rewards will be worth the effort.

Suppose that we have carried out an experiment that involved measuring the growth (dry matter at harvest) of plants treated with 10 different concentrations of mineral supplement. The experiment was carried out on two different genotypes, one cloned from material obtained from an arid habitat and a second from a moist habitat. The data are shown in Table 2.1. We are interested in a whole set of questions about the form of the relationship between growth and mineral concentration (is it linear or

Table 2.1 The data show mineral concentration ($g\,m^{-2}$) and plant dry weight (g) for 10 individuals from each of two plant genotypes. The data are in pairs, first the value of the explanatory variable, x (mineral concentration), then the plant dry weight at harvest associated with that level of mineral supplementation (the response variable, y). The data file glex1.dat contains the x and y data in 20 rows (10 plants × 2 genotypes), with the data for genotype A followed by the data for genotype B

Genotype A		Genotype B	
x	y	x	y
1	2.8215	1	5.2357
2	2.3590	2	5.4479
3	3.0912	3	6.1229
4	2.5297	4	4.4046
5	3.4753	5	10.823
6	3.6493	6	9.0844
7	6.6999	7	8.1000
8	8.5834	8	11.029
9	5.3210	9	14.502
10	10.1300	10	12.414

non-linear?), about the values of the parameters (is the slope greater than zero?), about the error variation in the data (is there a trend in the variance with changing values of the mean?) and about differences between the genotypes (are the regression slopes the same for the two genotypes?).

The analysis will be carried out in GLIM with very little comment. The idea is simply to demonstrate:

1 how to use GLIM;
2 the most important GLIM directives;
3 the kind of output produced by GLIM.

Detailed, worked examples of all phases of the analysis will be provided in later chapters. Experience has demonstrated that the best (some would say the only) way for ecologists to learn about statistics is to work with them, using data that look and feel relevant. This is the approach adopted here. Throughout the book, the expectation is that you will work through the exercises as you read, learning a new directive here, an unfamiliar test there, while all the time becoming progressively more familiar with the GLIM language. All the data are provided in machine-readable form on the disk provided at the back of the book, and transcripts of all of the exercise sessions are also given in full.

2.1 Getting started

The GLIM code is probably stored in a directory called glim on drive C of the hard disk. The first step is to enter the directory with the DOS command:

```
cd\glim
```

Then, to run GLIM, you simply type glim and press the Return key. If all is well, you will get the banner shown on p. 5, followed by a question mark. In subsequent chapters the ? is omitted from lines with GLIM commands, but it is included at this stage so you see exactly what is on the screen.

We start by telling GLIM the size of the data set. There are 20 values of the response variable (harvest dry weight), 10 for each genotype.

```
? $units 20 $
```

The explanatory variable (mineral concentration) is to be called x and the response variable (plant dry weight) y.

```
? $data x y $
```

The data are read from a file called glex1.dat (a mnemonic for 'GLIM example 1 in a data file') stored in directory glim on drive C (the current directory) using the **dinput** directive (this stands for data input); the number 6 is called the channel number (see Section 3.6).

```
? $dinput 6 $

File name? glex1.dat
```

If the data were on a floppy disk in drive A, then you would type:

```
a:glex1.dat
```

after the File name? prompt. In GLIM 4 we can type the file name directly, instead of the channel number, as part of the **dinput** directive.

2.2 Data exploration

When the data have been successfully read into GLIM, another ? will appear. The first step in the analysis of the data is to see whether the experiment of changing mineral supplementation has produced a response in plant dry weight. We do this by plotting a graph of dry weight y against mineral concentration x (notice that y comes first in the pair of variable names after the **plot** directive):

```
? $plot y x $
```

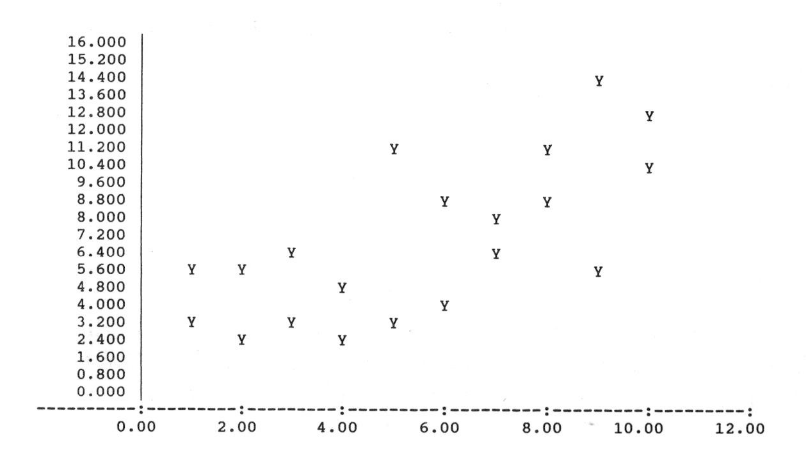

It appears that mineral supplementation has indeed had an effect, and the next question is whether the response is different for the two genotypes. We create a variable G to denote the genotype in question; this will take the value 1 for the genotype from the arid site and 2 for the genotype from the moist habitat. This is achieved using a built-in function (%gl) to generate the levels of the factor in question (the function generates 10 ones followed by 10 twos). This saves us the effort of typing in all 20 of the factor levels. We then declare G to be a *factor* with *two levels*:

```
? $calc g=%gl(2,10) $

? $factor g 2 $
```

GLIM 4 has a more powerful feature than this, and the **gfactor** directive generates the levels and declares the factor in a single line.

Some summary statistics are in order. To find the mean value of plant dry weight for each of the genotypes we simply use the **tabulate** directive (usually abbreviated to **tab**, but it can be abbreviated all the way to **t**):

```
? $tab the y mean for g $
```

and GLIM responds by printing:

```
        1       2
[ ]  4.866   8.716
```

The plants from the moist site (site 2) are an average of nearly 4 g heavier than the plants from the arid site (site 1). The square brackets [ ] at the left-hand end of the row mean that there is no label for this row.

The **tab** directive is an immensely powerful way of producing summary tables and contingency tables of counts that can then be used for subsequent statistical analysis. So important is the **tab** directive that it gets the whole of Chapter 6 to itself.

To look at the two variances, we write:

```
? $tab the y variance for g $
```

and GLIM responds with:

```
       1      2
[ ]  7.51  11.73
```

so it looks as if their variances differ by a factor of 1.56. To see whether the response of plant weight to mineral supplementation differs between the genotypes we can plot each genotype separately like this:

```
? $plot y x 'ab' g $
```

so GLIM plots two graphs, using a to denote the points from genotype A and b to denote genotype B (the two plotting symbols are enclosed in single quotation marks in the desired order; in this case, the first and second levels of factor G) (see p. 10).

2.3 Model-fitting

Now we can do the statistics. First we declare the response variable (the *y*-variable) to be plant dry weight (y):

```
? $yvar y $
```

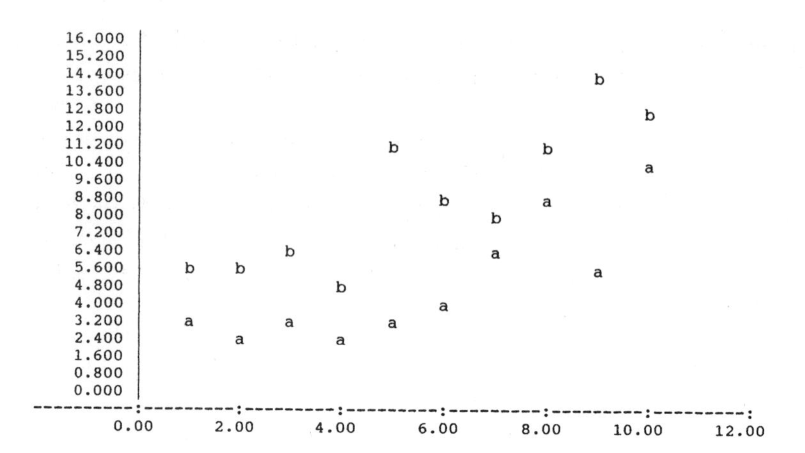

Next we fit the overall mean value of plant dry weight to the model, ignoring both mineral supplementation and genotype:

```
? $fit $
```

to which GLIM responds with:

```
deviance = 247.27
    d.f. = 19
```

which means that the total sum of squares is 247.27 with 19 degrees of freedom. To assess the relationship between plant dry weight and mineral supplementation, we add mineral supplementation to the model by writing:

```
? $fit +x $
```

Note the plus sign in front of the x. GLIM responds with:

```
deviance = 121.15 (change = −126.1)
    d.f. = 18     (change = −1)
```

This means that adding x to the model as an explanatory variable has produced a reduction of 126.1 in the deviance, leaving an error sum of squares of 121.15 (this change in deviance is highly significant). To see the parameter estimates for the slope and the intercept of the fitted line, we simply type:

```
? $disp e $
```

which stands for 'display the estimates', and GLIM responds with:

```
  estimate  s.e.    parameter
1 1.983     1.253   1
2 0.8743    0.2020  X
scale parameter taken as 6.731
```

Parameter 1 is the intercept (1.983) and parameter 2 (labelled X in the right-hand column) is the slope of the graph of plant dry weight against mineral supplementation (0.8743). The other two numbers are the standard errors of these parameters. We can do *t*-tests with these standard errors to see whether the parameter estimates are significantly different from zero. It looks as if the intercept is not significantly different from zero ($t = 1.58$) but that the slope is significantly different from zero ($t = 4.33$). The scale parameter is GLIM's jargon for the error variance ($s^2 = 6.731$).

To add genotype to the model, we write:

```
? $fit + g $
```

and GLIM responds with:

```
deviance = 47.025 (change = −74.12)
    d.f. = 17     (change = −1)
```

Not surprisingly, given the difference between the means that we noted earlier, adding genotype to the model produces another dramatic reduction (74.12) in the residual sum of squares. To see the new model we write:

```
? $disp e $
```

and GLIM prints:

```
  estimate  s.e.     parameter
1 0.05740   0.8853   1
2 0.8743    0.1295   X
3 3.850     0.7438   G(2)
scale parameter taken as 2.766
```

The intercept is now 0.0574, the slope of the graph is 0.8743 and the graph for genotype B is 3.85 units higher than the parallel graph for genotype A. It is possible, of course, that the lines are not parallel for the two genotypes, and plant weight might respond to mineral supplementation more steeply for one genotype than the other. To test this we fit separate regression lines for each genotype:

```
? $fit +g.x $
```

The '.' between G and x means 'the interaction between G and x'. GLIM responds with:

```
deviance = 45.147 (change = −1.878)
    d.f. = 16     (change = −1)
```

The interaction accounts for only 1.878 of the variance in plant dry weight and this is not significant. The interaction has an *F* ratio of only 0.6655, which is calculated by dividing the change in deviance (1.878) by the error variance (the scale parameter of 45.147/16 = 2.822) and comparing this with F tables with 1 and 16 degrees of freedom (5% tables give $F = 4.49$). To see how fitting the interaction has affected the parameter estimates, we type:

```
? $disp e $
```

and GLIM prints:

```
  estimate  s.e.    parameter
1 0.6442    1.148   1
2 0.7676    0.1849  X
3 2.677     1.623   G(2)
4 0.2134    0.2615  X.G(2)
scale parameter taken as 2.822
```

The intercept is a bit higher (0.6442) and the mean contribution of genotype B a bit lower (2.677) but nothing dramatic has happened. A t-test for the significance of the interaction term is not significant ($t = 0.2134/0.2615 = 0.816$).

2.4 Model simplification

Given that we should be looking for the *minimal adequate model*, there is no point in retaining non-significant terms in the model. Estimating unnecessary parameters also inflates the variance of the remaining parameters. To simplify the model, we remove the interaction term using the **fit minus** directive, and return to the model with parallel regression lines for the two genotypes:

```
? $fit -g.x $
```

and GLIM writes:

```
deviance = 47.025 (change = +1.878)
    d.f. = 17     (change = +1)
```

An F-test on this increase in deviance (above) shows that the model simplification produces a non-significant increase in the deviance, so the simplification appears to be justified. In general, the technique of model simplification is a better (more reliable) test than a t-test on the parameter values (although in this case the response is identical, because with normal errors F is equal to t^2; $0.6656 = 0.8158^2$).

2.5 Model criticism

Next, we test how well the model fits the data. We start by plotting the fitted values (%fv) and the data on the same axes. The fitted values are produced within GLIM, by evaluating the equation:

$$y = 0.0574 + 0.8743x$$

for genotype A, and:

$$y = 0.0574 + 0.8743x + 3.85$$

for genotype B, and the values are stored in a system vector called %fv (this stands for 'fitted values'). We can use lower-case letters for the data and upper-case letters for the fitted straight lines:

```
? $plot y %fv x 'abAB' g $
```

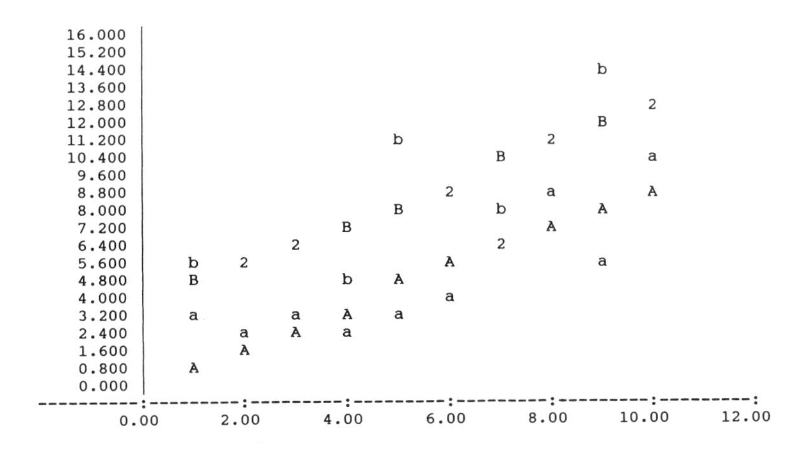

The number 2 appears where GLIM tries to print two symbols in the same place. The fit looks reasonably good, and the residuals look similar for the two genotypes. We check this out by plotting the residuals against the fitted values. We also need to check the assumed normality of the error distribution with a probability plot, in which the standardized residuals are ordered and plotted against the standard normal cumulative distribution function. Both these tasks are carried out by the model-checking macro, which is run from a file called mcheck.mac. You should put the disk that came with this book in drive A and type:

```
$input 8 $

File name? a:mcheck.mac
```

Then, after a short pause while the macro is read from disk, you will see a graph of the residuals against the fitted values (see p. 14) which shows no obvious pattern in the residuals, so we conclude that the residuals are well behaved (ideally, this kind of plot should look like the sky at night, with the points scattered randomly all over the place). In ecological work, the residuals often increase with the mean, which would show up as a fan-

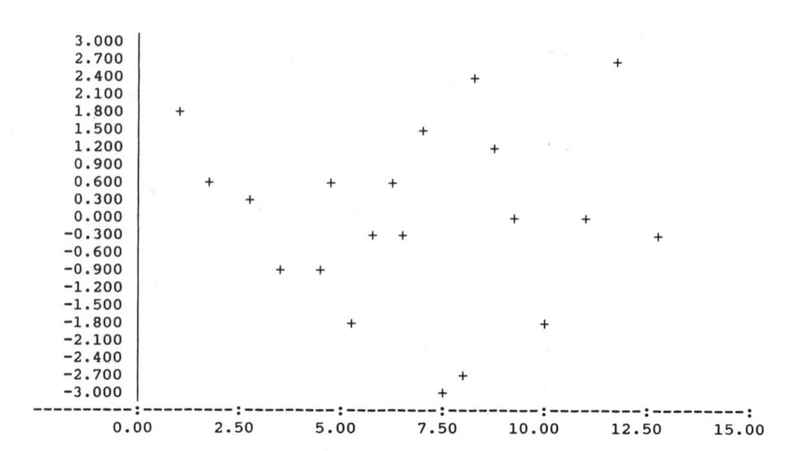

shaped pattern of increasing residuals with increasing plant weight. Such a pattern would require attention, perhaps by transformation of the response variable, by redefining the error structure, or by transforming one or more of the explanatory variables. If you press the Return key, you will now get a second graph, showing the probability plot:

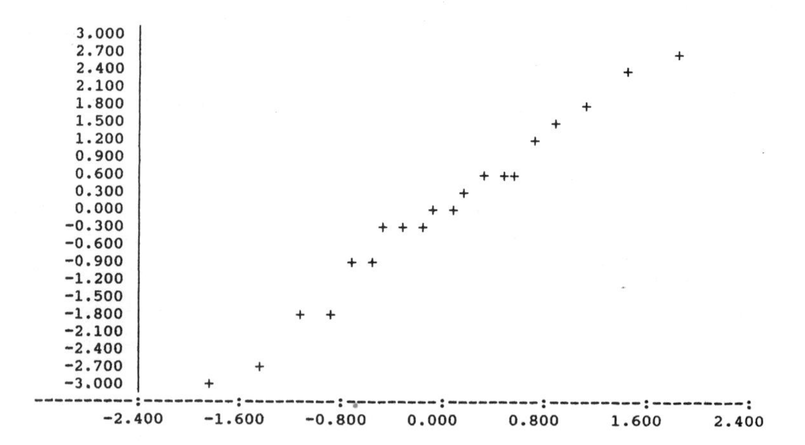

This is as good a straight line as we are ever likely to get, and we conclude that the assumption of normal errors was appropriate. If the plot had been markedly curved (S-shaped or J-shaped, for example), then we should need to respecify the model structure and repeat the **fit**.

To sum up: (i) there is a strong positive relationship between plant dry weight and mineral supplementation; (ii) the genotypes differ significantly in their mean dry weight at harvest; (iii) the slope of the relationship is not significantly different for the two genotypes; (iv) the residual variance is reasonably well behaved, and the errors are normally distributed; (v) the minimal adequate model for the relationship is $y = 0.0574 + 0.8843x$ for the genotype from the dry habitat, and the line is higher by 3.85 for the genotype from the moist habitat; (vi) the model accounts for 81% of the total sum of squares in plant dry weight ($r^2 = (247.27 - 47.025)/247.27$); and (vii) the standard errors are reasonably small, giving us some confidence in the values of the parameter estimates.

This gives the flavour of a GLIM analysis. Of course there is much more we could have done (e.g. we have made no serious attempt at model simplification; we might ask whether the inclusion of the intercept of 0.0574 is justified by the data, or whether a value of zero might not do equally well). Also, the example did not make use of GLIM's outstanding abilities to deal with non-normal errors and non-identity link functions. These topics are addressed in full in later chapters. For the time being, the important messages to learn from this first example are:

1 the **units** directive is vital (it shows the number of different y values; this is normally the number of rows in the data set);

2 the **data** directive lists the variable names that will be used (it also determines the order in which GLIM reads the data);

3 statistical analysis is carried out with the **fit** directive;

4 there is great flexibility in the **plot** directive;

5 data are conveniently read from external text files using the **dinput** directive;

6 programs are conveniently read from external text files using the **input** directive.

The statistical part of the GLIM output will almost certainly look difficult and unfriendly at this stage. Do not worry about this. We shall deal with all the statistics in a detailed, step-by-step manner in subsequent chapters. ●

CHAPTER 3

The GLIM language

The GLIM language consists of a series of *directives* to do such things as calculate, read data, fit models and so on, with a set of built-in functions to evaluate logs, square roots and so forth. It also has powerful facilities for creating and saving summary tables and plotting graphs and histograms. Its most important features, however, involve the specification and fitting of linear models. In particular, GLIM allows the specification of an *error structure* and a *link function* (you will have used normal errors and the identity link in regression and analysis of variance). The structure of the model is embodied in the *linear predictor* as specified in the *fit* directive. Once the user has specified the *response variable* GLIM can fit continuous variates (like height, weight or temperature) or factors (discrete categories like low, medium and high) in an attempt to explain the variation in the response variable.

3.1 Why is GLIM so unfriendly?

GLIM's legendary unfriendliness is due to the fact that it requires you to tell it absolutely everything that you want it to do. At first this is quite daunting, but we shall approach the task of learning the GLIM language in a simple, step-wise manner. Another component of the unfriendliness is due to the way in which GLIM presents the results of its labours. In fact, the output is exceptionally general and highly efficient once it has been mastered. But there is no getting away from it, GLIM output does take a considerable amount of getting used to.

All GLIM directives must begin with a dollar sign $ (or with the back slash \, which has the advantage of being a lower-case character and hence quicker to type). This is tedious, but it does mean that many GLIM directives can be placed on the same line (the $ symbol is the directive separator). It is a good idea to end each line of code with a $ sign as well, because then you do not have to remember which directives need a terminal $ sign and which do not.

A GLIM session usually begins with a **$units** directive and ends with **$stop**. The **$units** directive tells GLIM how many rows there are in the data matrix; this will usually be the number of measurements in the whole experiment (e.g. the number of points on a graph, or the total number of samples in an analysis of variance (ANOVA)). The **$stop** directive terminates the GLIM session and returns control to the operating system.

3.2 GLIM as a calculator

One of the most important directives in GLIM is the **calculate** directive. This is used in transformation, in creating category variables to describe factor levels, as a scratch-pad for doing intermediate calculations during modelling, and for a host of other purposes. Throughout this book we shall abbreviate the directive to **$calc** (although, strictly, all we need to type is **$ca**; the minimal abbreviation of directives is explained in the index and in the GLIM Reference Guide in the GLIM Manual.

For example, if you want to calculate a *t*-ratio during a modelling session you will have a difference (say 2.345) and a standard error of the difference (say 1.124); to work out the value of *t* you simply type:

```
$calc 2.345/1.124 $
```

and GLIM prints the answer 2.086 immediately beneath.

Lots of the calculations you will want to do involve mathematical functions. These are listed in Table 3.1. The thing to remember is that all GLIM's functions must be preceded by a percentage sign. Thus to calculate the natural log of 0.569 we use the %log function like this:

```
$calc %log(0.569) $
```

and GLIM prints −0.5639 beneath. You should familiarize yourself with

Table 3.1 Mathematical functions in GLIM

Name	Meaning	Comments
%log(x)	Logarithm	Base e
%exp(x)	Antilog	e^x
%ang(x)	Arcsine $\sqrt{x}$	Angular transformation (radians) x must be between 0 and 1
%sqrt(x)	$\sqrt{x}$	x must not be negative
%sin(x)	Sine x	x in radians
%cu(x)	Accumulate	See Section 3.2
%tr(x)	Truncate	The integer part of x
%sr(0)	Random fraction	$0 < \text{real} < 1$
%sr(n)	Random integer	Integers between 0 and n
%np(x)	$\Phi(x)$	Normal probability integral from $-\infty$ to x
%nd(x)	$\Phi^{-1}(x)$	Normal deviates x between 0 and 1
%nd(%sr(0))	Random normal deviates	Mean = 0, s.d. = 1

all the functions, and use your hand calculator to check the results (e.g. does GLIM use degrees or radians for its arcsine function?).

We often want to save the results of calculations. Normally we will create a whole vector of numbers, and this vector will be of the length defined in the **units** directive. But on other occasions we may want to save a single value as a scalar. In GLIM vectors are given names as in any computer language. You can write the variable names as long as you like to help legibility, but GLIM will recognize only the first eight characters. You will need to take care if you are using GLIM 3.77, because only the first four characters are recognized, so variable names like data1 and data2 would both be read as data by GLIM 3.77. You are allowed to use up to 26 scalars in GLIM 3.77; these are recognized as a single letter preceded by a percentage sign. Thus a, b and c would be vectors and %a, %b and %c would be scalars. GLIM 4 allows unlimited scalar names to be defined by the **number** directive.

A very important function used in **calc** statements is %cu. The function name stands for 'cumulative' and it adds together all the elements of a vector, storing the sub-totals as it goes. If %cu is used with a constant as its argument, then the result is a linear series, beginning with the constant, and increasing by steps equal to the constant. This sounds confusing, but an example should make it clear. Suppose we want to create a variable called x which goes from 1 to 10 in steps of 1. There are 10 numbers in this vector so, unless **units** is already set to 10, we need to declare x to be a variable of length 10 using the **var** directive. Then we use %cu(1) to generate the values for x, like this:

```
$units 10 $

$var 10 x $

$calc x = %cu(1) $
```

To see what GLIM has done, we can inspect the values of x in one of two ways. If we want numbers printed in a row across the screen we use **print** x, while if we want the numbers in a column down the screen we use **look** x. You should try both to see what happens. The machine has generated the numbers 1, 2, 3, 4, 5, 6, 7, 8, 9 and 10 in sequence. To recap, it has started at 1 and counted in increments of 1 because that was the constant specified in brackets in the %cu function. If we wanted to start at 7 and go up in steps of 1 we would need to write:

```
$calc x = 6 + %cu(1) $
```

Note that we write 6+ and not 7+ because we want the first number to be 7 not 8. If we wanted the 10 numbers to start at 500 and increase in steps of 50, we would write:

```
$calc x = 450 + %cu(50) $
```

In other cases, we can use %cu to do quite sophisticated calculations. Suppose that we have a vector y containing our data and we want the sum of the y values and the sum of the squares of the y values. First we enter the y values. There are several ways of doing this, but one of the simplest is to use the **assign** directive:

```
$assign y = 3, 6, 8, 3, 3, 5, 7, 8, 5, 4 $
```

This is one of the few cases in GLIM where the elements of a list are separated by commas; blanks are the usual separator. You do not need a **var** directive with **assign** because the length of the vector will be set automatically to the number of numbers in the list (10 in this case). Now the sum of the y values is obtained simply by writing:

```
$calc %cu(y) $
```

but this causes GLIM to list all 10 sub-totals up to 52.0. If we want to save the grand total for subsequent use, we might write:

```
$calc %a = %cu(y) $
```

where the name of our variable a is prefixed by a percentage sign to tell GLIM that it is a scalar. To see the total we need to write:

```
$look %a $

52.0
```

If we had omitted the % sign in front of the a, GLIM would save a vector of length 10 with the sub-total of %cu(y) in each element and the grand total in the 10th element. To get the sum of squares of y we first calculate a new vector containing the squares (say y2), then use %cu to calculate the total:

```
$calc y2 = y * y $

$calc %b = %cu(y2) $

$look %b $

306.0
```

Again, if we did not include the **look** directive, GLIM would calculate the sum of squares, but would not print it.

We can use the %cu function embedded within other parts of a calculation. For example, here is a single line formula to calculate the variance of y and save it in a scalar called %v:

```
$calc %v=(%cu(y**2)-%cu(y)**2/%nu)/(%nu-1) $
```

```
$look %v $

3.956
```

Notice the use of ** to mean 'to the power of' and the use of the *system scalar* %nu which contains the number of units (in this case, 10). Note that this formula for calculating variance is dangerous for general use on computers because errors will creep in when the values of y are very large or very small. In general it is better to calculate the mean %m in advance:

```
$tab the y mean $

5.200

$calc %m=5.2 $
```

and use:

```
$calc %v=%cu((y-%m)**2)/(%nu-1) $
```

We can also use %cu to do multiplications. The trick is that multiplications can be expressed as additions of logarithms. Thus, if we wanted to compute factorials in GLIM we might write a simple program as follows. To compute 5! (which is $5 \times 4 \times 3 \times 2 \times 1 = 120$) we first create a vector of length 5, then generate the numbers 1 to 5 using %cu(1). Then we take the logs (using the %log function) of these numbers (calling the vector, say, logn) then use %cu(logn) to add up the logs. The answer is found by calculating the antilog (using the %exp function) of the total as follows:

```
$var 5 n $

$calc n = %cu(1) $

$calc logn = %log(n) $

$calc %f = %cu(logn) $

$calc %exp(%f) $
```

when GLIM will print 120.0.

If you want more than the default of four decimal places, you can change this by the **accuracy** directive:

```
$accuracy 7 $
```

If you want to suppress the printing of decimals for whole numbers, then you can use the ***i** option in the print statement. Suppose you wanted to write 'with 4 degrees of freedom' in a **print** directive, and the number 4 was stored in a scalar called %d. You would write:

```
$calc %d=4 $
```

```
$print 'with ' *i %d ' degrees of freedom' $
```

To free space during a GLIM session, we can delete unwanted vectors and macros with the **delete** directive. For example,

```
$del x y z $

$del poisson $
```

will delete the vectors x, y and z and the macro called poisson. It is important to use **del** to get rid of variables that have been defined inside macros because, unless you do this, you may have problems caused by mixed vector lengths the next time the macro is invoked. A good general rule is that you should delete all the variables defined in a macro in the last statement before the **endmacro** directive (see Section 3.12).

If you forget which variable names are currently assigned, how many levels a factor has or what macros are currently accessible, you can use the **environment** directive, **env d**, to obtain a full listing:

```
$env d $
```

•

Other important **env** directives are explained below.

3.3 Data editing

Extensive data editing is best done in a full-screen editor like a spreadsheet or a word-processor rather than in GLIM. Simple editing tasks are most conveniently carried out using the **calc** directive. For example, if we want to change the eighth element of x from 18.5 to 22.1 we just type:

```
$calc x(8)=22.1 $
```

which writes over the old value of 18.5. The **edit** directive is not much use, but it does allow several successive values to be replaced. Thus

```
$edit 12 15 x 0.2 0.1 0.7 0.5 $
```

will do the same as

```
$calc x(12)=0.2 : x(13)=0.1 : x(14)=0.7 : x(15)=0.5 $
```

where the **colon** ':' is a very useful piece of shorthand meaning 'repeat the last directive' (i.e. repeat the **calc** directive in this case).

3.4 Transformation

Transformation is carried out using the **calc** directive. The commonest transformations are logs using the %log function to obtain natural logs (GLIM does not have a function for logs to the base 10, but if you should need them all you do is multiply natural logs by 0.4343), square root using the %sqrt function, and arcsine using the %ang function.

3.4.1 Logarithmic transformation

Taking logs is probably the most familiar transformation in ecology. The idea is that badly behaved data can be tamed by logging them. Specifically, the object is:

1 to reduce heterogeneity of variances;
2 to improve normality of errors;
3 to introduce additivity.

While transformation of the response variable is an important option in modelling with GLIM, using GLIM allows a greater variety of responses, and encourages more thought about why and how to transform the response variable. The important thing to bear in mind is that transformation changes more than one attribute of the response variable. For example, when we take logs of y we alter lots of things about the model; the errors in $\ln y$ are now assumed to be normal (which means that the raw data y are log-normally distributed), the parameter estimates are differences between logs, the fitted values are logs (i.e. treatment effects are multiplicative rather than additive), and the residuals are logs. The *link function* in GLIM differs from transformation in that the *errors are still additive in the linear predictor* (see Chapter 10). Thus, transformation will alter the assumptions about error distribution in important ways.

$$\ln y = a + bx$$

On the transformed scale, the model represents variation in the mean of the normally distributed transformed variable. But, on the original scale, the variation is in the *median* of the variable:

$$\text{median}(y) = \exp(a + bx)$$

The mean value on the original scale is larger than the median; the higher the variance in $\log y$, the bigger the difference:

$$\text{mean}(y) = \exp\left(a + bx + \frac{\sigma^2}{2}\right)$$

The variance on the original scale is not the antilog of the variance on the log scale, but the much more complicated expression:

$$\text{var}(y) = [\exp(\sigma^2) - 1] \exp[2(a + bx) + \sigma^2]$$

In summary, the additive model for the mean of log y is a multiplicative model for the median of y, but the intercept is changed by $\sigma^2/2$. It is important to understand that the variance of y is not constant, but goes up with the antilog of twice the linear predictor.

To carry out back-transformation of fitted values following log transformation in GLIM we would put:

```
$calc %exp(%fv) $
```

to get the median response, but

```
$calc %exp(%fv + %sc/2) $
```

to get the mean. The system scalar %sc is calculated automatically within GLIM as

```
%sc = %dv/%df
```

where %dv is the residual deviance (the error sum of squares in the case of normal errors and the identity link) and %df is the residual degrees of freedom. Both are system scalars with GLIM and are assigned values as soon as a **fit** directive is executed (see Section 11.2).

3.4.2 Arcsine transformation

In the old days before people had access to computers, the arcsine transformation was routinely applied to percentage data. Transformation was important when any of the percentages were less than 20 or greater than 80 because percentage data are bounded (they cannot be greater than 100 or less than 0). This means that standard errors of untransformed data might include nonsensical values like $92 \pm 14\%$. The transformation works like this. First the percentages must be converted to proportions by dividing by 100. Then GLIM works out the square root of the proportions, and finds the angle, in radians, that has this value as its sine. For example, suppose we have the percentages 2, 4, 10, 90, 92 and 99 in a vector called perc:

```
$assign perc= 2, 4, 10, 90, 92, 99 $
```

Then we can print the arcsine-transformed percentages and the raw data as follows:

```
$calc p = perc/100 $

$calc t = %ang(p) $

$look perc t $
```

and GLIM will print:

```
      PERC    T
1    2.000  0.1419
2    4.000  0.2014
3   10.000  0.3218
4   90.000  1.2490
5   92.000  1.2840
6   99.000  1.4706
```

The values of t are angles in radians. To convert them to degrees you would need to multiply them by 57.296 (because a right angle, 90°, is $\pi/2$ radians). Thus, 4% has been converted into 11.54° and 99% into 84.26°. Notice that the transformation turns small percentages into larger degrees, and large percentages into smaller degrees.

The arcsine transformation is often criticized because it produces such an uninterpretable model. Because GLIM can do a better job with models that have much more straightforward ecological interpretation, the arcsine transformation is rather little used today. Better ways of dealing with percentage data are described in detail in Chapters 9 and 15.

3.5 Random numbers

Another important function is the *random number generator*, %sr. This generates rectangularly distributed random numbers between 0 and 1. To get 20 random fractions, just type:

```
$units 20 $

$calc rn = %sr(0) $

$look rn $
```

You may be puzzled as to why you always get the same string of random numbers after you first switch on the machine; how can they be random, if they always come in the same sequence? The reason is that computers cannot generate really random numbers because they use a recipe to produce them. If the same recipe is followed, then the same numbers are bound to be generated. If you want to change the string of random numbers you must alter the *random number seed* as follows:

```
$sseed 2345 $
```

where the directive **sseed** stands for 'set seed' and the number is an integer between 1 and 4095. The reason that the seed is not set differently in each session (e.g. by reading the computer's date or clock) is that when you are testing a simulation model you actually *want* the same string of 'random' numbers so that you can check that your program is working properly.

To obtain random whole numbers (integers) we could use the truncating function, %tr; this knocks off all the decimal places. If we want to round to the nearest whole number we must add 0.5 to the data before using %tr. Thus, to get five random integers between 20 and 30 we would write:

```
$var 5 a $

$calc a = 20+%tr(%sr(0)*11) $
```

We multiply by 11 because there are 11 possible values for the random number (20, 21, 22,..., 29, 30). The %tr part of the function produces numbers between 0 and 10 because the largest possible value of %sr(0)*11 is 10.9999 and this would truncate to 10. It would be simpler, however, to write:

```
$calc a = 20 + %sr(10) $
```

which does the same thing in a simpler way. A non-zero term, n, in brackets after %sr tells GLIM to select at random from the $n+1$ integers between 0 and n. ●

3.6 Data input

For small sets of data it is convenient to use the **assign** or **read** directives for data input, as in the examples above. For longer sets of data, however, it is good practice to create the data file in a full-screen editor (e.g. a spreadsheet or a word-processor), and then to read the edited data into GLIM from an ASCII file using the **$dinput** directive (this stands for 'data input'). The **dinput** directive must be preceded by a **data** directive which lists the variables in the order in which they appear in the data file. The amount of data required is determined by the product of **units** and the number of variables listed in the **data** directive. Thus:

```
$units 9 $

$data xv yv $
```

requires 18 numbers (because there are two variables in the **data** list, xv and yv, and nine rows in the data matrix). To read values into the two variables from a file called glex37.dat, we would type:

```
$dinput 6 $

File name? glex37.dat

$look xv yv $
```

The 6 following dinput is called the channel number, and is explained below. In GLIM 4 we could simply write: $dinput glex37.dat $. At this point, GLIM should read the 18 values from the file. You will know that the data have been read successfully when the familiar ? reappears (reading large data sets from a floppy disk can take quite a long time on older machines). If something goes wrong during data input, you will get an error message. The commonest reasons for failure of data entry from an external file are:

1 there are fewer data in the file than were required to fill the data matrix defined in the **units** and **data** directives;

2 your data table is wider (has more columns) than GLIM can read;
3 there are non-numeric characters in the file;
4 the channel number you defined is not operative, or has not been rewound since its last use.

3.6.1 Insufficient data

The number of rows in the data matrix is defined by the **units** directive (in our example this is nine). The number of columns is defined by the number of variables named in the **data** directive (two, in our case, xv, and yv). Thus, the matrix has $9 \times 2 = 18$ elements. If the data file had fewer than 18 numbers in it, the **dinput** directive would fail. This kind of error is a major problem only with large data files, when you might miscount the number of rows. It is a useful tip to read your data file into a spreadsheet or word-processor before reading it into GLIM in order to count the rows automatically. This is then the number that you write into the **units** directive. It also allows you to see whether any text or other non-numeric characters are in the file, and gives you the opportunity to edit them out, prior to reading the file into GLIM.

Another common error is that the input goes ahead successfully, but data are allocated to xv and yv in the wrong way. For example, if the data file contained 18 numbers, but the first nine were the xv values and the second nine were the yv values, there would be no error message. But the subsequent analysis would be totally meaningless, because the second value in the data file would have been allocated to yv(1) instead of xv(2). This happened because the **data** directive specified that the numbers should be read in pairs: an xv value then a yv value.

If your data file were to contain all the xv values in sequence, followed by all the yv values, then you would read the data as follows:

```
$units 9 $

$data xv $

$dinput 8 $

File name ? glex37.dat

$data yv $

$dinput 8 $

$look xv yv $
```

Note that you are not prompted for a file name on the second **dinput** directive. This is because GLIM knows that input channel 6 is associated with the file name glex37.dat, and the end-of-file marker has not been encountered yet (since only half of the 18 numbers have been read so far). This example demonstrates the great flexibility of input in GLIM:

you can read little bits of data from lots of different files in any one session; each file has its own channel number, and the file name is defined on the first occasion that the channel number appears in an **input** or **dinput** directive. Note, however, that if there is more than one number per line in the data file, and input ends before the end of the present line is reached, then the next **dinput** begins from the start of the next record (i.e. after the next 'hard return'), and some numbers will be skipped. The simplest solution is to put only one number on each line, then no numbers will be missed when multiple **dinput** directives are employed.

3.6.2 Data in wide format

Unless told otherwise, GLIM will look for data only in the first 80 columns of the data table. If your table is much wider than this, then GLIM will read only the first 80 columns, and will therefore fail to find sufficient numbers by the time it gets to column 80 in the last row of the table (as defined by the **units** directive). The simplest way round this is to ensure that your data can squeeze into fewer than 80 columns. If they cannot, then you can inform the program of your wide data table by means of an optional parameter after the channel number in the **dinput** directive, as follows:

```
$dinput 6 132 $
```

which specifies that input will come from channel 6 (the left-hand number) and is up to 132 columns wide (the second number). This is the maximum width allowed (the minimum is 30), and most installations have 80 as the default width.

3.6.3 Non-numeric data

Many computer packages produce output files that contain the names of the variables in the first row. Other packages produce a non-numeric character in place of missing values (e.g. SX writes M and MINITAB writes *) when writing text files. GLIM will not read these files without special provision (see Format in the GLIM Manual). The simplest solution is to ensure that data files contain nothing but numeric characters, decimal points or minus signs. Individual data entries can be separated by one or more blanks, by commas, or by other specified separators. Non-separated data in contiguous, formatted columns can be read using the **format** directive, but the practice of using tightly packed data files is to be discouraged in everyday use, simply because the files are so difficult to check for data-entry errors.

3.6.4 Channel number problems

The notion of channel numbers is an initially confusing aspect of GLIM. As with other features of GLIM, the confusion arises from the great flexibility of the language. Briefly, the program reserves a number of channels for its own use (the primary input channel is from the keyboard, the primary output channel is to the screen, and there is a channel number for output to the primary printer), leaving others for user-defined purposes. You can see what the channel numbers are used for on your system by typing:

```
$env c $
```

which is the **environment** directive for channel information. It lists the settings for current input, output, dump and library channels. A good idea when beginning with GLIM is to pick a channel number that works on your machine, and stick with it (we shall use number 6 throughout the present book, with larger numbers when we have more than one file open during a given GLIM session; note that in GLIM 4 you should use number 7 or larger). •

3.7 Factors

GLIM deals with two different kinds of explanatory variables: continuous variables and factors. Factors are category variables like light = 'high, medium or low', colour = 'black, yellow, white or red', or sex = 'male or female'. In GLIM these factor levels are described by numbers rather than names, so light would have the levels 1, 2 and 3 and sex would have levels 1 and 2. We tell GLIM that they are factors rather than continuous variables with the **factor** directive:

```
$units 12 $

$factor light 3 colour 4 sex 2 $
```

which defines the maximum number of levels of each, and says that in this case light has three levels, colour has four levels and sex has two. GLIM will check to ensure that none of the values for colour is larger than 4 or sex is larger than 2. Factors are not allowed to have the value 0.

We could type the numbers to represent the different factor levels, but there is a much quicker way using GLIM's function %gl. This stands for *generate levels*, and is one of the most frequently used directives in GLIM. The idea is to produce sequences of 1's, 2's and 3's to represent, say, the light level associated with every value of response variable. The following material is hard to understand on first acquaintance, and may need to be read several times.

The argument of the %gl function contains two terms: the *maximum* and the *repeat*. Thus to obtain four 1's followed by four 2's followed by four 3's (a total of 12 numbers), we would write:

```
$calc light = %gl(3,4) $
```

where 3 is the maximum and 4 is the repeat. If we wanted the sequence of numbers 1, 2, 3, 4 repeated three times, it would be:

```
$calc colour = %gl(4,1) $
```

Consider a more complicated example. The data below come from an experiment with three treatments: irrigation, planting density ('stand') and fertilizer. There are two levels of irrigation (irrigated and not), three plant densities (high, medium and low) and three fertilizer regimes (high, medium and low). The whole experiment was replicated in four blocks, making a total of 72 yield measurements. The preamble, therefore, is:

```
$units 72 $

$data yield $

$dinput 7 $

File name? glex19.dat
```

The way we use %gl to generate factor levels is determined by the order in which the yield data appear in the file. In the present case, all the data from block 1 are entered first, then all the data from block 2, and so on. There are four blocks (the maximum) and 18 numbers in each block (the repeat). The repeat is 18 because there are two irrigation levels, three stand levels and three fertilizer levels in each block. We calculate the block factor levels like this:

```
$calc b=%gl(4,18) $
```

Within block 1, the nine non-irrigated samples come first, then the nine numbers from the irrigated samples. To generate the levels for the irrigation factor, therefore, we want nine 1's followed by nine 2's (then this big pattern repeated four times). The maximum is 2 and the repeat is 9:

```
$calc i=%gl(2,9) $
```

Within each irrigation the three low density measurements come first, then the three medium density, then the 3 high density data; the maximum is 3 and the repeat is 3, so we write:

```
$calc s=%gl(3,3) $
```

Finally, within each density treatment, the three fertilizer treatments — low, medium and high — appear in sequence; the maximum is 3 and the repeat is 1:

```
$calc f=%gl(3,1) $
```

To recap, we can look at all four statements together:

```
$calc b=%gl(4,18) $
```

```
$calc i =%gl(2,9) $
$calc s=%gl(3,3) $
$calc f =%gl(3,1) $
```

The first of the two arguments (the maximum) is the number of levels of each factor (4 for blocks, 2 for irrigation, 3 for stand and 3 for fertilizer). The second numbers (the repeats) show how the data file is put together. Going from bottom to top, the configuration demonstrates that fertilizer with its repeat of 1 is nested within stand (3), stand within irrigation (9), and irrigation within blocks (18). The procedure is completed by declaring b, i, s and f to be factors:

```
$factor b 4 i 2 s 3 f 3 $
```

This whole procedure can be performed in a single step in GLIM 4, once **units** have been declared, alleviating the need for the four %gl directives:

```
$gfactor b 4 i 2 s 3 f 3 $
```

The **gfactor** directive generates the levels and declares the variables as factors in a single statement. GLIM assumes a nesting from left to right, so for blocks the repeat must be (units = 72)/(levels = 4) = 18. The two levels of irrigation are nested within this 18 so they must have a repeat of 18/2 = 9. And so on. The full statistical analysis of this split-plot experiment is described in Exercise 8.5.

To see what has been created in this data set, turn on the **page** facility (this stops the screen from scrolling) and then look through the data matrix. Press the Return key to get the next screen-full of information.

```
$page $
$look yield b i s f $
```

After you have finished, turn off the **page** by repeating the directive:

```
$page $
```

To view the data in tabular form we could use the **tprint** directive as follows:

```
$tprint yield i;s;f;b $
```

Note that the list of classifying factors is separated by semicolons.

3.8 Logical operators

GLIM has a very elegant set of logical operators (Table 3.2) that are used in calculations of various sorts. They are especially useful when we want to operate on only some of the numbers in a vector (e.g. in weighting

Table 3.2 Logical operators in GLIM; these are used in **calc** directives to return the value true (=1) or false (=0) as appropriate; examples of logical calculations are given in Exercise 3.1

Logic	GLIM code	Meaning
Equals	`x == y`	x equals y returns 1; x not equal to y returns 0
Not equals	`x /= y`	x not equal to y returns 1; x equals y returns 0
Less than	`x < y`	x less than y returns 1; x greater than or equal to y returns 0
Greater than	`x > y`	x greater than y returns 1; x less than or equal to y returns 0
Less than or equals	`x <= y`	x less than or equal to y returns 1; x greater than y returns 0
Greater than or equals	`x >= y`	x greater than or equal to y returns 1; x less than y returns 1
AND	`x & y`	Returns 1 if both x and y are not equal to 0; returns 0 if either x or y, or both x and y, are equal to 0
OR	`x ? y`	Returns 1 if either x or y, or both x and y, are not equal to 0; returns 0 when both x and y are equal to 0
NOT	`/x`	Returns 1 if x is equal to 0; returns 0 if x is not equal to 0

certain values out of an analysis). Suppose that yield(54) is to be left out of the analysis in order to assess its influence on the parameter estimates. We would create a weight vector w as follows:

```
$calc w=1 $

$calc w(54)=0 $

$weight w $
```

More generally, we might want to remove from the analysis all data points with certain attributes. For example, to weight out all of the yield values >125 we could write:

```
$calc w=(yield<=125) $
```

which has 1's (logical 'true') for yields less than or equal to 125, and 0's (logical false) for all yields greater than 125.

Another useful feature of logical operators is that they allow a simple means of *counting* the number of elements in a vector that have certain attributes. Thus, to count how many yields out of the total of 72 were less than or equal to 125, we just write:

```
$calc %t = %cu(w) $

$look %t $

64.00
```

In other circumstances we may want to make a copy of the vector that has certain of its values replaced by other values. In model simplification, for example, we sometimes want to replace the values of an explanatory variable by 0's in order to alias them (see Section 12.5). Suppose that in the present case we wish to replace all the yield values lower than 80 by 0's in a new vector called q. Here we keep the values of yield by multiplying them by 1 (true, when yield ≥ 80) and replace them by 0's by multiplying them by 0 (false, when yield < 80).

```
$calc q=yield * (yield >= 80) $
```

To see how many yields were <80 we could count the 0's in the vector q:

```
$calc %t = %cu(q==0) $

$look %t $

8.00
```

so eight of the yields were <80. Note the use of the double equals sign to denote logical equality.

Although these logical functions are confusing on first acquaintance, it is worth practising with them, because they are extremely powerful and you will meet them again and again in transformation (see Section 3.4) and model simplification (see Section 12.5). There is ample opportunity to practise logical calculations in Exercise 3.1. ●

3.9 The logical function %if

Most people find this function rather difficult to use. As ever in GLIM, its generality is the cause of the difficulty. It is like an IF statement in any of the familiar computer languages, except that it is used within the **calc** directive. It says: if expression A is true, then the function takes the value of expression B. If expression A is false, then the function takes the value of expression C. It looks like this:

```
$calc z = %if(a,b,c) $
```

Some examples should help. Suppose we have the x vector

```
$assign x=5,8,1,3,2,5,7 $
```

and the conditional expression A is ($x > 2$); this is true when x is greater than 2. When $x > 2$ we want to set z equal to $x^2 + 3$, and otherwise we want to set z to 0. To do this we would write:

```
$calc z = %if(x>2, x**2+3, 0) $look x z $

      X       Z
1  5.000  28.00
2  8.000  67.00
3  1.000   0.00
4  3.000  12.00
5  2.000   0.00
6  5.000  28.00
7  7.000  52.00
```

The x values 1 and 2 have been replaced in z by 0's.

To set a weight variable k to 0 if the value in z is 0 and to 1 otherwise, we write:

```
$calc k = %if(z==0,0,1) $
```

To replace x with its absolute value, we write:

```
$calc x = %if(x<0,-x,x) $
```

Efficient use of the **%if** directive means that program branching can be minimized (there are no 'go to' statements in GLIM).

3.10 Sorting, shuffling and ranking

Note that because there is no 'unsort' directive, sorting is irreversible and the **sort** directive is potentially risky because it reorders some but not all of the data. This means that data could be separated from their subscripts, with potentially dire consequences. The **sort** directive in GLIM can take three different forms. The one-vector form is the easiest to understand. Do not execute this command or we shall lose the unsorted vector:

```
$sort x $
```

would replace the values in x with the sorted values in x from smallest to largest. So where x had been 5 8 1 3 2 5 7 to begin with, it would now be 1 2 3 5 5 7 8.

The second form of **sort** has two vectors. The second vector in the list is sorted and the result is stored in the first vector (the order of the variables in the list is therefore very important). After the sort directive has been executed, the first vector is unaltered. So if we had the unsorted x vector as before, then the command:

```
$sort y x $look x y $
```

produces a sorted y vector, while x has remained unaltered.

```
       X       Y
1  5.000   1.000
2  8.000   2.000
3  1.000   3.000
4  3.000   5.000
5  2.000   5.000
6  5.000   7.000
7  7.000   8.000
```

The three-vector form of the **sort** directive is more complicated, but very powerful. The command:

```
$sort z y x $

$look z y x $

     Z       Y       X
1 3.000   1.000   5.000
2 5.000   2.000   8.000
3 5.000   3.000   1.000
4 1.000   5.000   3.000
5 7.000   5.000   2.000
6 8.000   7.000   5.000
7 2.000   8.000   7.000
```

can be read as follows. Take the values in x and sort them. Use this ordering to sort the values of y. Store the ordered values of y in the new vector z. Make sure that you see how this operates by working out how the values of z were obtained.

There are three things you can do with the three-vector **sort**: (i) shuffling, which is useful in generating random numbers for experimental layout; (ii) ranking, which is useful in non-parametric correlations; and (iii) lags, which are useful in calculating growth increments and in analysis of population dynamics and life-table data on sequential mortality factors. ●

3.10.1 Shuffling

The first step is to set the number of **units** to the number of digits to be shuffled (say 10). Then we generate a vector of 10 random numbers r which will form the basis of the randomization:

```
$units 10 $

$calc r=%sr(0) $
```

Finally, use the three-vector sort with a 1 in place of the second vector:

```
$sort s 1 r $
```

```
$look s r$

      S        R
 1    1.000   0.03490
 2    5.000   0.86215
 3    9.000   0.76471
 4    4.000   0.17966
 5    3.000   0.11659
 6    7.000   0.86985
 7    2.000   0.76850
 8    8.000   0.86867
 9    6.000   0.13067
10   10.000   0.87919
```

after which the vector s will contain the shuffled integers 1 through 10. Placing a 1 in place of the second vector is the same as creating an ordered vector using %cu(1).

The advantages of shuffling rather than simply generating a series of 10 random numbers are that no numbers are repeated, and all 10 numbers appear in the shuffled list. This is very useful for allocating treatments to positions in a field experiment, or plant pots to positions on a greenhouse bench.

3.10.2 Ranking

Suppose we want to replace a set of data by their ranks. To do this in GLIM requires just two steps. We sort the data in r in the same way as in the previous example, and then we repeat the sort on the ordered list s. It is worth working through the two steps in detail, to see precisely what is happening:

```
$sort s 1 r $

$look s r $

      S       R
 1    3.000   0.5889
 2    9.000   0.6870
 3    4.000   0.1397
 4    6.000   0.1922
 5    5.000   0.2799
 6    1.000   0.2607
 7    8.000   0.9467
 8    2.000   0.6375
 9   10.000   0.1528
10    7.000   0.8238
```

The largest value in r (which will get rank 10 eventually) is 0.9467 and the smallest (which should get rank 1) is 0.1397. Look carefully at the row numbers (the ranked numbers 1 to 10 on the left) and the values in s. In the ranked set, we want the 1 to be placed in position 3, against the smallest value 0.1397. We want the 2 to be placed in position 9 against the 0.1528, and the 10 in position 7 against the 0.9467. If you think of the row number as the rank, then the value in s is the position in which the rank should be placed (i.e. number 1 in row 3, number 2 in row 9, number 3 in row 4, etc.). To do this we just type:

```
$sort s 1 s $

$look s r $

        S       R
 1   6.000  0.5889
 2   8.000  0.6870
 3   1.000  0.1397
 4   3.000  0.1922
 5   5.000  0.2799
 6   4.000  0.2607
 7  10.000  0.9467
 8   7.000  0.6375
 9   2.000  0.1528
10   9.000  0.8238
```

and we see that s now contains the ranks of the data in r.

GLIM does not deal with tied ranks in a predictable way. If it is important for your application to deal with ties in a consistent way, you will need to write your own macro to do the sorting.

3.10.3 Lags

An integer m in the third position of the **sort** directive means shift the contents of the vector by $(m-1)$ places downwards (if $m>0$) or upwards (if $m<0$). If $m=1$ there is no shift. In working with time series, it is useful to be able to calculate the change in the response variable that occurred between one time period and another (i.e. we want to subtract the previous value of y from the present value of y). Suppose that the following data are in a vector a:

```
$assign a=3,9,4,6,5,1,8,2,10,7 $
```

and we want to create a new vector b that contains the same data lagged by one time period. We use the three-vector **sort** directive with −2 in the third position, as follows:

```
$sort b a -2 $
```

```
$look b a $

        B        A
 1    9.000    3.000
 2    4.000    9.000
 3    6.000    4.000
 4    5.000    6.000
 5    1.000    5.000
 6    8.000    1.000
 7    2.000    8.000
 8   10.000    2.000
 9    7.000   10.000
10    3.000    7.000
```

Notice that all the data in a have been pushed up by one row and that the number that was in a(1) is now in b(10). This is an artefact that we could do without, and we would want to exclude the 10th row in computing the change from one period to the next. Since we can only compute nine increments from 10 numbers, we are not losing any information by this.

```
$calc diff=b-a $

$calc wt=(%cu(1)/=10) $

$weight wt $

$look diff wt $
```

where the **weight** directive removes the 10th difference from any subsequent calculations (the weight vector is set to wt = 1 for all rows except the 10th for which wt = 0).

If we wanted the array b to contain the contents of a advanced by one increment (i.e. moved downwards), we would simply type:

```
$sort b a 2 $

$look b a $
```

•

3.11 GLIM log

During each session of GLIM, all the information on the computer screen is written to an ASCII file called glim.log. This *transcript file* is very useful because it means you can obtain a hard copy of your session from another machine (you do not need to have a printer connected to the machine you are working with). All you do is copy the file called glim.log on to a floppy disk, take it to a machine with a printer, and list the file. Alternatively, you can read the file into your word-processor and cut out any of the output that you do not want to keep, then print only the bits that are required.

It is important to note that on many machines glim.log from the previous session is erased as soon as a new session begins, so that if you plan to do several sessions, it is good practice to copy glim.log to a file with a unique name at the end of each session (i.e. after **$stop**).

Unless you tell it otherwise, GLIM will echo all of your data to glim.log following the **dinput** directive, and with large data files this is most inconvenient. You can suppress the transcription of data to glim.log by the **trans** directive, as follows:

```
$units 9 $

$data x y $

$trans $

$dinput 8 $

File name: glex2.dat

$plot y x $

$trans i w f h o $

$plot y x $
```

The first **trans** directive on line 3 cuts off transcription to glim.log, then the data are read in from the file called glex2.dat. Transcription is restored by the **trans** directive on line 7. The letters following **trans** are the letters that appear in square brackets at the beginning of each line in glim.log, namely:

```
[i]  input
[w]  warning messages
[f]  fault messages
[h]  help messages
[o]  ordinary output
```

When you inspect glim.log for this session you will find only one copy of the plot of y against x because output transcription was switched off when the first plot directive was issued (note, however, that this plot did appear on the screen). •

3.12 Macros

GLIM is a powerful computing language as well as a statistical package. Programs are just sets of GLIM commands saved in files. Sections of code that are used repeatedly can be stored as macros, and these can be used in GLIM for a variety of tasks:

1 for eliminating repetitive typing tasks;

2 for statistical work (e.g. tests of normality);
3 for simple computing tasks like performing *loops*;
4 more complex sub-routines (e.g. jacknife residuals).

Once you have mastered the elements of the GLIM language, it is easy to write your own GLIM programs using macros, but we shall defer detailed discussion of this to the end of the book (see Chapter 19 and the exercises therein).

At this stage we shall work through two simple examples that will be useful from the beginning, and look briefly at a third example that gives a taste of how a GLIM program works. The first example is text substitution, the second is a test for normality. In all cases, note that there must be a blank space between the end of the macro name and the $ sign.

3.12.1 Text substitution

Suppose that we are carrying out a simulation study in which we have to generate lots of sets of normally distributed random numbers. As we have already seen, this can be done by typing:

```
$calc x = %nd(%sr(0)) $
```

Repeatedly typing %nd(%sr(0)) gets tedious very quickly, but help is at hand. We can define the oft-repeated piece of text as a macro called n (standing for normal random numbers), like this:

```
$macro n %nd(%sr(0)) $endmac $
```

then each time we want to generate a new set of random numbers we just use the *text substitution symbol* # to write:

```
$calc y = #n $

...

$calc z = #n $
```

and so on. You can write a different macro for each complicated piece of text that you need to type. For example, substitution macros are very useful for specifying complicated model formulas (see Section 14.8), and to save repeatedly writing:

```
$fit :+a*b*c*d-a.b.c.d-b.c.d $
```

it would be a good idea to begin by defining the model structure as a macro:

```
$macro m :+a*b*c*d-a.b.c.d-b.c.d $endmac $
```

Then each time we wanted to fit this particular model, we need only write:

```
$fit #m $
```

3.12.2 Program macros

This example assumes that you have already fitted a model to your data (you can skip to 3.12.3 on first reading). The following code generates a plot of raw residuals against the standard normal cumulative distribution function (see Section 5.12). You can write program macros within a GLIM session, but it is much more efficient to write them outside GLIM in your word-processor, and store the macros as ASCII files that can be read into GLIM as required. The macro is called, say, rplot (for residual plot), and might look like this: ○

```
$macro rplot $

$calc resid=%yv−%fv $

$sort resid $

$calc n=%cu(1) $

$calc norm=%nd((n−0.375)/(%nu+0.25)) $

$plot resid norm '*' $

$endmac $
```

All macros must begin with **macro** and end with **endmac**. The macro name (rplot in this case) appears on the first line. The purpose of the macro illustrated here is to calculate the residuals from the present model, and then compare the distribution of the residuals with the pattern that would be expected if the residuals were normally distributed. On line 3 the residuals are sorted into ascending order. Line 4 creates a vector n containing the ranks 1, 2, 3, up to %nu, and line 5 calculates normal deviates associated with each of the fractions n/%nu (using the correction terms 0.375 and 0.25). Finally, line 6 plots the ranked residuals on the *y*-axis against the ranked normal deviates. If the residuals are normally distributed, then the plot will be linear. Non-normality in the residuals will show up as a curve in one or more tails of the plot (see Section 13.3). If the macro had been stored in a file, then the directive **$return** would need to be added after the **$endmac** directive. If the file was called resplot.mac, we could use it like this:

```
$input 9 $

File name: resplot.mac

...

$use rplot $
```

3.12.3 Using macros

There are four different ways of executing macros within GLIM. The simplest is unconditional and employs the **use** directive. The next two are particularly important in programming: the **while** directive executes the macro repeatedly until a logical condition becomes false, while the **switch** directive selects one macro from a list of macro names, depending on the value of a scalar. The fourth method involves text substitution by the use of the # symbol, as described above. ○

Unconditional

```
$use rplot $
```

causes the macro called rplot to be executed, then control of the program moves on to the next GLIM directive.

*Repeated (**while**)*

```
$calc %a=1 $

$while %a update $
```

means that the macro called update will be executed again and again, until the value of the scalar %a becomes 0 (i.e. false). The macro should contain a statement which sets the value of %a to 0 when the task is complete, otherwise the macro will be executed for ever! For example:

```
$macro update $

$calc %z1=%z1+1 $

...

...

$calc %a=%if(%z1<100,1,0) $

$endmac $
```

*Branching (**switch**)*

Many programming applications call for one macro to be executed under one set of circumstances, but a different macro in others. The scalar %a controls which of a list of macro names is executed, so:

```
$calc %a=2 $

...

$switch %a eggs larvae pupae adults $
```

would cause the macro called larvae to be executed because %a is 2 and larvae is second in the list of macro names. If %a is zero, or is larger than the length of the list (say 6 in this case), then none of the macros is executed.

For conditional execution of a macro, you can use a single macro list:

```
$switch %a update $
```

which will be executed if %a is true (i.e. %a = 1) but not if it is false (%a = 0).

3.12.4 A simple program

Just as a taster, this is how you write a program loop in GLIM. The object is simply to print the numbers 10, 9, 8,..., 0 in descending sequence. We need to write a piece of code that creates a loop that goes round 11 times, subtracting 1 from the previous value and printing the result. This is achieved by writing a macro called loop which looks like this:

```
$macro loop $

$calc %x=%x−1 $

$print %x $

$calc %s=(%x>0) $

$endmac $

$return $
```

The working part of a macro is sandwiched between **$macro** and **$endmac** directives. The second line reduces the value of the scalar %x by 1, and the third line prints the new value. The fourth line checks to see whether the job is finished. The program will continue to loop so long as the logical scalar %s is true, so we set %s to false immediately after printing the final number of our required sequence (%x = 0). The **return** directive sends control back from the file to the point in the main program where the file was **input**.

The loop is controlled from the main program as follows:

```
$calc %x=11 $

$calc %s=1 $

$while %s loop $
```

The initial value of %x is set to 11, so that the first value printed inside loop will be 10 as required. The second line sets the logical scalar %s to 1 (true). This controls the execution of the loop in line 3. The **while** directive will continue to invoke the macro called loop so long as %s is true. This is why it is necessary to recompute the value of %s inside the macro, and to set %s to false (0) when the macro is finished (i.e. after we have printed the final 0).

When each execution of the macro is completed, the program returns to the **while** directive that invoked it. When %s = 0 the program passes to the next line after the **while** directive which, in this case, is the end of the program.

Suppose that the macro called loop is stored in a file called count.mac, then we would write:

```
$input 10 $

File name? count.mac

$calc %x=11 $

$calc %s=1 $

$while %s loop $
```

●

Remember that macros need to appear *before* they are invoked by GLIM (in some languages, sub-routines appear at the end of the program).

3.12.5 Using library macros

When you buy GLIM you are supplied with a library of useful macros like the Box–Cox transformation (see Section 13.6) and programs for assessing the distribution of errors (see Section 13.1). GLIM users also exchange custom-built programs with one another. All you need to do to use a library macro is to read the code from file like this: ○

```
$input %plc boxcox $
```

where %plc is the primary library channel that your machine is configured with (you do not need to know this channel number, but you can see what is it by typing **$env c**; this gives the width and height of all the primary and current channels) and boxcox is the name of the file containing the macro(s) you require. Once the macros have been read you can invoke them directly by the **use** directive:

```
$use boxcox $
```

or conditionally with the **while** or **switch** directives, as above.

It is a good idea to load as few library macros as possible, because space is limited within GLIM and several of the library macros are quite space-consuming. Examples of macros that you might like to add to your library are given in Chapter 19, and by Aitkin *et al.* (1989). When you write macros of your own, it is a good idea to put an exclamation mark !

at the end of each statement, because material after the ! is not read into GLIM. Remember that even the blanks at the end of a line will take up storage space. Also, the space after the ! is useful for documenting the code, to remind the user what each line of code is supposed to be doing.

There are lots of examples of the use of macros in Exercise 19.1.

Users of GLIM 4 should note the following points.

1 OWN models are defined differently than in this book; the logic is the same, but the familiar equations for the linear predictor, fitted values, variance function and deviance increment are defined as optional parameters of the **link** directive (%fv as a function of %lp, and the derivative of the link function, %dr) and **error** directive (variance function %va and deviance increment %di); new directives with OWN options include **initial**, **load** and **method** (see *The GLIM System: Release 4 Manual* (Francis *et al.*, 1993), pp. 728, 736 & 742).

2 The lowest channel number that can be used for input from files is 7 (not 6 as used throughout this book).

3 All the standard probability distributions are accessible both as deviates (e.g. %chd) and as probabilities (e.g. %chp for χ^2).

4 Model-fitting of polynomials and contrasts is much more sophisticated (see Francis *et al.*, 1993, pp. 156–183 for details).

5 The **$end** directive now means only **$endmacro** and has been replaced by **$newjob** to finish one GLIM session and begin another.

General users should note the following tips.

1 The backslash \ can be used as a lower-case replacement for the $ symbol in separating GLIM directives.

2 Typing mistakes in complicated directives can be edited using the F1 and F3 keys; F1 produces the last command line, character-by-character from the left, while F3 produces the entire last line, with the cursor at the right.

3 Solid bullet (•) means: type **$stop** before starting the next exercise (or **$newjob** in GLIM 4).

4 Open bullet (○) means: do not type these commands into GLIM (they are not set in the necessary context and would cause error messages to be produced).

CHAPTER 4

Introduction to experimental design

4.1 Experimental design

Whole library shelves of books have been devoted to the subject of experimental design. Some valuable examples are Fisher (1954), Cox (1958), Cochran & Cox (1957) and Mead (1989). My sole object here is to provide a few notes that are relevant to using GLIM.

The most important concepts to master are *replication*, *randomization*, *initial conditions* and *blocking*. It does not matter very much if you cannot do your own advanced statistical analysis. If your experiment is properly designed, you will often be able to find somebody to help you with the stats. But if your experiment is not properly designed, or not thoroughly randomized, or lacking adequate controls, then no matter how good you are at stats, some (or possibly even all) of your experimental effort will have been wasted. No amount of high-powered statistical analysis can turn a bad experiment into a good one. GLIM is good, but not that good.

There is always a trade-off between:

1 including a wide range of conditions, in an attempt to make the experiment general; and

2 restricting the set of conditions, so as to reduce variability and increase the likelihood of reaching firm conclusions.

4.2 Pulse and press experiments

We need to distinguish between experiments that consist of a single kick to the system (pulse experiments) and those that involve a constant push in the back (press experiments; see Bender *et al.*, 1984). We should recognize, however, that the short-term response (say, over 1–5 years) to either kind of experiment may exhibit transitory dynamics that are quite atypical of equilibrium or long-term behaviour. The long-term consequence of the experiment may be impossible to predict from the short-term dynamics (e.g. a species may increase in abundance in the first year following a disturbance, but that same species may disappear completely in the long run).

4.3 The principle of parsimony (Occam's razor)

An important theme running through this book concerns model simplification. The principle of parsimony is attributed to the 14th century English nominalist philosopher William of Occam, who insisted that, given a set of equally good explanations for a given phenomenon, *the*

correct explanation is the simplest explanation. It is called Occam's razor because he 'shaved' his explanations down to the minimum. In statistical modelling, the principle of parsimony means that:
1 models should have as few parameters as possible;
2 linear models should be preferred to non-linear models;
3 experiments relying on few assumptions should be preferred to those relying on many;
4 models should be pared down until they are *minimal adequate*;
5 simple explanations should be preferred to complex explanations.
The process of model simplification is an integral part of hypothesis testing in GLIM (see Section 11.4). In general, a factor is retained in the model only *if it causes a significant increase in deviance when it is removed from the maximal model*.

4.4 Observation, theory and experiment

There is no doubt that the best way to solve ecological problems is through a thoughtful blend of observation, theory and experiment. In most real situations, however, there are constraints on what can be done, and on the way things can be done, which mean that one or more of the trilogy has to be sacrificed. There are lots of cases, for example, where it is ethically or logistically impossible to carry out manipulative experiments. In these cases it is doubly important to ensure that the statistical analysis leads to conclusions that are as critical and as unambiguous as possible.

4.5 Strong inference

One of the most powerful means available to demonstrate the accuracy of an idea is an experimental confirmation of a prediction made by a carefully formulated hypothesis. There are two essential steps to the protocol of *strong inference* (Platt, 1964):
1 formulate a clear hypothesis;
2 devise an acceptable test.
Neither one is much good without the other. For example, the hypothesis should not lead to predictions that are likely to occur by other extrinsic means. Similarly, the test should demonstrate unequivocally whether the hypothesis is true or false.

A great many ecological experiments appear to be carried out with no particular hypothesis in mind at all, but simply to see what happens. While this approach may be commendable in the early stages of a study, such experiments tend to be weak as an end in themselves, because there will be such a large number of equally plausible explanations for the results. Without contemplation there will be no testable predictions; without testable predictions there will be no experimental ingenuity; without experimental ingenuity there is likely to be inadequate control; in short, equivocal interpretation. The results could be due to myriad plausible causes. As Hairston (1989) is fond of saying, 'Nature has no stake in being understood by ecologists'. We need to work at it. Without replication, randomization and good controls we shall make little progress.

4.6 Weak inference

The phrase weak inference is used (often disparagingly) to describe the interpretation of observational studies and the analysis of so-called 'natural experiments'. It is silly to be disparaging about these data, because they are often the only data we have. The aim of good statistical analysis is to obtain the maximum information from a given set of data, *bearing the limitations of the data firmly in mind.*

Natural experiments arise when an event (often assumed to be an unusual event, but frequently without much justification about what constitutes unusualness) occurs that is like an experimental treatment (a hurricane blows down half a forest block; a landslide creates a bare substrate; a submarine volcano produces a new island, etc.).

> The requirement of adequate knowledge of initial conditions has important implications for the validity of many natural experiments. Inasmuch as the 'experiments' are recognized only when they are completed, or in progress at the earliest, it is impossible to be certain of the conditions that existed before such an 'experiment' began. It then becomes necessary to make assumptions about these conditions, and any conclusions reached on the basis of natural experiments are thereby weakened to the point of being hypotheses, and they should be stated as such. (Hairston, 1989)

4.7 How long to go on?

Ideally, the duration of an experiment should be determined in advance, lest one falls prey to one of the twin temptations:

1 to stop the experiment as soon as a pleasing result is obtained;

2 to keep going with the experiment until the 'right' result is achieved (the 'Gregor Mendel effect').

In practice, most experiments probably run for too short a period, because of the idiosyncrasies of scientific funding. This short-term work is particularly dangerous in ecology, because the kind of short-term dynamics exhibited after pulse experiments may be entirely different from the long-term dynamics of the same system (see Tilman, 1988, for a detailed discussion of transitory dynamics in plant ecological experiments). Only by long-term experiments of both the pulse and the press kind will the full range of ecological dynamics be understood. The other great advantage of long-term experiments is that a wide range of weather patterns ('kinds of years') is experienced.

4.8 Degrees of freedom

Many people find the concept of degrees of freedom hard to grasp. Formally, degrees of freedom (d.f.) is defined as *the number of measurements minus the number of parameters estimated from the data*:

$$\nu = n - p$$

where ν (pronounced 'new') is the d.f., n is the number of measurements and p is the number of parameters estimated from the data.

What this means is most clearly seen by a numerical example. Suppose we have five numbers and their mean value is 4. This implies that the sum of the numbers must be $4 \times 5 = 20$. Now we ask: how many different values could the first number take? Clearly, it could be anything at all. Let's say it was 7. Then we ask: how many values could the second number take? Again, it could be anything at all. Let's say 3. The third number could take any value as well. Let's say it is another 7. The fourth value, too, could be anything; let's say it was -5. Now the numbers so far total 12. The grand total must be 20. Thus we have no choice at all in selecting the fifth and final number. It must be 8. Thus, with five numbers, we have four degrees of freedom in choosing values. The fifth number is constrained because the sum of the numbers must be 20.

When we estimate the variance of a sample, we calculate $\Sigma(y - \bar{y})^2$. We cannot calculate the variance until we have estimated the parameter $\bar{y}$ from the data. This constrains Σy and so the sample variance has $n - 1$ degrees of freedom.

When we estimate the error variance in a linear regression, we calculate $\Sigma(y - a - bx)^2$. We cannot calculate *SSE* until we have estimated the two parameters a and b from the data. Thus, if there are n points on the graph, we have $n - 2$ degrees of freedom for error.

In a contingency table with r rows and c columns the grand total is fixed at n. Take a 2×2 table (where $r = c = 2$). If there are f_c individuals in the first column, there must be $n - f_c$ in the second column. Similarly, if there are f_r individuals in the first row, there must be $n - f_r$ in the second row. Here we have 1 d.f. for rows and 1 d.f. for columns, giving 1 d.f. overall. Degrees of freedom for an r by c contingency table is simply the product of the row and column degrees of freedom; $\nu = (r - 1)(c - 1)$.

In GLIM, if you add a factor with four levels to a model, this will bring about a change of -3 in the residual degrees of freedom. If you add an interaction between this factor and another factor with five levels, the residual degrees of freedom will decline by $(4 - 1) \times (5 - 1) = 12$.

If a factor is wholly or partially aliased (see Section 4.20) then the change in degrees of freedom associated with the addition of the factor will be less than (levels – 1). If the factor is completely aliased (as a main effect is *intrinsically aliased* once an interaction term involving that factor is included), then adding the factor will cause no change at all in the residual degrees of freedom. Missing values or 0's can also cause a factor to be aliased (such factors are said to be *extrinsically aliased*).

4.9 Factorial designs

A factorial design is an experiment whose object is to test for, and to measure the strength of, statistical interactions. Interaction effects occur when *the response to one factor depends upon the level of another factor*. A simple way of thinking about interactions is to consider a graph of the response variable against the levels of factor A (Fig. 4.1). If the graphs for the different levels of factor B are parallel with one another, then there is no interaction. Interactions show up as departures from parallelness. One objective of a factorial experiment is to test whether any observed lack of parallelism is statistically significant.

Interaction effects are scientifically exciting because they mean that in order to predict the response to one factor you need to know the level of another. For example, in a factorial experiment on parasitoid egg-laying, we might find that host density and female parasite density show a statistically significant interaction in their impact on egg-laying rate. This means that we cannot say 'this is the effect of host density on egg-laying' but rather 'the response to host density depends upon the number of female parasites; with low numbers of females we observe a straightforward functional response, but with high numbers of females the response is less pronounced and non-asymptotic'.

4.10 Fractional factorials and confounding

Confounding means having less than one replicate of every treatment in a block. This sounds like a bad idea at first, but sometimes it is forced upon us. Suppose that we have time to handle only two treatments in a day, but there are six treatments. We do not want to make a three-day period into a block because we use different batches of animals on every day. Efficient use of confounding requires a careful analysis of its advantages and disadvantages. The advantages of confounding are:

1 reduction in experimental error arising from the use of smaller, and therefore internally homogeneous, blocks;

2 reduced size of the whole experiment;

while the disadvantages are:

1 reduction in replication of the confounded treatment comparisons;

2 increased complexity of calculations.

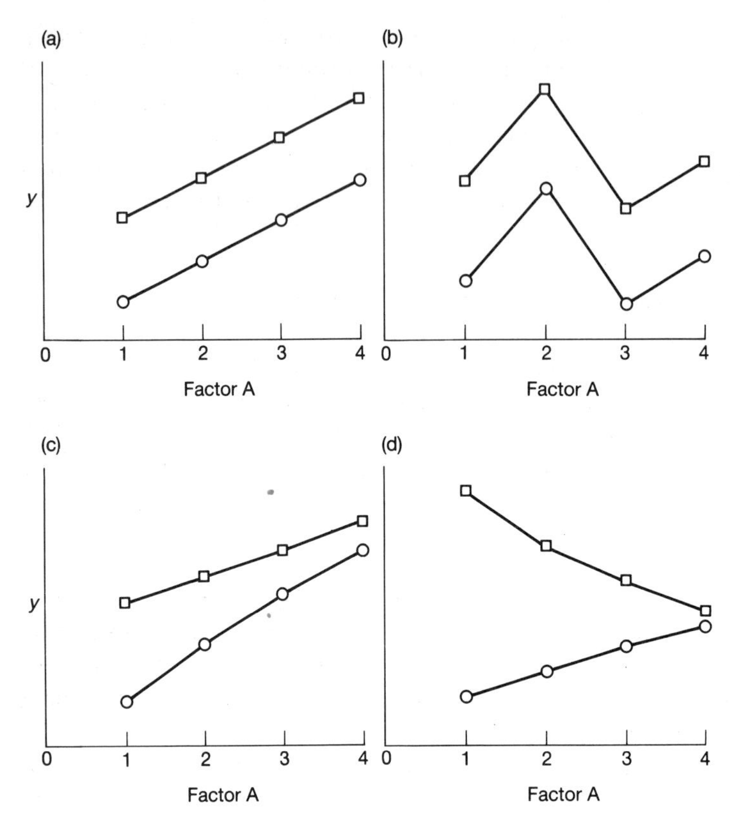

Fig. 4.1 Statistical interaction occurs when a graph of the response variable against factor A is not parallel to similar graphs for other levels of factor B. (□) B(1); (○) B(2). Cases (a) and (b) show no interaction, and the response to factor B does not depend upon the level of factor A. Case (c) shows interaction, because high levels of A produce smaller responses to factor B. Case (d) shows an extreme form of interaction, where the opposing slopes cancel out. In this case, the main effect of factor A would appear to be insignificant.

The question that needs to be addressed is whether the reduction in variance per unit that comes from having smaller blocks more than compensates for the loss of replication due to confounding (Cochran & Cox, 1957).

Factorial experiments with fractional replication

A large factorial experiment (say with six or seven different factors, each at two levels) may be beyond the resources of the experimenter, or it may be unnecessarily big, in the sense that it would give more precision of the estimates than was really needed. Under these circumstances, fractional replication may be appropriate (see Cochran & Cox, 1957). Thus, instead of having all eight of the treatment combinations from a 2^3 factorial in each block, one would have only four with half replication, or two with quarter replication. The difficulty is that experiments with fractional replication are open to misinterpretation because of aliasing. For instance, in an three-factor factorial with two levels of each factor, one half replicate might have only the three main effects (a), (b) and (c), and the three-way interaction (abc). Then the main effect of A would be calculated as:

$$A = (abc) + (a) - (b) - (c)$$

However, the two-way interaction AB would be given by:

$$AB = (abc) + (c) - (a) - (b)$$

which you will notice is precisely the same as the formula for the main effect of factor C. Thus C and AB are aliases (see Cochran & Cox, 1957, pp. 244 *et seq.* for details). The three-way interaction ABC cannot be estimated at all, because:

$$ABC = (abc) + (a) + (b) + (c) - (ab) - (ac) - (bc) - (1)$$

and we only have information on the terms with positive signs. In this particular half replicate we have lost all information on the three-way interaction, and all of the two-factor interactions are inextricably confounded with main effects. Thus, if the experiment suggests that the main effect of A is important, we do not know whether this is really due to the effect of A, the interaction between B and C or a combination of both.

4.11 Nested designs

It is tempting to assume that any experiment that has two or more factors can be analysed as a factorial design. There are several pitfalls for the unwary:

1 a factorial design must have replication for each of the interaction terms that needs to be estimated;

2 the treatment combinations must be independent;

3 the treatment combinations must be assigned at random.

The commonest mistake is to assume that the treatment combinations are independent when in fact they are not. Take a simple example involving an experiment on the effects of diet on insect growth at different temperatures. We have four controlled temperature (CT) rooms at 10, 15, 20 and 25°C and in each CT room insects are cultured on five different diets, with each diet repeated three times at each temperature. Now ask yourself the question: how many CT rooms would be needed if this were a factorial design? There are five diets and three replicates, so we would need 15 CT rooms at each temperature, so with four temperatures this is 60 CT rooms in all. Since we have only four CT rooms, it is pretty clear that the present experiment is not a factorial design. But the structure of the data does not tell us this, because we have 60 numbers, and we could easily make the mistake of analysing the results as if they had come from a factorial design. Failure to recognize split-plot and other nested designs is one of the commonest mistakes in ecological statistics. It is important because *it leads to the use of the wrong error term* in hypothesis testing, and causes the standard errors to be wrongly calculated if the correct design is not recognized.

4.11.1 Recognizing nested designs

The simple rule of thumb for recognizing a factorial experiment is this. Multiply together all your treatment combinations and replicates (4 temperatures $\times$ 5 diets $\times$ 3 replicates = 60). Then ask, do I have this many *independent* repeats of the experiment. If you do, then you have a full factorial. If you do not, then you have some kind of nested or split-plot design, and you will need to think carefully about how you specify the error term in the model.

In the present example, the repeats are not independent of one another because we have only four CT rooms, one at each temperature. Thus, if one of the CT rooms was odd in some way (perhaps someone sharing the CT room was using a volatile growth regulator), then *all* the results from that CT room would be affected (all the diet treatments and all the replicates of each diet treatment). In the present case, there are no degrees of freedom for temperature, because temperature and growth room are completely confounded. The way to analyse this experiment is to treat it as a split-plot design. The CT rooms are blocks, and we compare the diets by one-way ANOVA nested within blocks, using the block–diet interaction term (with its 12 degrees of freedom) to test the significance of differences between the diet means (i.e. we do not use the overall error degrees of freedom with its $4 \times 5 \times (3 - 1) = 40$ degrees of freedom). Differences due to temperature effects will emerge as the block sum of squares, but we shall not be able to estimate temperature–diet interactions, because these are confounded with differences between the CT rooms. Because they all occur in the same CT room, the 'replicates' are actually pseudoreplicates when it comes to comparing diets at different temperatures. The repeats in each CT room are not completely useless, of course, because they allow four independent comparisons of the five diets (one in each CT room, with $5 \times (3 - 1) = 10$ degrees of freedom for error in each room).

The skills involved in spotting pseudoreplication come only with practice. This, again, is not something that it is easy to pick up from reading about it (but see Section 4.14.1).

4.12 Efficient regression designs

Too little thought tends to be given to the precise nature of the question being asked in regression studies. Two contrasting examples should make this point. Suppose that we have the financial and human resources to carry out 14 experimental measurements. Should we gather seven replicates at just two extreme levels of the x-axis, or one replicate at each of 14 different levels of the explanatory variable? Should the levels of x be equally spaced or unequally spaced along the x-axis (see Fig. 4.2)? These are design questions that can be resolved only by thinking carefully about exactly what it is about the relationship between y and x that we wish to establish.

In the first case, let us assume that we suspect that there is a threshold level of x below which y shows no response (this might be a temperature threshold, below which there is no development in an insect). The question in this case is where, precisely, is the threshold located?

In the second case, let us suppose that we know that there is a smooth relationship between y and x, but we suspect that the relationship is not a straight line, and we wish to test whether there is evidence for a quadratic term in the model (i.e. do the data provide support for a model that

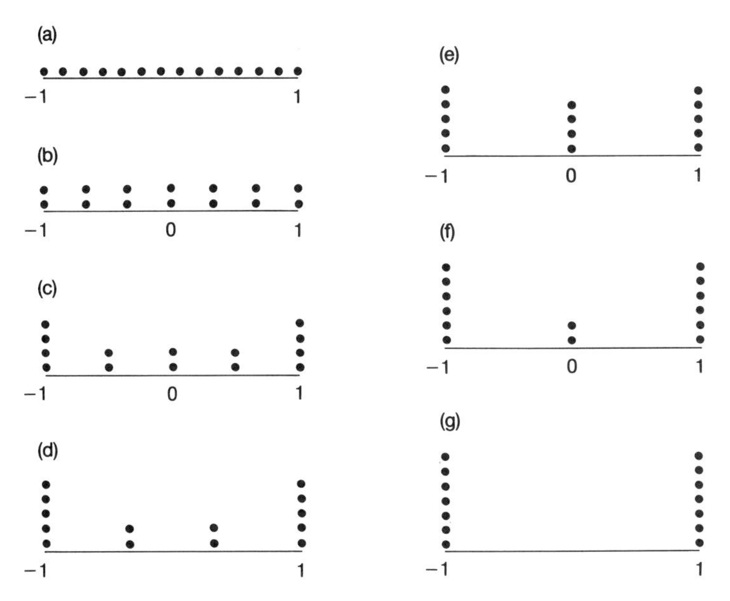

Fig. 4.2 Regression designs. The optimal distribution of sampling effort along the length of the x-axis depends upon the question in hand. To minimize the standard error of the slope, half of the data should be gathered at each extreme end of the x-axis (g). To detect the location of thresholds or non-linearities, the data should be evenly spaced along the whole axis (a). Most cases require some compromise between these extremes. (After Draper & Smith, 1981.)

contains an x^2 term in addition to a term for x?). This might occur if the value of y increased more slowly at higher levels of x than at low (as in a study of functional responses, where predator feeding rates were investigated at a range of prey densities).

We begin by considering some of the options. Figure 4.2 shows seven different ways that our 14 measurements might be distributed along the x-axis. Which of these is best? The answer, of course, depends upon the question. For the first case, we wish to locate the position of a threshold. Design (a) may be the best bet if we have no idea where the threshold lies, while design (g) is completely hopeless. But design (g) is not always hopeless, because it is the design that gives the lowest standard error for the slope. The standard error of a regression slope is $\sqrt{s^2/SSX}$, and so we always want to make SSX as large as possible. This is achieved by having lots of measurements at the extreme left- and right-hand ends of the x-axis, and fewer in the middle (see Section 7.5).

The solution in the second case is a little more subtle. The best design cannot be (g) because it can provide no evidence of non-linearity, since a straight line will always fit perfectly between two points. The number of degrees of freedom for non-linearity is given by $k-2$, where k is number of levels of x. In order to be able to test for non-linearity, therefore, we need at least three levels of x. The best design is a compromise between degrees of freedom allocated to non-linearity and degrees of freedom allocated to estimating pure sampling error (see Section 7.6). On balance, designs (c), (d) and (f) look best as a compromise between error estimation and detection of non-linearity. If we are reasonably sure that a quadratic term is the most complex model we need to fit, then perhaps choice (f) would be preferred, with choice (c) if we were less certain about the nature of the non-linearity (e.g. the curve might be sigmoid, in which case the mean of a treatment midway between the extremes of the x-axis could fall on the straight line joining the two extreme points, thus providing no evidence of non-linearity; see Draper & Smith, 1981).

4.13 Variables

There are two kinds of variables in linear modelling: the *response variable* and one or more *explanatory variables*.

4.13.1 Response variable

The response variable is the thing that we measured, or the observational data that we collected. The aim of statistical modelling is to understand the causes of variation in the response variable. A synonym for the response variable is the *y-variable*. This is because the response variable is always drawn on the y-axis of a graph rather than the x-axis; failure to observe this convention can lead to untold confusion.

The response variable is drawn on the y-axis because graphs are constructed in order to demonstrate *functional relationships*; thus 'y is a function of x' means that changing the level of x is associated with changes in the value of y. This matters, because ecological relationships are usually not symmetrical or reversible. Thus, we graph photosynthetic rate on the y-axis against light intensity on the x-axis because changing light intensity affects the rate of photosynthesis, but changing the rate of photosynthesis does not alter the incident light intensity. This is the origin of the old-fashioned name for the response variable; because y depends upon the level of x, but not the other way round, y was called the dependent variable and x was called the independent variable. This usage has been all but abandoned amongst statisticians, and we should follow suit. Lots of response variables in modelling do not strictly *depend* upon the level of x. More importantly, many x-variables are not at all independent, either of one another, or of the value of y (see Section 12.5).

The response variable can take one of several forms (see Table 4.1), and knowing which form is appropriate to the data in hand is important in the choice of GLIM model you should use.

You should aim to become thoroughly familiar with this way of classifying variables. A useful exercise is to write down four examples of category, proportion, count and continuous variables from your own area of interest. The chances are that you will make at least two mistakes at your first attempt.

Category variables are extremely important in GLIM because they define the factors that are used in statistical design and in analysis of variance (see Section 3.7). Thus, in a well-planned experiment, the initial material is divided up into blocks on the basis of as many category

Table 4.1 Types of response variable

Continuous	Count	Proportion	Category
Body weight	Young born per female	Death rate (a number r died out of a cohort of n)	Sex
Shoot length	Ringed animals found dead	Fraction breeding	Experimental treatment (e.g. irrigated or not)
Feeding time	Colonies on an agar plate	Sex ratio	Geographic origin
Dose	Deaths through lightning strikes	Percentage parasitism	A statistical block
Water volume	Days with frost in May	Infection rate	Genotype

variables as it is sensible and cost-effective to incorporate, and then the experiment is repeated in each block.

4.14 Replication

As the statistician with a sense of humour once said: it's the *n*'s that justify the means. The requirement for replication arises because if we do the same thing to different individuals we are likely to get different responses. The causes of this heterogeneity in response are many and varied (genotype, age, sex, condition, history, substrate, microclimate and so on). The object of replication is to assess the variability that is found within the same treatment. Only if the variation of observations about the treatment means is small compared with the variation about the grand mean will it be legitimate for us to conclude that the treatment has caused significant differences. The object of replication is to provide more precise estimates of mean and effects. To qualify as replicates, the repeated measurements:

1 must be independent;

2 must not form part of a time series (data collected from the same place on successive occasions are not independent);

3 must not be grouped together in one place (aggregating the replicates means that they are not spatially independent);

4 must be of an appropriate spatial scale;

5 ideally, one replicate from each treatment ought to be grouped together into a block, and each treatment repeated in many different blocks;

6 repeated samples (e.g. from the same plant or the same quadrat) are not replicates (this is probably the commonest cause of pseudoreplication in ecological work). The analysis of repeated measures is discussed by Gurevitch & Chester (1986) and Crowder & Hand (1990).

Positive covariance between experimental units would occur if some samples were grouped together in good habitat and others grouped in poor habitat. Under these conditions, the sample means would differ for reasons that had nothing to do with the treatments that were applied to them.

4.14.1 Spotting pseudoreplication

Pseudoreplication is rife amongst ecological studies (see Hurlbert, 1984). There are two reasons for this. The first is that lots of ecologists cannot recognize it. The second reason is more fundamental. A good rule of thumb in ecology is that *everything varies*. What this means in terms of experimental design is that it is exceptionally difficult, and frequently impossible, to find identical experimental conditions in which to replicate a given treatment. An obvious example is years. Carrying out the same experiment once in each of three years does not give three replicates, because the years will have differed from one another in countless ways. Another example is spatial heterogeneity. Carrying out the same experiment in three different places does not give three replicates because the places will differ in soil, microclimate, access by natural enemies and so on (although the closer the places are together, the more like replicates the experimental plots are likely to become). In some circumstances, of course, it may be physically impossible to replicate (e.g. studies on global climate change, or in biogeographic studies involving whole islands).

Pseudoreplication is generally quite easy to spot. The question to ask is this. How many degrees of freedom for error does the experiment really have? If an ecological field experiment appears to have lots of degrees of freedom, it is probably pseudoreplicated. Take an example from pest control of insects on plants. There are 20 plots, 10 sprayed and 10 unsprayed. Within each plot there are 50 plants. Each plant is measured five times during the growing season. Now this experiment generates $20 \times 50 \times 5 = 5000$ numbers. There are two spraying treatments, so there must be one degree of freedom for spraying and 4998 degrees of freedom for error. Or must there? Count up the replicates in this experiment. Repeated measurements on the same plants (the five sampling occasions) are certainly not replicates. The 50 individual plants within each quadrat are not replicates either. The reason for this is that conditions within each quadrat are quite likely to be unique, and all 50 plants will experience more or less the same unique set of conditions, irrespective of the spraying treatment they receive. In fact, there are 10 replicates in this experiment. There are 10 sprayed plots and 10 unsprayed plots, and each plot will yield only one independent datum to the response variable (the proportion of leaf area consumed by insects, for example). Thus, there are 9 degrees of freedom within each treatment, and $2 \times 9 = 18$ degrees of freedom for error in the experiment as a whole. It is not difficult to find examples of pseudoreplication on this scale in the literature (Hurlbert, 1984). The problem is that it leads to the reporting of masses of spuriously significant results (with 4998 degrees of freedom for error it is almost impossible *not* to have significant differences). The first skill to be acquired by the budding ecological experimenter is the ability to plan an experiment that is properly replicated.

4.14.2 Sample covariance

Negative covariance between individuals might occur in a competition experiment where, if one of the individuals grew large, its neighbours would necessarily be small.

When there is positive covariance between samples, the measurements should be added together and the statistics carried out on the average values. In the case of negative covariance, then it may be legitimate to analyse the individual responses (as in individual weights in a competition experiment) (H. M. Wilbur in Hairston, 1989).

4.14.3 How many replicates?

The usual answer is 'as many as you can afford'. There are ways of working out the replication necessary for testing a given hypothesis, but these all require a good deal of information about the sampling variance and the degree of between-block heterogeneity. Often we know nothing about any of these values when we are planning an experiment. Experience is important. So are pilot studies. These should give an indication of the variance between initial units before the experimental treatments are applied, and also of the approximate magnitude of the responses to experimental treatment that are likely to occur.

Sometimes it may be necessary to reduce the scope and complexity of the experiment, and to concentrate the inevitably limited resources of manpower and money on obtaining an unambiguous answer to a simpler question. It is immensely irritating to spend three years on a grand experiment, only to find at the end of it that the response is significant only at $\alpha = 0.08$. A reduction in the number of treatments might well have allowed an increase in replication to the point where the same result would have been unambiguously significant.

4.14.4 Replicates or blocks?

There is always a trade-off between replication and blocking, because we always have limited resources (time, space, pairs of hands). The question always arises, therefore, as to whether it is better to have lots of replicates in a small number of blocks, or lots of blocks with no replication. It is impossible to make a cast-iron generalization, but in many ecological cases it is better to go for blocks rather than replicates, because ecological material is so variable, and ecological conditions are so heterogeneous (both in time and in space), that attempts at replication are often futile. On the other hand, replication within blocks does allow us to estimate treatment–block interactions, and to obtain estimates of pure sampling errors (see Section 8.2). As always, the best solution will depend on the nature of the problem in hand.

4.15 Controls

No controls, no conclusions.

4.16 Randomization

Randomization is something that everybody says they do, but hardly anybody does properly. Take a simple example. How do I select one tree from a forest of trees, on which to measure photosynthetic rates? I want to select the tree at random in order to avoid bias. For instance, I might be tempted to work on a tree that had accessible foliage near to the ground, or a tree that was close to the lab. Or a tree that looked healthy. Or a tree that had nice insect-free leaves. And so on. I leave it to you to list the biases that would be involved in estimating photosynthesis on any of those trees.

Now one common way of selecting a 'random' tree is to take a map of the forest and select a random pair of coordinates (say 157 m east of the reference point, and 72 m north). Then pace out these coordinates and, having arrived at that particular spot in the forest, select the nearest tree to those coordinates. But is this really a randomly selected tree?

If it was randomly selected, then it would have *exactly the same chance of being selected as every other* tree in the forest. Let us think about this. Look at Fig. 4.3, which shows a plan of the distribution of trees on the ground. Even if they were originally planted out in regular rows, accidents, tree-falls and heterogeneity in the substrate would soon lead to an aggregated spatial distribution of trees. Now ask yourself how many different

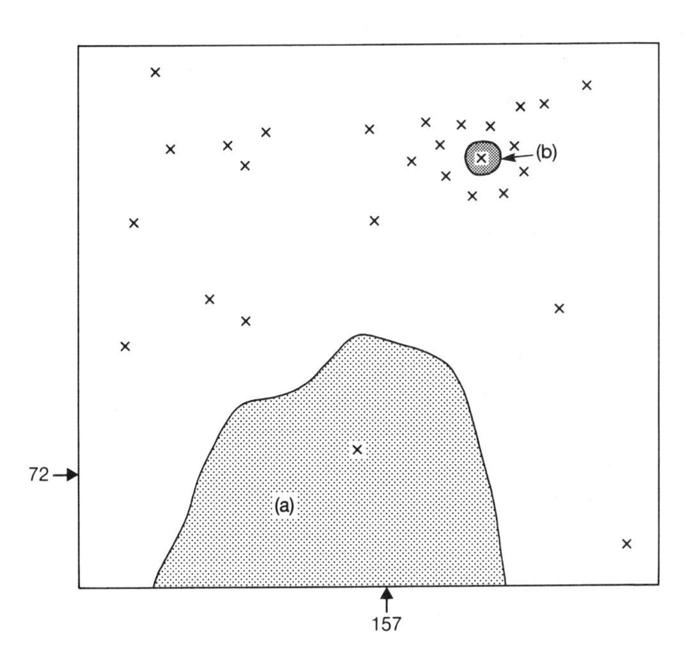

Fig. 4.3 Selection of a random individual. The nearest individual to a randomly placed point is not randomly selected, because in aggregated populations isolated individuals and individuals on the edges of clumps stand a much greater chance of being selected than individuals in the centre of clumps. Random sampling requires that each individual has the same probability of being selected. The shaded area shows the probability of selection for (a) an isolated individual and (b) an individual in the centre of a group. The isolated individual is about 30 times more likely to be selected.

random points would lead to the selection of a given tree. Start with tree (a). This will be selected by any points falling in the large shaded area. Now consider tree (b). It will be selected only if the random point falls within the tiny area surrounding that tree. Tree (a) has a much greater chance of being selected than tree (b), and so *the nearest tree to a random point is not a randomly selected tree*. In a spatially heterogeneous woodland, isolated trees and trees on the edges of clumps will always have a higher probability of being picked than trees in the centre of clumps.

The answer is that to select a tree at random, every single tree in the forest must be numbered (all 24 683 of them), and then a random number between 1 and 24 683 must be drawn out of a hat. There is no alternative. Anything less than this is not randomization.

Now ask yourself how often this is done in practice, and you will see what I mean when I say that randomization is a classic example of 'Do as I say, and not as I do'. As an example of how important proper randomization can be, consider the following experiment that was designed to test the toxicity of five contact insecticides by exposing batches of flour beetles to the chemical on filter papers in Petri dishes. The animals walk about and pick up the poison on their feet. The *Tribolium* culture jar was inverted, flour and all, into a large tray, and beetles were collected as they emerged from the flour. The animals were allocated to the five chemicals in sequence; four replicate Petri dishes were treated with the first chemical, and 10 beetles were placed in each Petri dish. Do you see the source of bias in this procedure?

It is entirely plausible that flour beetles differ in their activity levels (sex differences, differences in body weight, age, etc.). The most active beetles might emerge first from the pile of flour. These beetles all end up in the treatment with the first insecticide. By the time we come to finding beetles for the last replicate of the fifth pesticide, we may be grubbing round in the centre of the pile, looking for the last remaining *Tribolium*. This matters, because the amount of pesticide picked up by the beetles will depend upon their activity levels. The more active the beetles, the more chemical they pick up, and the more likely they are to die. Thus, the failure to randomize will bias the result in favour of the first insecticide because this treatment received the most active beetles.

What we should have done is this. Fill $5 \times 4 = 20$ Petri dishes with 10 beetles each, adding one beetle to each Petri dish in turn. Then allocate a treatment (one of the five pesticides) to each Petri dish at random, and place the beetles on top of the pretreated filter paper. We allocate Petri dishes to treatments most simply by writing a treatment number on a slip of paper, and placing all 20 pieces of paper in a bag. Then draw one piece of paper from the bag. This gives the treatment number to be allocated to the Petri dish in question. All of this may sound absurdly long-winded but, believe me, it is vital.

The recent trend towards 'haphazard' sampling is a cop-out. What it means is that 'I admit that I didn't randomize, but you have to take my word for it that this did not introduce any important bias'. You can draw your own conclusions.

4.17 Initial conditions

Many otherwise excellent ecological experiments are spoiled by a lack of information about initial conditions. How can we know whether something has changed if we do not know what it was like to begin with? It is often implicitly assumed that all the experimental units were alike at the beginning of the experiment, but this needs to be demonstrated rather than taken on faith. One of the most important uses of data on initial conditions is as a check on the efficiency of randomization. For example, you should be able to run your statistical analysis to demonstrate that the individual organisms were not significantly different in mean size at the beginning of a growth experiment. Without measurements of initial size, it is always possible to attribute the end result to differences in initial conditions. Another reason for measuring initial conditions is that the information can often be used to improve the resolution of the final analysis through analysis of covariance (see Section 9.1).

4.18 Fixed effects and random effects

Traditional statistical analyses distinguish between experimental designs where you, the experimenter, impose treatments upon subjects (fixed effects) and experiments where you go looking for different places in which to repeat the same experiment (random effects). The statistical methods appropriate to the first case are called Model I ANOVA and to the second, Model II (the reasons for the Roman numerals are lost in the mists of time). In many ecological experiments, the two may be mixed together. It is commonplace in field experiments, for example, to select initially different plant communities to serve as blocks, and then to apply all the treatments at random within each block. Thus the blocks are sometimes said to be random effects, and the treatments fixed effects. The important difference is that, with fixed effects, we think we know that the cause of variation can be attributed to our experimentally imposed treatments, whereas with random effects we know things differ but we have no idea *why* they differ. In other cases, we may well know the cause of random effects (e.g. soil fertility trends between blocks) but we want to treat them as instances from a distribution, rather than specific fixed effects.

4.19 Taylor's power law

One of the most robust empirical generalizations in ecology is known as Taylor's power law (Taylor, 1961). It states that as the sample mean increases, so the variance increases, such that on a graph of log variance against log mean, the slope of the graph is approximately equal to 2.0. Statisticians refer to this as the *power-of-the-mean model* (Carroll & Ruppert, 1988, p. 89).

Since one of the most important assumptions underlying regression and ANOVA is that sample variance is constant, you will see at once that ecologists have a major problem. There are two ways out of this dilemma. The traditional way is to take the log of the response variable (or keep taking logs repeatedly) until the variance stabilizes. GLIM offers a more attractive alternative, because it can deal directly with data that have a slope of 2 on Taylor plots (i.e. data with a constant coefficient of variation) by means of gamma errors (see Section 17.1).

4.20 Aliasing

This section will be hard to understand on first reading, but it is so important as a background to interpreting GLIM output that you should reread it as often as necessary until the penny drops.

Aliasing occurs when certain combinations of parameters in the model cannot be distinguished from each other because there is no independent information about the components. *Intrinsic aliasing* occurs when this is due to *the structure of the model. Extrinsic aliasing* occurs when it is due to *the nature of the data.*

4.20.1 Intrinsic aliasing with continuous variables

Suppose we are modelling the density of pin-galls on leaves, and we have measurements of the length, breadth and area of each leaf. We might propose to fit the model:

$$\text{galls} = a + b.\text{log length} + c.\text{log breadth} + d.\text{log area}$$

with four parameters (a, b, c and d). It turns out that we can estimate only three of the four parameters because:

$$\text{log area} = e + \text{log length} + \text{log breadth}$$

It is worth working through this example, in order to understand the concept of aliasing. Let the explanatory variables be x_1, x_2 and x_3 so:

$$y = a + bx_1 + cx_2 + dx_3$$

and because leaf area is proportional to length times breadth we can write:

$$x_3 = e + x_1 + x_2$$

where e is the log of the leaf-shape constant. Now we replace x_3 in the equation for the linear predictor:

$$y = a + bx_1 + cx_2 + d(e + x_1 + x_2)$$

Multiplying through by d and gathering the terms for x_1 and x_2 we get:

$$y = (a + d.e) + (b + d)x_1 + (c + d)x_2$$

so that the model will be able to estimate only three separate quantities from the data:

$$a + d.e \qquad b + d \qquad c + d$$

The fourth parameter is said to be intrinsically aliased.

As an exercise, you should convince yourself that if leaf breadth is proportional to leaf length (i.e. $x_2 = f + x_1$), then we can estimate only two parameters rather than three.

4.20.2 Intrinsic aliasing with factors

If we had a factor with four levels (say none, light, medium and heavy shading) then we could estimate four means from the data, one for each factor level. But the model looks like this:

$$y = m + a_1x_1 + a_2x_2 + a_3x_3 + a_4x_4 \qquad (4.1)$$

where x_1 to x_4 are dummy variables having the value 0 or 1 (see Section 8.5). Clearly there is no point in having five parameters in the model if we can estimate only four independent terms. One of the parameters must be intrinsically aliased.

There are innumerable ways of dealing with this, but three equally logical options are as follows:

1 set the grand mean m to 0, so that a_1 to a_4 are the four individual treatment means;

2 set the first term a_1 to 0 so that m is the mean of the first group and a_1 to a_4 are the differences between the first group mean and the other group means;

3 set the sum of a_1 to a_4 to 0 so that m is the grand mean and each a is a departure from the grand mean.

Although it sounds like the most complicated option, the default in GLIM is (2), where the mean of the first treatment is aliased. Because a full understanding of this idea is so important to your ability to interpret GLIM output, you should work slowly through the following numerical example.

Suppose that the grand mean was 10, and that the individual treatment means were 6, 9, 12 and 13. Then the components of the linear predictor under the different constraints of (1), (2) and (3) on equation (4.1) are as shown in Table 4.2.

Table 4.2 Components of linear predictor

Term	Symbol	(1)	(2)	(3)
Grand mean	m	0	6	10
None	a_1	6	0	−4
Light shade	a_2	9	3	−1
Medium shade	a_3	12	6	2
Heavy shade	a_4	13	7	3

When we display the results of an analysis of variance in GLIM, the mean of the first treatment is aliased (it is called 'parameter 1'), and *all the other parameters are differences between means* (option (2), above). The advantage of option (2) is that is has only four estimates (cf. option (3) which has five) and that, because the estimates are differences between means (cf. option (1) where the parameters are treatment means), hypothesis testing is extremely straightforward; means are significantly different from the mean of treatment 1 if their parameter estimates are significantly different from zero.

In two-way tables, the question of aliasing is a little more complicated. If we have replication for every treatment combination then the calculations are reasonably straightforward. Let us say the two factors are nitrogen fertilizer (two levels: none and regular application rate) and potassium fertilizer (also two levels). This design defines a 2×2 table, and the linear model we want to fit is:

$$1 + N + K + N.K$$

where N.K represents the *interaction* between nitrogen and potassium. There are four cells in the table and hence four means to be estimated. But the linear predictor looks like this:

$$\eta_{ij} = \mu + \alpha_i + \beta_j + \gamma_{ij}$$

so, in the case of a 2×2 table, there are nine terms in the model (the overall mean, two α, two β and four γ terms). Since we can estimate only four quantities, five of the nine parameters must be aliased. We shall need *five constraints on the estimates* in order to produce uniqueness. As with the one-way example, above, there are lots of ways of dealing with this. The default option in GLIM is that *any parameter with either of its subscripts equal to 1* is aliased, and μ is set to the mean of cell (1,1). These five constraints leave four estimatable quantities: the mean yield in cell (1,1), a *main effect term* for nitrogen (measured by α), a main effect for potassium (measured by β) and an interaction term (measured by γ).

If the mean yields under the different treatments were as in Table 4.3,

Table 4.3 Mean yields

	Treatment	
Treatment	None	K added
None	3	4
N added	4.5	6

and the model terms were as in Table 4.4,

Table 4.4 Model terms

	Treatment	
Treatment	None	K added
None	μ	$\mu+\beta$
N added	$\mu+\alpha$	$\mu+\alpha+\beta+\gamma$

then the estimated parameters in GLIM would be as in Table 4.5.

Table 4.5 Parameters

	Treatment	
Treatment	None	K added
None	$\mu=3$	$\beta=1$
N added	$\alpha=1.5$	$\gamma=0.5$

Again, check the tables carefully so that you can see where each of the terms has come from. Notice that to calculate the mean for the treatment receiving both nitrogen and potassium, you need to add the nitrogen effect ($\alpha=1.5$), the potassium effect ($\beta=1$) and the interaction effect (γ=0.5) to the 'intercept' (the mean of the 'none : none' treatment combination); $3+1.5+1.0+0.5=6.0$ as required.

In two-way tables with no replication the logic is a little more difficult, because it is hard to see where GLIM obtains its estimate for the expected value of cell (1,1). There is a fully worked example in Exercise 8.2, but for the present simply note that if we had r levels of factor A and c levels of factor B, we should have $n=rc$ numbers in total because there is no replication. This would enable us to estimate only $r+c-1$ quantities, as we can see by an example. Suppose that we have three rows ($r = 3$) and four columns ($c = 4$); see Table 4.6,

Table 4.6 Two-way ANOVA tables in GLIM are made up of three components: i is the intercept (parameter 1); c are the differences in the column means (compared with the mean of row 1, column 1); r are the differences in the row means (compared with the mean of row 1, column 1); see text for details on how the value of the intercept is calculated by GLIM

	c_1	c_2	c_3	c_4
r_1	i	$i + c_2$	$i + c_3$	$i + c_4$
r_2	$i + r_2$	$i + r_2 + c_2$	$i + r_2 + c_3$	$i + r_2 + c_4$
r_3	$i + r_3$	$i + r_3 + c_2$	$i + r_3 + c_3$	$i + r_3 + c_4$

where i stands for the 'intercept' (the fitted value for cell 1,1). As before, GLIM aliases any parameter with either of its subscripts equal to 1. Thus, all the row parameters in the top row of the table are aliased, and so are all the column parameters in the left-hand column.

So how does GLIM estimate the fitted value for the cell (1,1)? The answer is not quite as easy as you might imagine. With replication (as in the earlier example) GLIM would simply calculate the average of the replicates from cell (1,1). But this does not work in the present case, because there is just one number in each cell. In order to calculate the fitted value for any given cell you need to add to the intercept:

1 the value of the difference between the row mean and the mean of row 1, plus

2 the value of the difference between the column mean and the mean of column 1.

The terms are laid out in full in Table 4.6. You will see that the intercept occurs in every cell (12 of them in this case), r_2 occurs in four cells, c_3 occurs in three cells and so on.

Now the grand mean is the sum of the values in all the cells divided by the number of cells. In the present case, this is:

$$\mu = \frac{12(i) + 3(c_2) + 3(c_3) + 3(c_4) + 4(r_2) + 4(r_3)}{12}$$

To determine the value of the intercept, we must first work out the grand mean (μ) and the means for the rows and columns that have subscripts greater than 1. Now work out the differences between the row means and the mean of row 1 (r_i, above) and the column means and mean of column 1 (c_j, above). Substitute these quantities in the equation, and rearrange to solve for i (check your answer against the worked example in Exercise 8.2). Notice that the intercept *is not* the average of the means of row 1 and column 1 (a common mistake).

4.20.3 Examples of aliasing in GLIM

In general, an *aliased* parameter is a parameter whose value cannot be estimated because either the model or the data contain no information on it. When the **aliased** message appears unexpectedly during a GLIM analysis (e.g. in the standard error column following **disp e**), it is because *you are trying to estimate more parameters than the structure of the data will allow*. There may be no information about the parameter because either: (i) the necessary values of the response variable are missing from the data set (extrinsic aliasing); or (ii) all the available information has already been used up in estimating parameters in earlier **fit** directives (intrinsic aliasing). Consider the following examples.

1 Suppose that, in a factorial experiment, all of the animals receiving level 2 of diet (factor A) and level 3 of temperature (factor B) have died accidentally as a result of fungal pathogen attack. This particular combination of diet and temperature contributes no data to the response variable, so the interaction term $A(2).B(3)$ cannot be estimated (it is extrinsically aliased, and its parameter estimate is set to zero).

2 If one continuous variable is a multiple of another variable that has already been fitted to the data, then the second term is aliased and adds nothing to the model; if, say, `x2=%a*x1` then **fit** `x1 + x2` will lead to `x2` being intrinsically aliased and given a zero parameter estimate (see the example of pin-galls on leaves, above).

3 If all the values of the explanatory variable are set to zero for a given level of a particular factor, then that level is intentionally aliased. This sort of aliasing is a useful programming trick when we wish a covariate to be fitted to some levels of a factor but not to others (see Section 12.5.4). Sometimes, an entire vector of zeros is fitted to the model, so that the whole explanatory variable is aliased. This is done when the user specifies the entire model in an **offset** directive, and wishes to fit the model to the data only for the purposes of estimating the residual deviance (i.e. there are no parameters left to be estimated and one is interested only in the explanatory power of the fully specified model; see Section 12.5.5). The need to set up a vector of zeros to fit a null model is another manifestation of GLIM's unfriendliness.

4.21 Orthogonal designs and non-orthogonal observational data

The data in this book fall into two distinct categories. In the case of planned experiments, all of the treatment combinations are equally represented and, barring accidents, there are no missing values. Such experiments are said to be *orthogonal*. In the case of observational studies, however, we have no control over the number of individuals for which we have data, or over the combinations of circumstances that are observed. Missing treatment combinations are commonplace, and the data are said to be non-orthogonal.

This makes an important difference to our statistical modelling because, in orthogonal designs, the deviance that is attributed to a given factor is constant – it does not depend upon the order in which that factor is removed from the model. In contrast, with non-orthogonal data, we find that the deviance attributable to a given factor *does* depend upon the order in which the factor is removed from the model. We must be careful, therefore, to judge the significance of factors in non-orthogonal studies, when they are *removed from the maximal model* (i.e. from the model including all the other factors and interactions with which they might be confounded).

Missing values may arise in any kind of study, and the more missing values there are, the more the value of the experiment is diluted. As we shall see, GLIM is very good at dealing with missing treatments and unequal replication, but only at a cost. Degrees of freedom are lost, and standard errors are inflated, thus reducing the likelihood of detecting significant differences.

CHAPTER 5

Understanding data: graphical analysis

5.1 The importance of knowing your data

It is a common failing to rush headlong into statistical analysis without first obtaining a thorough understanding of the data. This attitude is encapsulated in the immortal words of the student who, on first being introduced to a statistical computer package, wanted to know 'Which button do I press for significance?' GLIM encourages good habits of data exploration by providing powerful functions for graphical and tabular output. The object of the exercise is to:

1 understand the distribution of the response variable;
2 look for trends with the explanatory variables;
3 consider the need for transformation;
4 look for potentially influential observations;
5 find errors that have occurred during data entry;
6 test the assumptions of the statistical models you intend to employ.

This chapter is concerned with graphical analysis, while the tabular summary of data is explained in Chapter 6.

5.2 Graphs

The graphics within GLIM are not sophisticated, but they are quick, powerful and very general. GLIM 4 has high-resolution graphics, but here we shall deal only with the simple graphs that can be produced by Version 3.77 as well as Version 4. These graphs also have the virtue that hard copy can be obtained on even the simplest dot-matrix printer.

The basic command is the **plot** directive. This is central to almost all preliminary data analyses. **Plot** can be used to plot *y*-variables against continuous variables (see Regression in Chapter 7), or against the levels of different factors (see ANOVA in Chapter 8). As usual, we shall work with an example. Read in the set of data from the file glex4.dat, which shows stomatal density (y) against leaf thickness (x) for 15 individual plants of the same species from three different soil types (45 pairs of numbers in all).

```
$units 45 $

$data y x $

$dinput 6 $

File name? glex4.dat
```

5.3 One *y*-variable

The simplest plotting command is just:

```
$plot y x $
```

which produces the output:

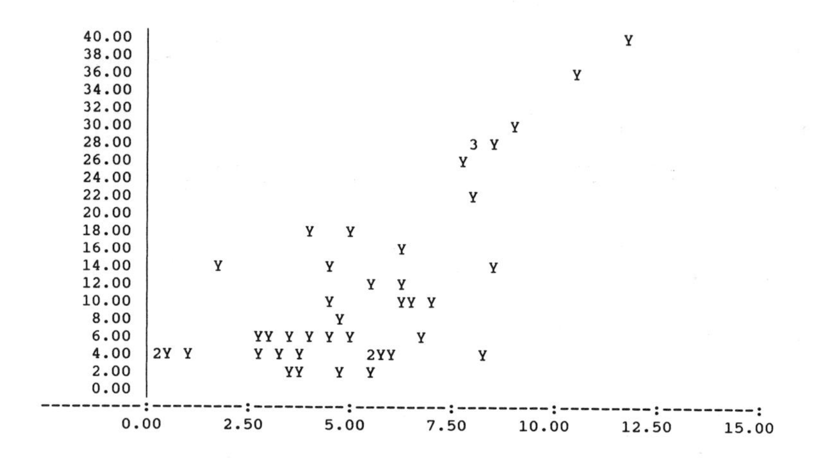

The first variable in the list is the *y*-axis and the second is the *x*-axis. There are three things to notice here.

1 The plotting symbol is the first letter of the name of the *y*-variable (simply Y in this case).

2 If more than one point on the graph is printed in a given position, then GLIM prints a numeral to show how many points are printed in a given place (9 is printed if there are more than nine points in a given position).

3 The scaling of the axes is rather odd (e.g. the *x*-axis goes from 0 to 15 with divisions of 2.5).

5.4 Changing the plotting symbol

To plot a symbol like + instead of the first letter of the name of the *y*-variable, we put the required symbol within single quotes after the name of the *x*-axis variable (see p. 71):

```
$plot y x '+' $
```

5.5 Scaling the axes

To get a consistent scaling of the axes (e.g. when we want to make a visual comparison of several different plots) we use the **ylimit** and **xlimit**

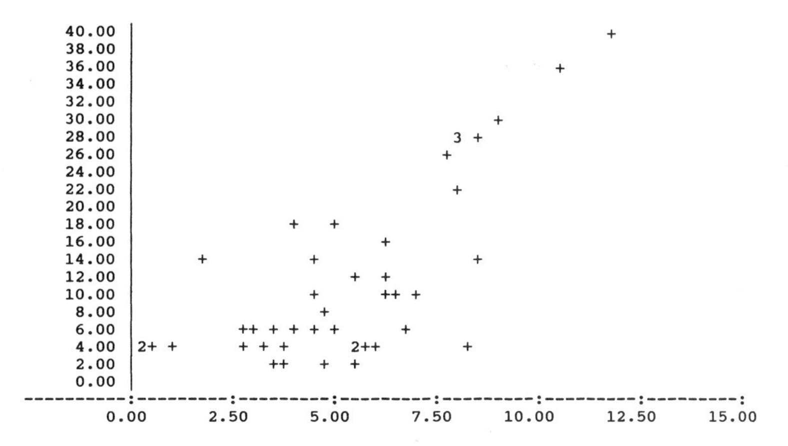

options (these can be abbreviated to **y** and **x**). Thus, to have y from 0 to 50 and x from 0 to 20, we would put:

```
$plot (y = 0,50 x = 0,20) y x '+' $
```

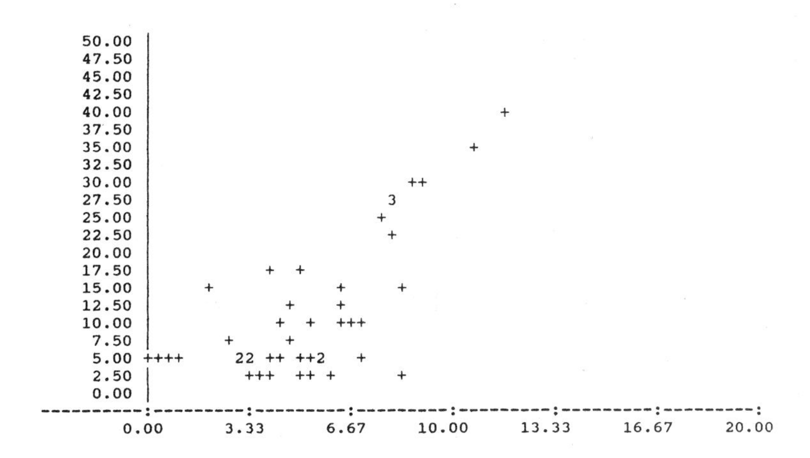

5.6 Plotting factor levels separately

When there are two or more levels of a factor, it is often informative to use different plotting symbols for each factor level. We generate the three levels of soil type, and declare soil as a factor:

```
$calc t=%gl(3,15) $

$factor t 3 $
```

then use the plotting style:

```
$plot y x 'abc' t $
```

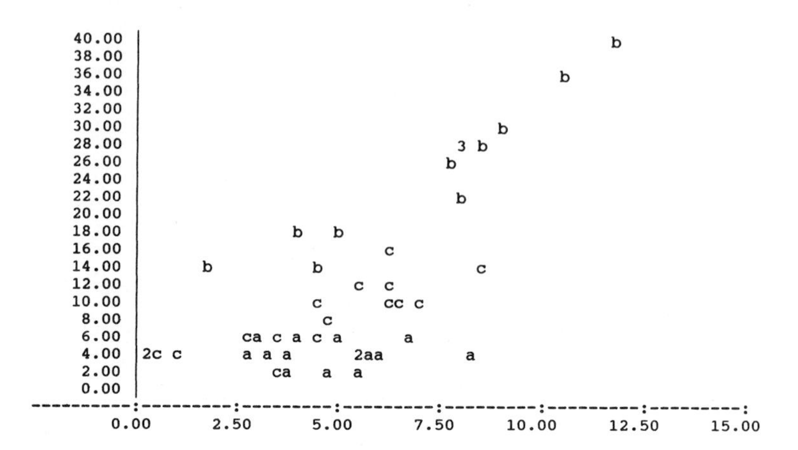

which means, plot y against x for each level of T, using the first symbol in the list 'a' for data coming from the first soil (T = 1), the second symbol 'b' for data from the second soil (T = 2), and so on. Note that T must be a factor and the length of the symbols-list must match the declared size of the factor T (i.e. there must be three characters in this case because there are three soil types).

5.7 Multiple *y*-variables

If we are analysing the fit of a model, we often want to plot the observed and expected values of y on the same axes. This means we have two *y*-variables, and each, obviously, should be given a different plotting symbol. Thus if we calculate the fitted values of a linear regression of stomatal density against leaf thickness for the present data, we would write:

```
$yvar y $

$fit x $
```

and then plot the data (+) and the fitted values (*) on the same axes, like this (see p. 73):

```
$plot y %fv x '+*' $
```

GLIM will understand this to mean that there are two *y*-variables (y and the fitted values %fv) to be plotted against x, and that the first symbol in the symbols-list '+' will represent the first *y*-variable in the list (y), and the second symbol '*' will represent %fv.

5.8 Multiple graphs with multiple factor levels

GLIM plots reach their pinnacle of complexity when we plot multiple

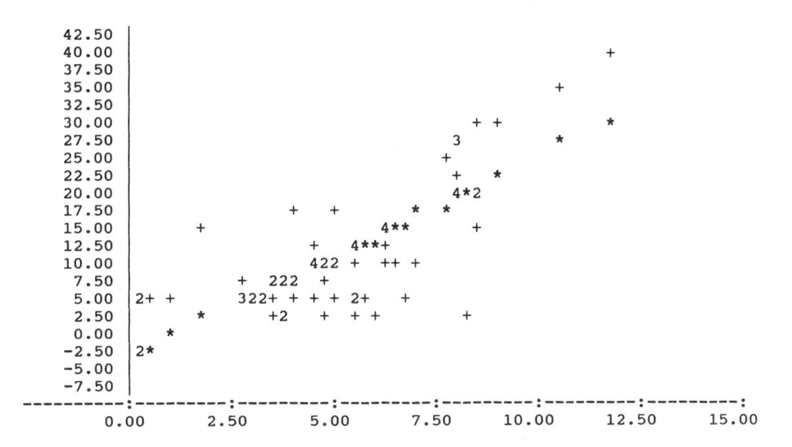

y-variables, each with multiple factor levels. Thus, we might want to plot observed and expected values of stomatal density for each of three soil types separately. First, we need to fit three separate regression lines, one for each soil type. This is simple; we just type:

```
$fit x*t $
```

Because there are two *y*-variables and three levels of T we must provide 2 x 3 = 6 symbols in the symbols-list. Then, to use a, b and c for the three soil types as before, and '+' for the three sets of fitted values, we write:

```
$plot y %fv x 'abc+++' t $
```

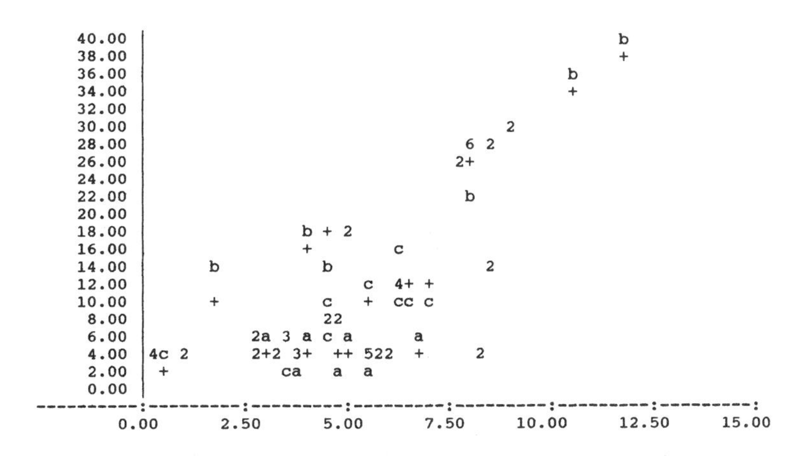

where the first three symbols in the plotting list 'abc' refer to the first *y*-variable (y) and the second three symbols '+++' to the second *y*-variable (the fitted values which are stored in the system vector %fv).

5.9 Restricted plotting

Often, we want to plot a sub-set of the data rather than all of it. In the present example, we might want to look at the graph for soil type 2 on its own. For this we use the %**re** directive, which stands for 'restrict plotting'.

```
$calc %re=(t==2) $
```

Note the use of the 'logical equals' function (see Section 3.8). Now,

```
$plot (y=0,40 x=0,15) y %fv x '*+' $
```

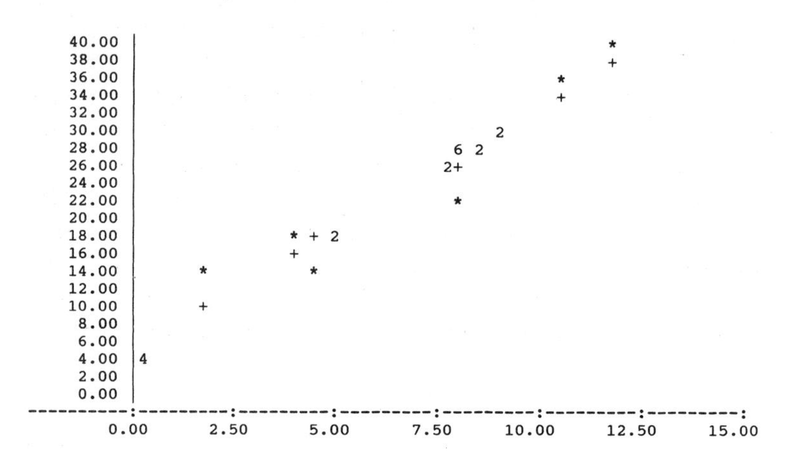

```
$stop $
```

●

5.10 Post-modelling plots

Up to this point, the plots have been of the kind we would do during data exploration. After the model has been fitted to the data there is a new set of plots to be inspected, which have to do with:

1 residuals;
2 error distribution;
3 leverage.

These are explained in detail in Chapters 13 and 18, but they use the same kind of **plot** directives as we have used so far.

5.11 Residual plots

Raw residuals are the differences between the observed and fitted values of the response variable:

```
$calc r=%yv−%fv $
```

Standardized residuals allow for the nature of the variance function, and for the magnitude of any prior weights (see Section 13.1). Residuals should be plotted in a number of ways.

1 Against the order of measurement (or against time), to look for systematic changes (like the experimenter becoming more experienced or becoming less attentive as time goes on).
2 Against the fitted values, to look for non-constant variance or for systematic inadequacy of the model (as when the model fits well at low values of *y* but poorly at high values).
3 Against each of the explanatory variables in turn; these may be non-linear functions of the response variable, in which case the fit might be improved by transformation of one or more of the explanatory variables (e.g. fitting log population density to the model rather than population density).
In each case, the hope is that the residuals will be distributed like the sky at night (as in Fig. 5.1a). Often, however, the residuals will show pronounced patterns of one sort or another, which are indicative that one or more of the assumptions about model structure or error distribution are wrong. The variance of the residuals may increase with *y*, giving a fan-shaped pattern like a V on its side (this is often dealt with by weighted regression; see Section 10.2). The size of the residuals may increase with *x* (this shows systematic failure of the model to account

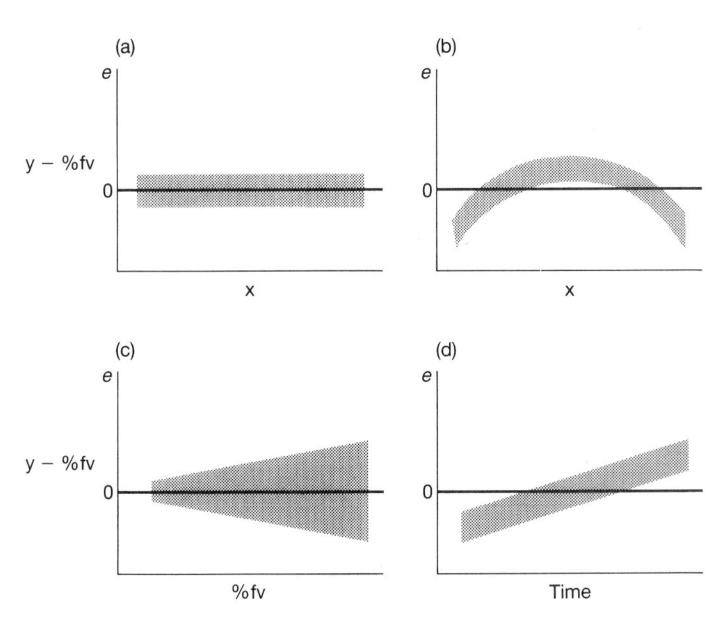

Fig. 5.1 Patterns of residuals. To test the assumptions that the residuals are normally distributed and the variance is constant, it is good practice to plot the residuals against the fitted values, the explanatory variables and the order of data collection. A fan-shape like a V on its side shows the residuals increasing in magnitude and suggests non-constant variance (c). Curved or n-shaped patterns suggest the need for transformation of one or more of the explanatory variables, or an alteration of the link function (b).

adequately for the values of y that are measured when x is large). Finally, the residuals may be a curvilinear function of x (this suggests that y is a non-linear function of x). These problems may be rectified by transformation, or by judicious changes to the link function, error distribution or weights, as described in Chapter 13.

5.12 Testing for normal errors

We begin by calculating 30 values for x from 1 to 30, then compute the straight line relationship $y = 2 + 3x$. Next, we generate 30 normally distributed, random errors with a standard deviation of 5.0, and add these on to the y-values. Then we carry out a regression of y1 against x:

```
$units 30 $
$calc x=%cu(1) $
$calc y=2+3*x $
$calc e=5*%nd(%sr(0)) $
$calc y1=y+e $
$yvar y1 $
$fit x $
```

Next, calculate the residuals, r, sort them and plot against the ordered standard normal residuals:

```
$calc r=%yv-%fv $
$sort r $
$calc n=%cu(1) $
$calc f=(n-0.375)/(%nu+0.25) $
$calc z=%nd(f) $
$plot r z '*' $
```

The simplest way to get a quick impression of the error distribution is to type:

```
$hist r '*' $
```

but this does not produce convincing tests with small sample sizes. A better test is obtained by plotting the ranked residuals against a cumulative normal distribution (see the macro written for this purpose in Section 3.12). If the line is curved, or has pronounced kinks in it, then the residuals are not normally distributed.

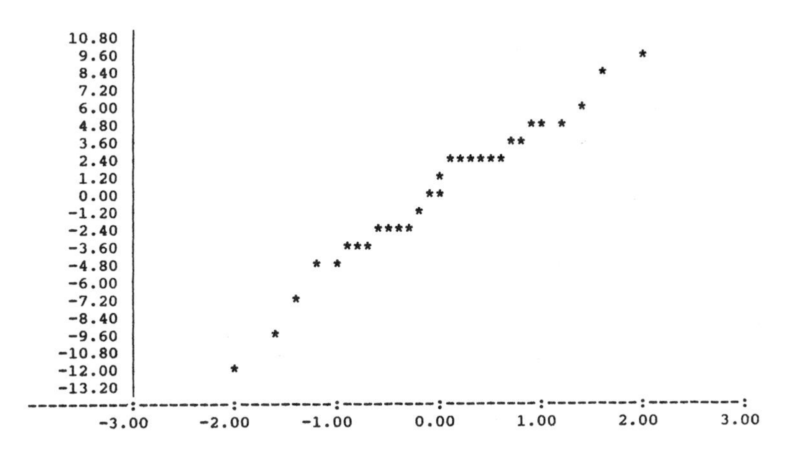

The line is reasonably straight, demonstrating that the errors were normal (as, indeed, we created them). Next, we create a set of non-normal errors:

$calc e2=(0.5−%sr(0))*3 $

Add these new errors to y to get a new variable called y2, and repeat the exercise:

$calc y2=y+e2 $

$yvar y2 $

[w] -- model changed

$fit x $

$calc r=%yv−%fv $

$sort r $

$plot r z '*' $

```
    1.600 |
    1.440 |                                             *   *
    1.280 |                                       * * *
    1.120 |
    0.960 |                                      *
    0.800 |                                    **
    0.640 |                                 ***
    0.480 |
    0.320 |                               **
    0.160 |
    0.000 |                             **
   -0.160 |                            **
   -0.320 |                           *
   -0.480 |
   -0.640 |                       ****
   -0.800 |                      *
   -0.960 |                    **
   -1.120 |                 * *
   -1.280 |         *   * *
   -1.440 |
   -1.600 |
----------:---------:---------:---------:---------:---------:---------:
        -3.00     -2.00     -1.00      0.00      1.00      2.00      3.00
```

This produces a distinctly S-shaped graph, demonstrating that the errors are not normally distributed. In particular, it shows that the largest negative residuals (in the bottom left-hand corner) are nowhere near large enough (the trend suggests that some residuals should be about −1.6). Similarly, the largest positive residuals are not large enough (there is none larger than 1.5). This is only to be expected, of course, since we generated the errors, and none of the residuals could be larger than 1.5 [i.e. $(0.5-0)\times 3$] or smaller than −1.5 [i.e. $(0.5-1)\times 3$].

5.13 What to do about non-normality

If inspection of the residual plots suggests non-normality, then we need to determine whether the non-normality was due to kurtosis (as in the previous example with rectangular errors) or to skewness (more likely, perhaps, in real ecological data). We may be able to specify the errors more precisely by using one of GLIM's own error structures (Poisson, binomial, gamma, etc.). Alternatively, we might try transforming the *y*-variable, say by taking logs, in order to improve the normality of the errors. At this stage, it is sufficient to be alerted to the fact that the errors might not be normal, and that plotting has an important role to play in checking for mis-specification of the error structure. These matters are discussed in detail in Section 13.3.

5.14 Leverage

Some points on a graph are more important than others. The most important points are often those at the extremes of the *x*-axis, i.e. the *y*-values associated with the very large and the very small values of *x*. The importance of individual points is further exaggerated if they lie a long way from their closest neighbouring values on the *x*-axis. An insidious fact is that *the most influential data often have the smallest residuals*, so inspection of the residuals is not enough. It is easy to see this with a simple example. Take a 'circle' of data showing no trend at all, say:

```
x 2 3 3 3 4
y 2 3 2 1 2
```

and then add a new datum at the right-hand end of the *x*-axis, say at the point (7,6). Now do a linear regression with and without this point, and look at the residuals.

```
$units 6 $

$assign x=2,3,3,3,4,7 $

$assign y=2,3,2,1,2,6 $

$plot (y=0,8 x=0,8) y x '*' $
```

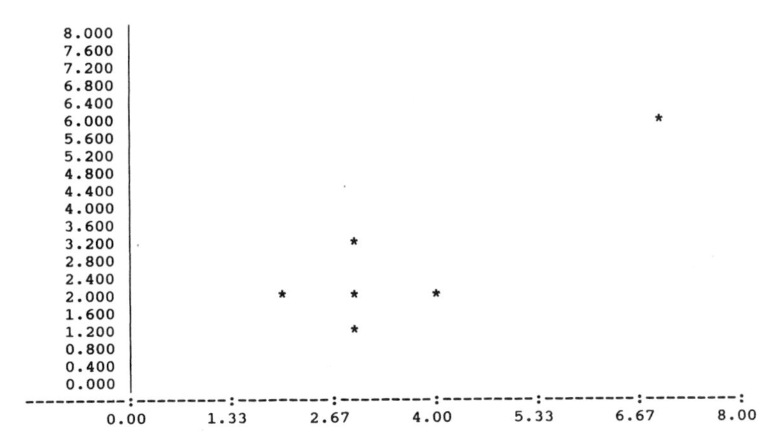

We determine the parameter estimates as follows:

```
$yvar y $

$fit x $

deviance = 3.7391
      d.f. = 4
```

and look at the parameter estimates:

```
$disp e $
```

	estimate	s.e.	parameter
1	−0.5217	0.9876	1
2	0.8696	0.2469	X

scale parameter taken as 0.9348

This shows a highly significant, positive effect of x on y, with a *t*-test of 3.52 (the slope of 0.8696 divided by its standard error of 0.2469), so the point at $x = 7$ is clearly of considerable influence. But let us look at the residuals, using the **disp r** directive:

```
$disp r $
```

unit	observed	fitted	residual
1	2.000	1.217	0.783
2	3.000	2.087	0.913
3	2.000	2.087	−0.087
4	1.000	2.087	−1.087
5	2.000	2.957	−0.957
6	6.000	5.565	0.435

Precisely because it is so influential, the point at unit 6 ($x = 7$) forces the regression line very close to it (in fact, it has the second *smallest* of all the residuals). Therefore, *looking at the size of the residuals is not a good way of assessing influence*. The statistical techniques for influence-testing are outlined by Atkinson (1985) and Cook & Weisberg (1982). It is good practice to check for influential values by doing selective deletions of points at the extreme ends of the axes of the explanatory variables. In the present example, we can use weighting to remove the value of y at x = 7, then refit the model, and compare the parameter values obtained. First calculate the weight vector, w, which has the value 1 for all units except those for which x is 7. These units get the value 0 (logical false):

```
$calc w=(x/=7) $

$weight w $

[w] -- model changed

$fit x $

deviance = 2.0000
d.f. = 3      from 5 observations

$disp e $

     estimate      s.e.     parameter
1  2.000         1.770    1
2  -8.327e-17  0.5774   X
scale parameter taken as 0.6667
```

The slope of −8.327e−17 is GLIM's attempt at zero. Removing the single, highly influential, point (7,6) has completely eliminated the relationship between y and x. The residuals now look like this:

```
$disp r $

unit  observed  fitted   residual
1     2.000     2.000     0.000
2     3.000     2.000     1.000
3     2.000     2.000     0.000
4     1.000     2.000    -1.000
5     2.000     2.000     0.000
6     6.000     2.000     0.000
```

Notice that the residual for the deleted observation is weighted to zero. When weights are employed to remove values temporarily from the working matrix, a residual of 0 appears in the row that has zero weight. The true residual for row 6 with the fitted value = 2 would obviously be $6-2=4$.

To obtain a warning about which points on the graph are likely to be highly influential we require a *leverage measure*. This is provided by:

$$h_i = \frac{1}{n} + \frac{(x_i - \bar{x})^2}{\Sigma(x_i - \bar{x})^2}$$

in which the denominator is *SSX* (see Section 7.5). Note that the leverage measure does not depend upon the numerical value of y; it depends only upon the relative isolation of the x-value. The value of h_i goes from a minimum of $1/n$ for x points that are exactly at $\bar{x}$, to higher values as x gets further away from the mean value of x. A value of h_i sufficiently large as to merit close attention is provided by the rule of thumb (Belsley *et al.*, 1980):

$$h_i > \frac{2p}{n}$$

where p is the number of parameters in the model. Thus for a two-parameter model of a relationship based on 10 points, we should think carefully about y-values that have x-values with leverage greater than $2 \times 2/10 = 0.4$.

A further use for leverage is in the analysis of residuals. Because outliers tend to force the model to pass close to them, they tend to have small residuals. This means that the variance of residuals is smaller for influential, outlying points. In particular,

$$\text{var } r_i = \sigma^2(1 - h_i)$$

This means that in order to find residuals with constant variance, it is necessary to weight them by leverage. This is how *standardized residuals* are calculated:

$$r_i' = \frac{r_i}{s\sqrt{(1 - h_i)}} = \frac{y_i - \hat{y}_i}{s\sqrt{(1 - h_i)}}$$

5.15 Deletion residuals

It is useful to know whether deletion of a single case, i, has an important effect on the model's parameter values. For example, we would like to know how close an observed y-value, y_i, is to the value of $\hat{y}_{(i)}$ that would be predicted by the model *if the value y_i was left out of the calculations*. If the point is not influential, the predicted value will be close to y_i. If y_i is influential, however, then removing it from the calculation will have a substantial effect on the parameter values of the model, and the predicted value will be a long way from y_i. The statistic r_i^*, called the *deletion residual*, is given by:

$$r_i^* = \frac{r_i}{s_{(i)}\sqrt{(1 - h_i)}} = \frac{sr_i'}{s_{(i)}}$$

The term $s_{(i)}$ is simply the standard deviation of the residuals, and r' is the standardized residual, calculated without the datum point (x_i, y_i).

Cook's statistic is an attempt to combine leverage and residuals in a single statistic. The absolute value of the deletion residual is weighted as follows:

$$C_i = |r_i^*| \left(\frac{n-p}{p} \cdot \frac{h_i}{1-h_i} \right)^{1/2}$$

A worked example is given in Exercise 5.1, showing how these leverage methods can be used in graphical analysis of data.

CHAPTER 6

Understanding data: basic statistics

GLIM contains an immensely powerful directive for carrying out basic statistical summaries of data. It is called the **tabulate** directive and is abbreviated to **tab** in this book (strictly, we could reduce it right down to **t**). Because it can do so many different things, the syntax of the **tab** directive takes a lot of getting used to. But you should persevere, because **tab** is probably the single most useful directive for preliminary (i.e. pre-modelling) analysis, and is vastly more flexible than the data-summary routines that are found in many other statistical packages.

6.1 Basic statistics

We can use the **read** directive to put 12 values into a variable y:

```
$units 12 $

$data y $

$read 3 4 6 4 5 2 4 5 1 5 4 6 $
```

6.1.1 Means and variances

An important preliminary to any statistical modelling is to obtain a thorough understanding of the properties of the variables we intend to analyse. For example, to obtain the mean value of the variable called y, we simply type:

```
$tab the y mean $
```

and obtain the answer 4.083. Similarly, to find out the variance of y we just type:

```
$tab the y variance $
```

after which GLIM prints 2.265; or the standard deviation:

```
$tab the y deviation $
```

which is 1.505. Other statistical descriptors can be obtained by using the **calc** directive after a set of **tab** directives. For example, the variance/mean ratio is 0.5547:

```
$calc 2.265/4.083 $
```

and the percentage coefficient of variation (100 × standard deviation/mean) is 36.86%:

```
$calc 100*1.505/4.083 $
```

6.1.2 *Percentiles of various kinds*

GLIM has a built-in 50 percentile command called **fifty** which prints the median value of the data:

```
$tab the y fifty $

4.000
```

but there is great flexibility because you can specify your own percentiles like this:

```
$tab the y percentile 95 $

6.000

$tab the y percentile 75 $

5.000

$tab the y percentile 25 $

3.500
```

This means that the central 50% of the data lie between 3.5 and 5.0. You should check the raw data carefully to ensure that you understand what GLIM has done in working this out (e.g. what is GLIM's policy on ties? what does the difference between an ending of 0.5 and 0.0 mean?).

6.1.3 *Maximum and minimum values*

It is extremely straightforward to obtain the maximum and minimum values of a vector. **Tab** uses the words **largest** and **smallest** for this (it cannot use maximum and minimum because they both begin with 'm' and since all these terms can be abbreviated down to their first letter, **m** has been reserved for **mean**).

```
$tab the y largest $

6.000

$tab the y smallest $

1.000
```

Using the maximum degree of abbreviation, and packing three directives on a line, we could find the mean, maximum and minimum of any data set just by typing:

```
$t t y m $t t y l $t t y s $
```

and GLIM would respond by printing:

```
4.083

6.000

1.000
```

6.2 Factors

Suppose that our 12 values of y can be classified by two factors. Soil type (A) has four levels and light intensity (B) has three. We assign the factor levels using %gl (see Section 3.7):

```
$calc a=%gl(4,3) $

$calc b=%gl(3,1) $

$factor a 4 b 3 $
```

or in GLIM 4 we could save two lines by writing:

```
$gfactor a 4 b 3 $
```

Our data matrix now contains the *y*-variable in column 1 with the two factors, A and B, in columns 2 and 3.

```
$look y a b $

        Y      A      B
 1  3.000  1.000  1.000
 2  4.000  1.000  2.000
 3  6.000  1.000  3.000
 4  4.000  2.000  1.000
 5  5.000  2.000  2.000
 6  2.000  2.000  3.000
 7  4.000  3.000  1.000
 8  5.000  3.000  2.000
 9  1.000  3.000  3.000
10  5.000  4.000  1.000
11  4.000  4.000  2.000
12  6.000  4.000  3.000
```

Notice how the subscript of factor B repeats within each level of factor A. Thus, the eighth row of the data matrix contains a y-value of 5 that was obtained with the third level of factor A and the second level of factor B.

Using factors makes many GLIM statements much more general. For example, we can organize the *y*-variable in different ways using the **tprint** directive:

```
$tprint y a;b $

B 1      2      3
A
1 3.000  4.000  6.000
2 4.000  5.000  2.000
3 4.000  5.000  1.000
4 5.000  4.000  6.000
```

Notice that the factor which appears first in the factor list (A;B) determines the rows and the second factor the columns. It is vital that the product (number of levels of A times the number of levels of B) is equal to the declared **units** of y (no table would be printed if A*B was not equal to 12 in this example).

The **tab** directive also becomes much more general by the incorporation of factors. We can obtain the different mean values of y for each level of A by writing:

```
$tab the y mean for a $
```

and GLIM prints:

```
     1      2      3      4
[ ]  4.333  3.667  3.333  5.000
```

or for factor B:

```
$tab the y mean for b $

     1      2      3
[ ]  4.000  4.500  3.750
```

where the square brackets [ ] just mean that the row is unlabelled.

6.3 Comparing variances

A very important part of the preliminary data analysis involves the comparison of the variances of y under different treatment combinations:

```
$tab the y var for a $

     1      2      3      4
[ ]  2.333  2.333  4.333  1.000

$tab the y var for b $

     1       2       3
[ ]  0.6667  0.3333  6.9167
```

This shows that the variance is unusually high in the third level of treatment A and in the third level of treatment B. This prompts us to look back at the raw data more critically. The value in y(3,3) is 1, much lower than the other values in the third row. Notice, also, that the third column B(3) contains the two largest and the two smallest values of y. These facts would need to be borne in mind in subsequent, more detailed, analysis.

A macro for carrying out Bartlett's test for equality of variances is demonstrated in Section 19.11.

6.4 Weighted statistics (tab with)

Assume that we have a vector *n* containing the sample sizes on which the values of *y* were based.

```
$data n $

$read 12 13 10 10 11 11 14 13 10 14 12 11 $
```

showing that the sample sizes vary between 10 and 14. Samples with higher replication are given more weight than samples with low replication, as we can see by calculating the overall weighted average using the **with** option:

```
$tab the y mean with n $

4.113
```

as compared with the unweighted average of 4.083 (above). We can also calculate weighted group means:

```
$tab the y mean with n for a $

      1      2      3      4
[ ]  4.229  3.656  3.541  4.973
```

as compared with the unweighted group means of 4.333, 3.667, 3.333 and 5.0.

6.5 Contingency tables (tab for)

The **tab** directive can also be used to create contingency tables. As so often, its simplicity belies its power. For tables of counts, we drop the **the** and use **for**, as follows:

```
$tab for n $

     10  11  12  13  14
[ ]  3   3   2   2   2
```

which counts how many values there are with each level of replication. This directive is extremely valuable with large, unbalanced data sets from observational studies, where we do not know how many cases there are of any given factor. Missing cells in the classifying table are filled with zeros.

This is particularly useful when we have tables with lots of zeros but we want to save data entry effort by typing in only the non-zero values.

Another use of **tab with** is to work out totals. For example, if we want to know the total of all the y-values at each level of A we can write:

```
$tab with y for a $

        1       2       3       4
[ ]  13.00   11.00   10.00   15.00
```

This is synonymous with $tab the y total for a $.

6.6 Saving summary tables (tab into)

A very important feature of GLIM is its ability to store summary tables for subsequent statistical analysis. The new commands we need are **into**, which names the variable where the new table will be stored, and **by**, which gives names to the classifying vectors. These need new names because the old factors will be too long. Because it is a summary table, the new table will have fewer than %nu units, and its classifying vectors will need to be the same length as the new table. Thus,

```
$tab the y mean for a into ym by a1 $
```

saves the means of y for each level of A into a new table called ym, classified by the factor A1. If we want to see the contents of the new table, we use **tprint** as follows:

```
$tprint ym a1 $

A1    1       2       3       4
YM  4.333   3.667   3.333   5.000
```

Note that if you tried to use $tprint ym a $ you would get an error message because the units of ym and A are inconsistent (ym has length 4 while A has 12). We can save a table of weighted means in exactly the same way:

```
$tab the y mean for a with n into wym by a1 $
```

These summary tables can by cut from glim.log and pasted directly into reports or documents in your word-processor. For high-quality graphs and histograms, you can export an ASCII file containing the summary data on y and x, and then import the numbers into a specialist graphics package to draw a publication-quality picture.

6.7 Tables with multiple entries in each cell (tprint)

In writing reports we often need tables with several values written in each cell. For example, we might want to print a table containing the mean, standard deviation and total sample size for each level of the factor A.

Multiple tables are assembled in two stages. First, we use a series of **tab into** directives to save all the necessary summary statistics and generate the factor levels. For practice, we shall use the fully abbreviated directives to calculate the mean m, standard deviation d and total sample size nt classified by am as follows:

```
$t t y m f a i m b am $

$t t y d f a i d b am $

$t t n t f a i nt b am $
```

Then we use **tprint** with the names of all the vectors to be printed in each cell separated by semicolons, like this:

```
$tp m;d;nt am $
```

and GLIM will print a table with three values in each cell:

AM	1	2	3	4
M	4.333	3.667	3.333	5.000
D	1.528	1.528	2.082	1.000
NT	35.00	32.00	37.00	37.00

The same method will work for multidimensional tables, but in that case the classifying factors also make a list separated by semicolons in the **tprint** directive. For example,

```
$tprint x;y;z a;b;c;d $
```
○

would produce a number of tables (depending upon the number of levels of A and B) with three numbers (x, y and z) in each cell. ●

6.8 Frequency distributions (tab for)

We shall use a different data set for this part of the exercise. This shows counts of mines per leaf for three trees at each of four sites (a total of 1440 leaves in all).

```
$units 1440 $

$data mines $

$dinput 6 $

File name? glex25.dat
```

An extremely valuable data summary is obtained with the **tab for** directive. This just lists the distinct values of a variable and counts how many times each different value appears in the data set. Thus, to get a summary of our leaf mine data we type:

```
$tab for mines $
```

and GLIM responds with:

```
        0.000     1.000     2.000    3.000    4.000  5.000
[ ]   735.000   488.000   145.000   40.000   29.000  3.000
```

which means that 735 of the leaves had no mines, and the maximum number of mines per leaf was five (three leaves had five mines).

6.9 Saving frequency tables (tab using)

In order to create new vectors containing the frequencies of different values of the numbers of mines per leaf, we employ the **tab using** directive:

```
$tab for mines using freq by value $
```

where **using** defines a new variable to contain the counts (freq) and **by** defines a new variable that will contain the different *y*-values (value). To view the result we would write:

```
$tprint freq value $
```

which gives the output:

```
VALU      0.000     1.000     2.000    3.000    4.000  5.000
FREQ    735.000   488.000   145.000   40.000   29.000  3.000
```

Notice how the **tab using** directive has automatically produced two summary vectors: the distinct values of mines per leaf (in value) and the frequency of occurrence of each value (in freq).

6.10 Histograms

We could obviously use data from the output of the **tab for** directive to draw a histogram by hand. But GLIM can produce its own histograms very simply with the **hist** directive:

```
$hist mines $
```

The output is more readily understood if we customize the axes. The trick is to specify the bars to be printed in the **ylimit** part of the directive and the number of bars in the **rows** part. So, if we want a bar for each of the values 0 through 6 (so that we have a zero at the right-hand end of the plot) we specify y=0,6 and, since this is 7 bars, we specify rows=7 like this:

```
$hist (y=0,6 rows=7 style=1) mines '*' $
```

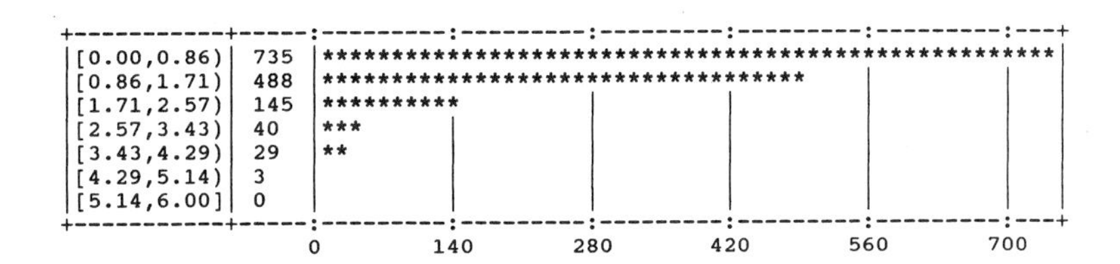

style=1 puts a box around the histogram and '*' changes the plotting symbol for the bars (the default is the first letter of the variable name of the vector containing the values; m in this case).

The rather odd-looking numbers on the left are the class intervals that GLIM has worked out, based on the **ylimit** and **rows** information we gave in the **hist** directive. Square left-hand brackets, [, mean *from and including* while round brackets, (, mean *from, but not including*. Right-hand square brackets,], mean *to and including* while right-hand round brackets,), mean *to, but not including*. If you had been working out the class boundaries for integer data you would probably have made the mid-points 0.5's and the intervals 1.0, so that the first bar included information on the 0's, that is to say the range [−0.5,0.5), and the last bar the 6's [5.5,6.5]. Can you see why GLIM's numbers are a bit odd?

If you specify the option **tails** = 1,1 in the list you will get the frequencies in the left- and right-hand tails of your distribution (i.e. from minus infinity up to your minimum value, and from your maximum value up to plus infinity). **Tails** = 1,0 would give you just the left-hand tail and 0,1 just the right-hand tail.

The histogram directive is especially useful when investigating mortality data. For data on time-to-death, t, the command:

```
$hist t $
```

gives a visual impression of the density function, and shows the importance of censoring (see Section 18.9).

CHAPTER 7

Regression

To demonstrate how the statistical part of GLIM works, and to show what kind of output it produces, we begin with two familiar statistical techniques: regression and analysis of variance. The aim is to work through a series of simple examples in sufficient detail that the calculations carried out within GLIM become apparent. In all these examples, we assume that the errors are normally distributed, and that the variances are homogeneous. Later on, we shall deal with cases that have non-normal errors and unequal variances.

7.1 The model

We begin with the simplest possible linear model: the straight line,

$$y = a + bx$$

where a *response variable* y is hypothesized as being a linear function of the *explanatory variable* x (sometimes called an independent variable, but this is unfortunate, since it need be neither independent nor a variate), and the two *parameters* a and b. In the case of simple linear regression, the parameter a is called the *intercept* (the value of y when $x = 0$), and b is the *slope* of the line (or the gradient, measured as the change in y in response to unit change in x).

Any errors are assumed to be confined to y, to be normally distributed, and to be independent of the level of x. Many ecological data are not like this at all; the errors are non-normal, the variance in y increases with the mean value of y, and there are also errors in x. We ignore these difficulties for the time being. Each of them is considered in detail in later chapters.

The aims of the analysis are as follows:

1 to estimate the values of the parameters a and b;

2 to estimate their standard errors;

3 to use the standard errors to assess which terms are necessary within the model (i.e. whether the parameter values are significantly different from zero);

4 to determine what fraction of the variation in y is explained by the model (the coefficient of determination, r^2) and how much remains unexplained (so called error variance).

7.2 Data inspection

The first step is to look carefully at the data. Is there an upward or downward trend, or could a horizontal straight line be fitted through the data? If there is a trend, does it look linear or curvilinear? Is the scatter of the data around the line more or less uniform, or does the scatter change systematically as x changes?

Consider the data in Fig. 7.1. They show how weight gain (mg) of individual caterpillars declines as the tannin content of their diet (%) increases. It looks as if there is a downward trend in y as x increases, and that the trend is roughly linear. There is no evidence of any systematic change in the scatter as x changes. This cursory inspection leads to several expectations:

1 the intercept a is greater than zero;
2 the slope b is negative;
3 the variance in y is constant;
4 the scatter about the straight line is relatively small.

It now remains to carry out a thorough statistical analysis to substantiate or refute these initial impressions, and to provide accurate estimates of the slope and intercept, their standard errors and the degree of fit.

7.3 Least squares

The technique of least squares linear regression defines the *best fit* straight line as the line that minimizes *the sum of the squares of the departures of the y-values from the line*. We can see what this means in graphical terms. The first step is to fit a horizontal line through the data, showing the average value of y (Fig. 7.2a). The scatter around the line defined by $\bar{y}$ is said to be the total variation in y. Each point on the graph lies a vertical distance d from the horizontal line ($d = y - \bar{y}$), and we define the total variation in y as being the sum of the squares of these departures:

$$SST = \Sigma(y - \bar{y})^2 \tag{7.1}$$

where SST stands for *total sum of squares*. Now fit a straight line through the data (Fig. 7.2b). There are two decisions to be made about such a best-fit line.

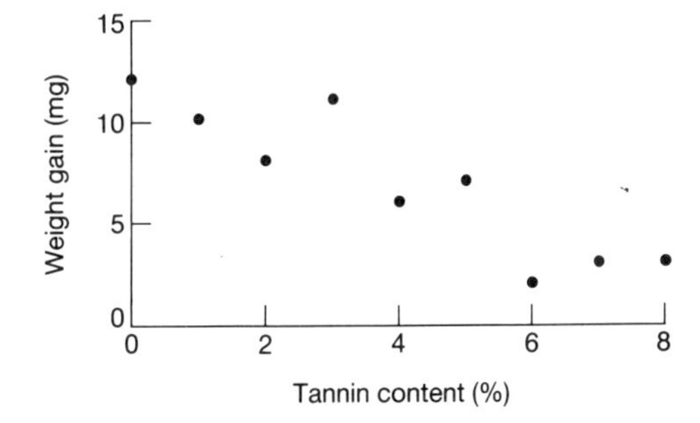

Fig. 7.1 Caterpillar weight gain (mg) as a function of dietary tannin (%). An example of linear regression (for calculations, see text).

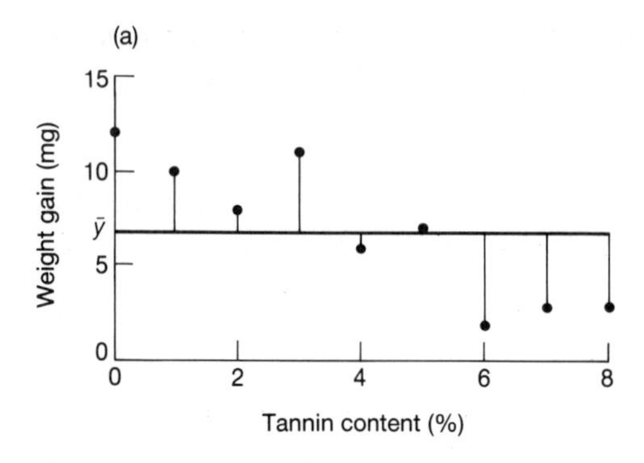

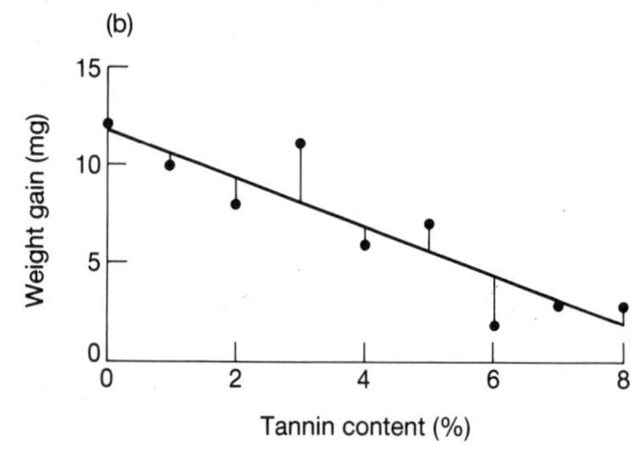

Fig. 7.2 Regression sums of squares. (a) The squared deviations from the overall average value of y give the total sum of squares, SST. (b) The squared deviations from the regression line $y = a + bx$ give the error sum of squares, SSE.

1 Where should the line be located?
2 What slope should it have?
Location of the line is straightforward, because a best-fit line should clearly pass through the point defined by the average values of x and y $(\bar{x},\bar{y})$. The line can then be pivoted at this point, and rotated until the best fit is achieved. The process is formalized as follows. Each point on the graph lies a distance $e = y - \hat{y}$ from the fitted line, where the predicted value $\hat{y}$ is found by evaluating the equation of the straight line at the appropriate value of x:

$$\hat{y} = a + bx$$

The location of the line is fixed by assuming that it passes through the point $(\bar{x},\bar{y})$, so that we can rearrange the equation to obtain an estimate of the intercept a in terms of the best-fit slope, b:

$$a = \bar{y} - b\bar{x} \tag{7.2}$$

We begin by replacing the average values of y and x by $\Sigma y/n$ and $\Sigma x/n$:

$$a = \frac{\Sigma y}{n} - b\frac{\Sigma x}{\text{n}} \tag{7.3}$$

The *best-fit* slope is found by rotating the line until the *error sum of squares*, *SSE*, is minimized. The error sum of squares is the sum of squares of the individual departures, e, shown in Fig. 7.2b:

$$SSE = \min \Sigma e^2$$

which, replacing each departure, e, by the distance between the measured y and the predicted $\hat{y} = a + bx$, becomes:

$$SSE = \min \Sigma(y - a - bx)^2 \quad (7.4)$$

noting the change in sign of bx.

This is how the best fit is defined. All that remains is to discover how to calculate the value of b that minimizes this function. The method is a simple application of calculus. We find the derivative of equation (7.4) with respect to b, then set this to zero, and find what value of b minimizes the function.

$$\frac{dSSE}{db} = \Sigma - 2x(y - a - bx)$$

because the derivative with respect to b of the bracketed term in equation (7.4) is $-x$, and the derivative of the squared term is 2 times the squared term. Thus, taking the constant -2 outside the summation, and multiplying through the bracketed term by x, gives:

$$\frac{dSSE}{db} = -2\Sigma(xy - ax - bx^2) \quad (7.5)$$

Now take the summation of each term separately, set the derivative to zero, and divide both sides by -2 to remove the unnecessary constant:

$$\Sigma xy - \Sigma ax - \Sigma bx^2 = 0$$

We cannot solve the equation as it stands because there are two unknowns, a and b. However, we already know the value of a in terms of b from equation (7.3). Also, note that Σax can be written as $a\Sigma x$; so, replacing a by equation (7.3) and taking a and b outside the summations gives:

$$\Sigma xy - \left(\frac{\Sigma y}{n} - b\frac{\Sigma x}{n}\right)\Sigma x - b\Sigma x^2 = 0$$

Now multiply out the central bracketed term by Σx to get:

$$\Sigma xy - \frac{\Sigma x \Sigma y}{n} + b\frac{(\Sigma x)^2}{n} - b\Sigma x^2 = 0$$

Finally, take the two terms containing b to the other side, and note their change of sign:

$$\Sigma xy - \frac{\Sigma x \Sigma y}{n} = b\Sigma x^2 - b\frac{(\Sigma x)^2}{n}$$

and then divide both sides by $\Sigma x^2 - (\Sigma x)^2/n$ to obtain the required estimate b:

$$b = \frac{\Sigma xy - \dfrac{\Sigma x \Sigma y}{n}}{\Sigma x^2 - \dfrac{(\Sigma x)^2}{n}} \tag{7.6}$$

Thus, the value of b that minimizes the sum of squares of the departures in Fig. 7.2b is given simply by:

$$b = \frac{SSXY}{SSX} \tag{7.7}$$

where $SSXY$ stands for the sum of the products x times y (the measure of how x and y co-vary), and SSX is the sum of squares for x, calculated in exactly the same manner as the total sum of squares SST, which we met in equation (7.1).

For comparison, these three important formulas are presented together:

$$\begin{aligned} SST &= \Sigma y^2 - \frac{(\Sigma y)^2}{n} \\ SSX &= \Sigma x^2 - \frac{(\Sigma x)^2}{n} \\ SSXY &= \Sigma xy - \frac{\Sigma x \Sigma y}{n} \end{aligned} \tag{7.8}$$

Note the similarity of their structures. SST is calculated by adding up y times y then subtracting the total of the y's times the total of the y's divided by n (the number of points on the graph). Likewise, SSX is calculated by adding up x times x then subtracting the total of the x's times the total of the x's divided by n. Finally, $SSXY$ is calculated by adding up x times y then subtracting the total of the x's times the total of the y's divided by n. Note that these formulas are not suitable for general computation, because they can suffer from severe rounding errors for very large or very small values of x or y.

It is worth reiterating what these three quantities represent: SST measures the total variation in the y-values about their mean (the sum of the squares of the d's in Fig. 7.2a); SSX represents the total variation in x (expressed as the sum of squares of the departures from the mean value of x), and is a measure of the range of x values over which the graph has been constructed; $SSXY$ measures the correlation between y and x in terms of the corrected sum of products ($SSXY$ is negative when y declines with increasing x, positive when y increases with x, and zero when y and x are uncorrelated).

7.4 Significance testing

Tests of significance are based on the *standard errors* of the parameters. These, in turn, depend upon the *error variance* and the *degrees of freedom*.

7.4.1 Standard error of the slope

Our confidence in the estimated value of the slope will be high when:

1 replication is high;

2 the maximum and minimum values of x are widely spaced (large SSX);

3 the error variance is low.

These considerations are combined in the formula for the standard error of the slope. Calculation of the standard error will be achieved in stages, but we begin with a graphical impression. Look at the relative sizes of the departures d and e in Figs 7.2a and 7.2b. Now, ask yourself what would be the relative size of Σd^2 and Σe^2 if the slope of the fitted line were *not significantly different from zero*? A moment's thought should convince you that if the slope of the best-fit line were zero, then the two sums of squares would be the same. If the slope were exactly zero, then the two lines would lie in exactly the same place, and the sums of squares would be identical. Thus, when the slope of the line is not significantly different from zero, we would find that $SST = SSE$.

Similarly, if the slope was significantly different from zero (i.e. significantly positive or negative), then SSE would be substantially *less* than SST. In the limit, if all the points fell exactly on the fitted line, then SSE would be zero.

Now we calculate a third quantity, SSR, called the *regression sum of squares*:

$$SSR = SST - SSE$$

This definition means that SSR will be large when the fitted line accounts for much of the variation in y, and small when there is little or no lincar trend in the data. In the limit, SSR would be equal to SST if the fit was perfect (because SSE would then equal zero), and SSR would equal zero if y was independent of x (because, in this case, SSE would equal SST).

These three quantities form the basis for drawing up the ANOVA (Table 7.1). The sums of squares are entered in the first column. SST is the corrected sum of squares of y, as given above. While we have defined SSE in equation (7.4), this formula is inconvenient for calculation, since it involves n different estimates of $\hat{y}$ and n subtractions. Instead, it is easier to calculate SSR and then to estimate SSE by difference. The regression sum of squares is simply:

$$SSR = b.SSXY \qquad (7.9)$$

Table 7.1 The ANOVA table for linear regression

Source	SS	d.f.	MS	F-ratio	Probability
Regression	SSR	1	SSR	SSR/s^2	Tables
Error	SSE	$n-2$	$s^2 = \frac{SSE}{n-2}$		
Total	SST	$n-1$			

so that:

$$SSE = SST - SSR \tag{7.10}$$

The second column of Table 7.1 contains the degrees of freedom. The estimation of the total sum of squares required that one parameter, the mean value of y, be estimated from the data prior to calculation. Thus, the total sum of squares has $n-1$ degrees of freedom when there are n points on the graph. The error sum of squares could not be estimated until the regression line had been drawn through the data. This required two parameters, the mean value of y and the slope of the line. Thus, the error sum of squares has $n-2$ degrees of freedom. The regression consists of the single estimated parameter, b, and therefore has one degree of freedom.

The third column contains two variances: the first row shows the regression variance and the second row the error variance. As usual, these variances are simply sums of squares divided by degrees of freedom. One of the oddities of analysis of variance is that the variances are referred to as *mean squares* (this is because a variance is defined as a mean squared deviation). The error variance, s^2 (also known as the *error mean square*, *MSE*), is the quantity used in calculating standard errors and confidence intervals for the parameters, and in carrying out hypothesis testing.

The standard error of the regression slope, b, is given by:

$$SE_b = \sqrt{\frac{s^2}{SSX}} \tag{7.11}$$

The standard error increases with the error variance, and decreases as the range of values over which x was measured is increased. This makes good intuitive sense; the larger the range of x values covered by the data, the greater will be our confidence in the estimate of the slope.

A 95% confidence interval for b is now given in the usual way:

$$CI_b = t_{\text{tables}} SE_b \tag{7.12}$$

where t is obtained from tables with $\alpha = 0.025$ (two-tailed) and $n-2$ degrees of freedom.

The standard error of the intercept, a, is given by:

$$SE_a = \sqrt{\frac{s^2 \, \Sigma x^2}{n \; SSX}} \tag{7.13}$$

which is like the formula for the standard error of the slope, but with two additional terms. Confidence in predictions made with linear regression declines with the square of the distance between the mean value of x and the value at which the prediction is to be made, i.e. with $(x-\bar{x})^2$. Thus, when the origin of the graph is a long way from the mean value of x, the standard error of the intercept will be large, and vice versa. As in the formula for the standard error of a mean, the standard error of the intercept declines as the number of points on the graph, n, increases.

A 95% confidence interval for the intercept, therefore, is:

$$CI_a = t_{\text{tables}} SE_a \tag{7.14}$$

with the same two-tailed t-value as before.

In general, for any predicted value of $\hat{y}$, the standard error is given by:

$$SE_{\hat{y}} = \sqrt{s^2\left(\frac{1}{n} + \frac{(x - \bar{x})^2}{SSX}\right)} \tag{7.15}$$

and the formula for the standard error of the intercept is just the special case of this for $x = 0$ (you should check the algebra of this result as an exercise).

An important significance test concerns the question of whether or not the slope of the best-fit line is significantly different from zero. If it is not, then the principle of parsimony suggests that the line should be assumed to be horizontal. If this were the case, then y would be independent of x (it would be a constant, $y = a$).

7.5 An example of linear regression

Consider the data in Fig. 7.1, which show how weight gain (mg) of individual caterpillars declines as the tannin content of their diet (%) increases. You will find it useful to work through the calculations longhand as we go. The raw data, rounded for computational convenience are:

Tannin % (x)	0	1	2	3	4	5	6	7	8
Weight gain (mg)	12	10	8	11	6	7	2	3	3

Obviously, you would not round the data to whole numbers if the analysis were in earnest. The first step is to compute the five sums: $\Sigma X = 36$, $\Sigma X^2 = 204$, $\Sigma Y = 62$, $\Sigma Y^2 = 536$ and $\Sigma XY = 175$. Note that ΣXY is $0 \times 12 + 1 \times 10 + 2 \times 8 \ldots + 8 \times 3 = 175$. Now calculate the three corrected sums of squares, using the formulas in equation (7.8):

$$SST = 536 - \frac{62^2}{9} = 108.889$$

$$SSX = 204 - \frac{36^2}{9} = 60$$

$$SSXY = 175 - \frac{36 \times 62}{9} = -73$$

then the slope of the best-fit line is simply:

$$b = \frac{SSXY}{SSX} = -1.21666$$

and the intercept is:

$$a = \bar{y} - b\bar{x} = 11.755$$

where the mean value of y is $62/9 = 6.889$ and the mean value of x is $36/9 = 4.0$. The regression sum of squares is:

$$SSR = b.SSXY = -1.2167 \times -73 = 88.82$$

so that the error sum of squares can be found by subtraction:

$$SSE = SST - SSR = 108.89 - 88.82 = 20.07$$

Now we can complete the ANOVA table as shown in Table 7.2. The calculated F-ratio of 30.98 is much larger than the 1% value in tables with one degree of freedom in the numerator and seven degrees of freedom in the denominator ($F = 12.25$), so we can unequivocally reject the null hypothesis that the slope of the line is zero. Increasing dietary tannin leads to significantly reduced weight gain for these caterpillars. Recall that the F table is used by looking up the *column* headed by the degrees of freedom in the numerator (i.e. the first column in this case) and the *row* number defined by the degrees of freedom in the denominator (i.e. the seventh row in this case).

We can now calculate the standard errors of the slope and intercept using the formulas in equations (7.11) and (7.13):

$$SE_b = \sqrt{\frac{2.867}{60}} = 0.2186$$

$$SE_a = \sqrt{\frac{2.867 \times 204}{9 \times 60}} = 1.041$$

Table 7.2 The completed ANOVA table for linear regression

Source	SS	d.f.	MS	F-ratio	Probability
Regression	88.82	1	88.82	30.98	0.0008
Error	20.07	7	2.867		
Total	108.89	8			

The two-tailed value of Student's t from tables of 2.365 is found in the column headed by $\alpha = 0.025$, and the row number 7 (the error degrees of freedom), so the confidence intervals for the slope and intercept are given by:

$$CI_b = 2.365\sqrt{\frac{2.867}{60}} = 0.470$$

$$CI_a = 2.365\sqrt{\frac{2.867 \times 204}{9 \times 60}} = 2.463$$

and we could write our parameter estimates for a and b as follows:

$$a = 11.756 \pm 2.463$$

$$b = -1.217 \pm 0.470$$

The numbers give the parameter estimates and their 95% confidence intervals.

7.6 Degree of scatter

So far we have estimated the equation of the best-fit line, but made no explicit statement concerning the degree of scatter about the line. As Fig. 7.3 shows, it is possible to have data sets with identical best-fit equations, but with very different degrees of scatter. It is clearly not sufficient to state only what the best-fit equation is, and we need also to say what fraction of the variance in y is explained by the model. In Fig. 7.3a almost all the variation in y is explained by the best-fit line, while in Fig. 7.3b the proportion of the variation explained is very low. If the fit were perfect, then all the data points would fit exactly on the line, and 100% of the variation in y would be explained (Fig. 7.3c). If the fit were completely hopeless, then the regression line would be horizontal, and none of the variation in y would be explained by the model (Fig. 7.3d).

Formulated in this way, it becomes clear that we could use some of the quantities already calculated to derive an estimate of scatter. In particular, we require our measure to vary from 1.0 when the fit is perfect, to zero when there is no fit at all. Now, recall that the total variation in y is measured by SST. Our question can be re-expressed to

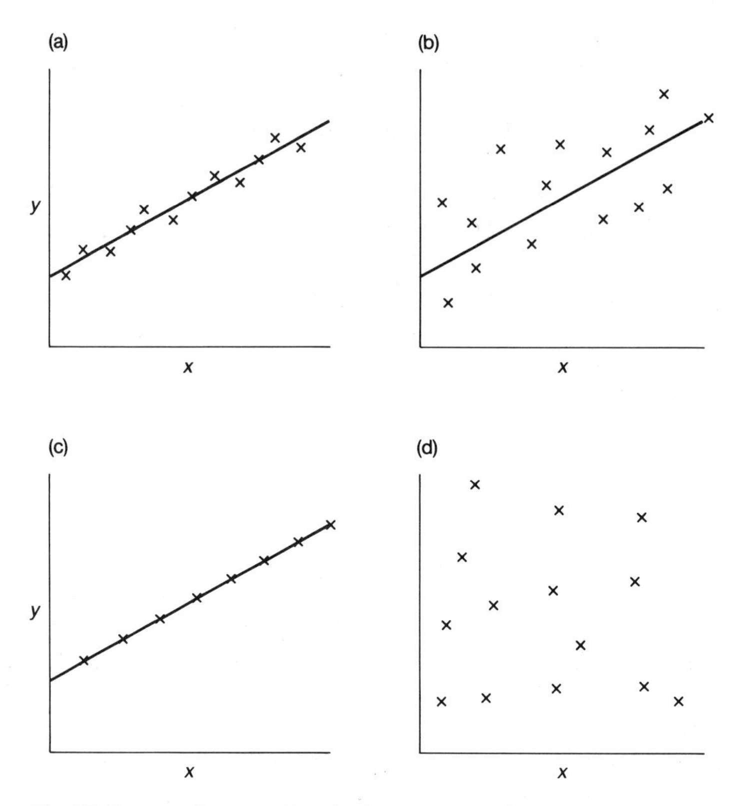

Fig. 7.3 Degree of scatter. Exactly the same regression equations can describe very different data sets: (a) degree of scatter low (r^2 is high); (b) high scatter (r^2 is low); (c) a perfect fit with no scatter at all ($r^2 = 1$); (d) scatter so great that there is no significant relationship between y and x ($r^2 = 0$). The coefficient of determination, r^2, measures the fraction of the total sum of squares in y that is explained by the regression line (see text for details).

ask: what fraction of SST is explained by the regression line? We defined the residual variation, once the line had been fit, as SSE. Thus, if the fit is perfect $SSE = 0$ and if it is hopeless, then $SSE = SST$. Since the regression sum of squares, SSR, is defined as $SST - SSE$, we note that $SSR = SST$ when the fit is perfect, and $SSR = 0$ when the fit is hopeless. A measure of scatter, therefore, is given simply as the ratio of SSR to SST. This ratio is called the *coefficient of determination* and is denoted by r^2:

$$r^2 = \frac{SSR}{SST} \tag{7.16}$$

where its square root, r, is the familiar *correlation coefficient*. Note that because r^2 is always positive, it is better to calculate the correlation coefficient from:

$$r = \frac{SSXY}{\sqrt{SSX.SST}} \quad (7.17)$$

You should check that these definitions of r and r^2 are consistent.

7.7 Using GLIM for regression

The procedure so far should have been familiar. Let us now use GLIM to carry out the same regression analysis. At this first introduction to the use of GLIM each step will be explained in detail. As we progress, we shall take progressively more of the procedures for granted. There are six steps involved in the analysis of a linear model:

1 get to know the data;
2 suggest a suitable model;
3 fit the model to the data;
4 subject the model to criticism;
5 analyse for influential points;
6 simplify the model to its bare essentials.

With regression data, the first step involves the visual inspection of plots of the response variable against one or more explanatory variables using the **plot** directive explained in Section 5.2. We need answers to the following questions.

1 Is there a trend in the data?
2 What is the slope of the trend (positive or negative)?
3 Is the trend linear or curved?
4 Is there any pattern to the scatter around the trend?

7.7.1 Data input

First, there are some formalities: (i) we must define the number of data points using the **units** directive; (ii) we must name the variables; and (iii) we need to read the data from an external text file.

The first two statements in any GLIM analysis should be the **units** and **data** directives, because between them they define the structure of the *data matrix*. **Units** is the number of rows (the number of points on the graph) and **data** defines the number of columns (the variables to be considered in the analysis are defined simply by declaring their names). For our regression example, we write:

```
$units 9 $

$data x y $
```

to show that the data matrix contained 18 elements (nine rows and two columns). Because the **data** directive has two entries, and they are in the order x, y, the computer will read numbers into the data matrix one row at a time, putting the first number into x(1), the second into y(1), the third into x(2), and so on until the 18th number is placed into y(9). The GLIM directive that reads the data from your prepared data file is:

```
$dinput 6 $
```

where 6 is the *channel number* (see Section 3.6.4). When the computer executes this directive, it will ask for the name of the ASCII file containing the data. You type in the name of the file, including the drive and path name if necessary, for example:

```
File name? glex37.dat
```

which will work so long as the *file* called glex37.dat is in the same *directory* as GLIM on *drive* C. If the file is on a diskette, you will need to type a:glex37.dat at the file name prompt.

If the machine produces an error message during its attempt to read data from file you will need to pause at this point. The commonest reasons for failure on input and their remedies are explained in Section 3.6. When the machine replies by printing a question mark, it means that the data have been successfully read from the file. It is a good idea to print out the data, to check that they have been read correctly, and that the file you read them from was the correct one.

```
$look x y $
```

7.7.2 Initial data inspection

The first priority is to get to know the data. This will typically involve the production of several graphs, in which we look for trends, for evidence of curvilinearity, and at various plots of residuals, and tables where we look at means and ranges. At this stage we shall simply plot y against x (more complex plots are explained in Section 5.7).

```
$plot y x $
```

Note that we put the *y*-variable first, then the *x*-variable (they should be separated by one or more blank spaces). The graph is crude, but it serves the basic purpose. Notice that the symbols are plotted using the first character of the name of the *y*-variable, and that the scaling of the *x*- and *y*-axes is rather odd. It appears from Fig. 7.1 that a linear model would be a sensible first approximation, as there are no massive outliers in the data, and there is no obvious trend in the variance of y with increasing x. We can use **tab** to find the mean value of y, and the maximum and minimum weight gains:

```
$tab the y mean $

6.889
```

```
$t t y l$t t y s $

12.00

2.000
```

This initial data exploration completed, we begin with the GLIM statistical analysis in earnest.

7.7.3 Regression statistics in GLIM

The first step is compulsory; we must state which of our variables is to be the response variable in the model. The is done by means of the **yvar** (the *y*-variable) directive. This is simple in our case since we have called the response variable (caterpillar weight gain) y:

```
$yvar y $
```

In principle, however, the *y*-variable could be any of the variables in our data list (or, indeed, other variables that we had calculated in the current GLIM session). Now there is a variety of optional directives. In more advanced applications, we will state the error structure of the *y*-variable at this point (it might be Poisson or binomial, for example). For the time being, let us assume that the errors in y are normally distributed. This is the default assumption in GLIM, and so we need to do nothing about it. We are now in a position to begin the statistical analysis.

The use of GLIM for statistical modelling centres around the **fit** directive. This directive takes the currently specified model, fits it to the data using maximum likelihood methods, then calculates standard errors for the estimates of the model's parameters. We intend to fit the linear regression $y = a + bx$. This has two parameters: the intercept, a, and the slope, b. There is a single explanatory variable x that we shall fit to the data in y. The procedure is straightforward. As a first step we fit the *null model* like this:

```
$fit $
```

The computer then produces the following output:

```
deviance = 108.89

      d.f. = 8
```

In order to see what this means, we need to look back to the hand calculations. The word 'deviance' has been written for the total sum of squares, SST, and the total degrees of freedom has been abbreviated to d.f. ($8 = 9$ points on the graph $- 1$ parameter (the mean value of y) estimated from the data).

To see the value of the parameter estimated by carrying out the **fit** directive, we simply type:

```
$disp e $
```

which stands for **display** the **estimate(s)**. The computer responds by printing:

```
     estimate   s.e.    parameter
1    6.889      1.230   1
scale parameter taken as 13.61
```

which you will recognize as the mean value of y (6.889) and the overall standard error of the mean ($1.230 = \sqrt{s^2/n} = \sqrt{13.611/9}$). The *scale parameter* is GLIM's way of saying the overall variance in y, which you can check by dividing *SST* (108.89) by its degrees of freedom (8), using the information in the ANOVA table we completed earlier.

What the **fit** directive has done is simply to fit the average value of y (the 'grand mean') to the data. We can see what this involved by plotting the *fitted values* estimated by GLIM along with the raw data.

```
$plot y %fv x '*+' $
```

using the symbol ⋆ to represent y-values and + to represent the fitted values.

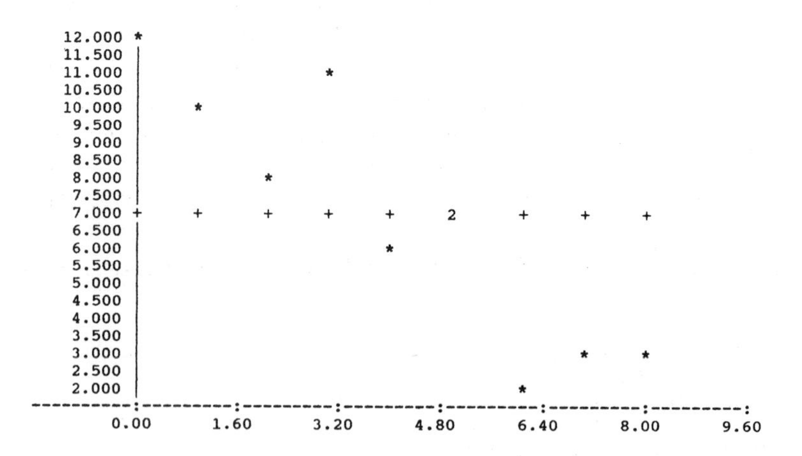

Within GLIM, the computer stores information that is calculated by the program during statistical analysis in vectors and scalars that have names beginning with the percentage symbol. Thus, %fv is the system vector containing the fitted values. Since we have fitted only the average value of y so far, all nine elements of %fv contain the same value (see the horizontal line of +'s, above). To see the numerical values that the model has fitted, we might type:

```
$print %fv $
```

which would produce the output:

```
6.889 6.889 6.889 6.889 6.889 6.889 6.889 6.889 6.889
```

To continue with the linear regression, we now fit the explanatory variable x to the data, by adding x to the model using the **fit**+ (**fit plus**) directive:

```
$fit +x $
```

which produces the output:

```
deviance = 20.072 (change = −88.82)
      d.f. = 7         (change = −1)
```

Deviance is now reduced to 20.072 and the degrees of freedom to 7. Reference to our earlier hand calculations shows that the new deviance is the error sum of squares, *SSE*, with error degrees of freedom (9 points on the graph − 2 (the two parameters estimated from the data being the mean value of y and the slope of the least squares regression line) = 7).

The terms in parentheses are the change in deviance and the change in degrees of freedom resulting from the **fit** + x directive just executed. Looking back at our hand calculations shows that the change in deviance of 88.82 is the regression sum of squares, *SSR*, and the change in degrees of freedom is 1 (the single additional parameter, b, has been estimated). So far, so good. The deviances printed immediately after the **fit** directives give us the sums of squares and degrees of freedom we need to fill in the first two columns of the ANOVA table (or now, more generally, the analysis of deviance table).

To see the parameter estimates and standard errors for a and b, we simply type:

```
$disp e $
```

and the computer prints:

```
     estimate   s.e.      parameter
1    11.76      1.041     1
2   −1.217      0.2186    X
scale parameter taken as 2.867
```

It is very important that you understand all the entries in this table. The rows of the table (1 and 2) are numbered in the leftmost column; the estimates of the parameters are found in the second column, with the estimate of the intercept, $a = 11.76$, in row 1, and the slope $b = -1.217$ in row 2; the standard errors of the parameters are in the third column, so that $SE_a = 1.041$ and $SE_b = 0.2186$; and in the final column are the names of the parameters as used by GLIM. The *scale parameter* of 2.867 is the error variance (or error mean square). You should check these values against those calculated by hand earlier on.

In order to see how well the model describes the data we can replot, showing the new fitted values as well as the data:

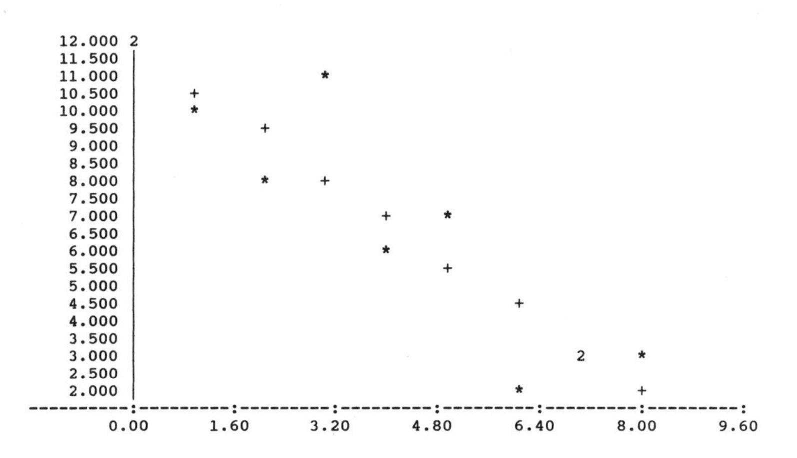

There is no evidence of curvature in the data, and the residuals appear to be reasonably well behaved (see below).

The only point requiring clarification regards the parameter names used by GLIM. While the explanatory variable x should be unambiguous (GLIM will print the first eight letters of whatever variable name we gave to the explanatory variable, but only four letters in version 3.77), the use of 1 for the intercept is a bit confusing. When we said **disp e** after the first **fit** directive, you will recall that parameter 1 was the grand mean $\bar{y} = 6.899$. After **fit** + x, parameter 1 is the expected value of y at $x = 0$ (the intercept = 11.76). If we were to fit more explanatory variables to the model, the meaning of parameter 1 would change again. If, say, we were to fit a second explanatory variable z to the data (i.e. **fit** + z), then the parameter 1 would be the expected value of y when *both* x and z were zero. And so on. In general, the first term in a GLIM regression model is the expected value of y when all the explanatory variables are equal to zero. Note that the interpretation of parameter 1 is slightly different in

ANOVA, where parameter 1 is the expected value of y for all subscripted factors set at level 1; this is explained in Section 4.21 and Chapter 8.

To recap, it is worth presenting the compete GLIM code for the modelling analysis up to this point (ignoring the **plot** and **tab** directives):

```
$units 9 $

$data x y $

$dinput 6 $

File name? glex37.dat

$yvar y $

$fit $

deviance = 108.89
      d.f. = 8

$disp e $

   estimate  s.e.    parameter
1  6.889     1.230   1
scale parameter taken as 13.61

$fit +x $

deviance = 20.072 (change = -88.82)
      d.f. = 7       (change = -1)

$disp e $

   estimate  s.e.      parameter
1   11.76    1.041     1
2  -1.217    0.2186    X
scale parameter taken as 2.867
```

The next question concerns the goodness of fit of the model. We can compute the residuals, and display their values like this:

```
$calc r = y - %fv $

$plot r %fv '+' $
```

or we can obtain a list of the residuals by using the display residuals directive:

```
$disp r $
```

which gives the following print-out:

unit	observed	fitted	residual
1	12.000	11.756	0.244
2	10.000	10.539	−0.539
3	8.000	9.322	−1.322
4	11.000	8.106	2.894
5	6.000	6.889	−0.889
6	7.000	5.672	1.328
7	2.000	4.456	−0.456
8	3.000	3.239	−0.239
9	3.000	2.022	0.978

The *unit* column contains the row number from the data table, the *observed* column contains the y-values (in general it will contain whatever variable is currently defined as being the **yvar**), the *fitted* column contains the %fv, obtained by evaluating the equation:

$$Y = 11.756 - 1.2167X$$

at the x-values corresponding to each of the observed y-values, while the *residual* column contains the raw residuals (the difference between the observed and fitted values, %yv−%fv). With Poisson or binomial errors, this column would show the standardized residuals (see Section 11.6).

In order to obtain the coefficient of determination r^2 from a GLIM analysis, we need to evaluate SSR/SST. These values are found in the first column of the analysis of deviance table, and

$$r^2 = \frac{SSR}{SST} = \frac{88.817}{108.89} = 0.8157$$

so that the correlation coefficient is $r = \sqrt{0.8157} = -0.9031$, bearing in mind the negative slope of the relationship.

In GLIM we can obtain r^2 simply by calculating the ratio of SSR to SST (i.e. the change in deviance following the **fit** + x directive divided by the deviance following the initial **fit** directive):

```
$calc 88.82/108.89 $

0.8157
```

so the line explains about 82% of the total sum of squares in caterpillar weight.

7.7.4 The scale directive

In our hand calculations, the significance of the slope parameter was assessed by dividing the regression mean square by the error variance to obtain an F-ratio of 30.97. This had one degree of freedom in the numerator and seven degrees of freedom in the denominator. A useful trick, especially in the analysis of more complex models, is to use the **scale** directive to convert the sums of squares in the analysis of deviance table directly into variances. Then, when parameters are added to or removed from the model, the change in scaled deviance can be used to compute F-ratios more simply.

The scale parameter in the present example was the error variance, 2.867. To alter the scale, we just type $scale 2.867 $, then repeat the model fitting as before:

```
$scale 2.867 $

-- model changed

$fit $

scaled deviance = 37.974
           d.f. = 8

$fit +x $

scaled deviance = 7.0000   (change = −30.97)
           d.f. = 7        (change = −1)
```

The effect of using the **scale** directive is that now, instead of deviance, the machine outputs *scaled deviance*. This is just a reminder that the scale parameter has been set to a non-default value (see Section 13.7). When the overall mean is fitted, the scaled deviance appears as 37.974. This is SST/s^2 and is not of any immediate interest.

However, when we **fit** + x the scaled deviance declines by 30.97 instead of by 88.82. The new value is the F-ratio we obtained earlier for testing the hypothesis that the slope equals zero. The 1% value of F in tables with 1 and 7 d.f. is 12.25, so the hypothesis is rejected with a high degree of confidence (there is much less than 1 chance in 100 that a slope this steep could arise by random sampling from a population where the true slope was zero). The advantage of using the **scale** directive is that it allows us to test the significance of a component of the model without carrying out extra calculations. Previously, in order to obtain the F-test statistic, we had to calculate the ratio of the regression deviance to the scale parameter (the error variance) in an extra step.

Note that following the use of the **scale** directive, the ratio of the residual scaled deviance (7.0000) to the residual degrees of freedom (7) is unity. Later on, when we use Poisson or binomial errors, the magnitude of this ratio will be an important component of model assessment (see Overdispersion, Sections 13.7 and 14.9). •

7.7.5 Summary

The steps in the regression analysis were: (i) data inspection; (ii) model specification; (iii) model fitting; and (iv) model criticism. We conclude that the present example is well described by a linear model with normally distributed errors. There are some large residuals, but no obvious patterns in the residuals that might suggest any systematic inadequacy of the model. The slope of the regression line is highly significantly different from zero, and we can be confident that, for the caterpillars in question, increasing dietary tannin reduces weight gain; and that, over the range of tannin concentrations considered, the relationship is reasonably linear. Whether the model could be used accurately for predicting what would happen with much higher concentrations of tannin would need to be tested by further experimentation.

CHAPTER 8

Analysis of variance

Instead of fitting continuous, measured variables to data (as in regression), many ecological experiments involve exposing experimental material to a range of discrete *levels* of one or more *factors*. Thus, a factor might be light intensity in which there were three levels: full sunlight, single shade-cloth or double shade-cloth. Alternatively, a factor might be a particular kind of mineral fertilizer, where the four levels represented four different nutrient mixtures. Factors are often used in experimental designs to represent statistical *blocks*; these are internally homogeneous units in which each of the experimental treatments is repeated. Blocks may be different fields in an agricultural trial, different genotypes in a plant physiology experiment or different growth chambers in a study of insect photoperiodism.

It is important to understand that regression and analysis of variance are not rigidly different approaches. The analysis of variance using factors often merges into regression using continuous variables. For example, it is a small step from having three levels of a shade factor (say light, medium and dark) and carrying out a one-way analysis of variance, to measuring the light intensity in the three treatments and carrying out a regression with light intensity as the explanatory variable. As we shall see later on, some experiments combine regression and analysis of variance by fitting a series of regression lines, one in each of several levels of a given factor (e.g. we might carry out separate regressions of respiration against temperature at each of three levels of light intensity; this is called analysis of covariance; see Chapter 9).

8.1 Statistical background

The emphasis in ANOVA has traditionally been on hypothesis testing. The aim of an analysis of variance in GLIM is to estimate means and standard errors of differences between means. You will be familiar with the technique of comparing two means by a t-test. This involves calculating the difference between the two means, dividing by the standard error of the difference, and then comparing the resulting statistic with the value of Student's t from tables. The means are said to be significantly different when the calculated value of t is larger than the value in tables. For large samples ($n > 30$) a useful rule of thumb is that a t-value >2 is significant.

It is much less easy to see how you can compare means by looking at variances. The underlying rationale of ANOVA is best seen by an example.

Assume that we have a single factor (say irrigation) with two levels (irrigated or not). We measure the mean plant growth by collecting shoot dry weight data from 20 irrigated quadrats and 20 similar quadrats that received only natural rainfall. We assume that the irrigation treatment was allocated at random to 20 of the 40 quadrats. It is likely that the mean shoot dry weights in each treatment will differ, and we need to establish whether this inevitable difference in their means could have come about by chance (i.e. we ask the question: is it likely that a difference as large as the one we have measured could have arisen by taking two random samples from populations that actually had the same means?). Let us draw a graph with the two sets of data (y_1 and y_2) on the same axes, showing the values simply in the order in which the measurements were taken (Fig. 8.1a).

First we calculate the overall mean, and draw this as a horizontal line through both sets of data (this is the total weight over all 40 quadrats,

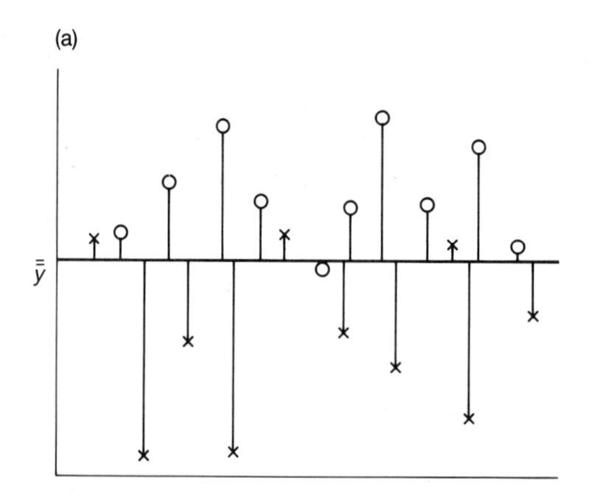

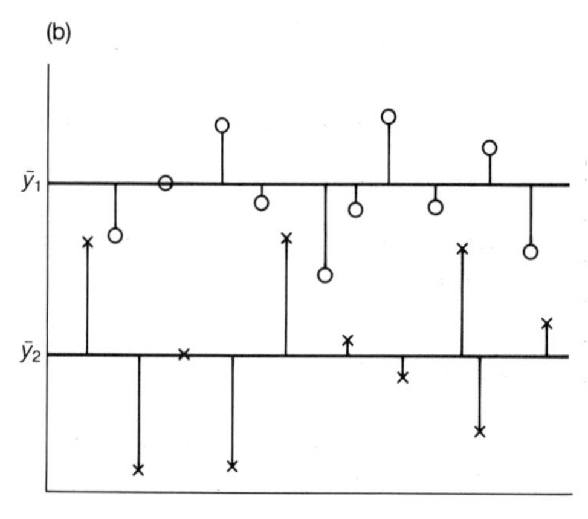

Fig. 8.1 ANOVA sums of squares. (a) The squared deviations from the overall average value of y give the total sum of squares, SST, just as in regression. (b) The squared deviations from the individual treatment means give the error sum of squares, SSE. When the treatments are significantly different from one another, SSE is small compared with SST.

irrigated and unirrigated, divided by 40). The individual quadrat yields differ from this overall mean by varying degrees. We can show each deviation by drawing a line from the point to the mean (Fig. 8.1a). Now the overall variation in the data is defined as the sum of the squares of these departures, the *total sum of squares*, *SST* (exactly the same as in regression in Chapter 7).

The next step is to calculate the individual means for each level of the factor (i.e. for each of the irrigation treatments). We then draw the two individual treatment means through the relevant data points. Each datum point departs from its own treatment mean by a certain distance, and we can represent this by drawing lines from the points to the treatment means (Fig. 8.1b). The sum of the squares of these departures now gives us the *error sum of squares*, *SSE*. In the previous chapter, *SSE* measured the variation from a fitted regression line. In ANOVA, the error sum of squares measures departures from individual level means within a factor (often called *treatment means*).

It is at this point that an understanding of ANOVA is won or lost. Ask yourself the following question: if the individual means really *aren't* different from one another, what is the relationship between *SST* and *SSE* (which is bigger, or are they equal)? Look at Figs 8.1a and b. After some head scratching you should be able to convince yourself that if the two means are the same, then *SSE* will be equal to *SST*, because the lines will actually lie on top of each other.

Now, if the means *are* significantly different, what will be the relationship between *SSE* and *SST*? This may take a little longer, and might be helped by a further, more extreme, example. Look at the data in Fig. 8.2. The value of *SST* will be large, because the overall mean lies halfway between the data from the two, very different, treatments. But what is the value of *SSE*? Because all the data lie exactly on their treatment means, *SSE* must be zero. Thus, *SST* is large and *SSE* is zero. In general, then, when the treatment means are significantly different from one another, *SSE* will be smaller than *SST*. The more different the means, the bigger will be the difference between *SSE* and *SST*.

To recap, we have discovered that *SSE is close to SST when means are not significantly different, but SSE is smaller than SST when the means are significantly different*. We can calculate the difference between *SST* and *SSE*, and use this as a measure of the difference between the treatment means; this is traditionally called the *treatment sum of squares*, and is denoted by *SSA*:

$$SSA = SST - SSE$$

The technique we are interested in, however, is analysis of variance, not analysis of sums of squares. We convert the sums of squares into variances

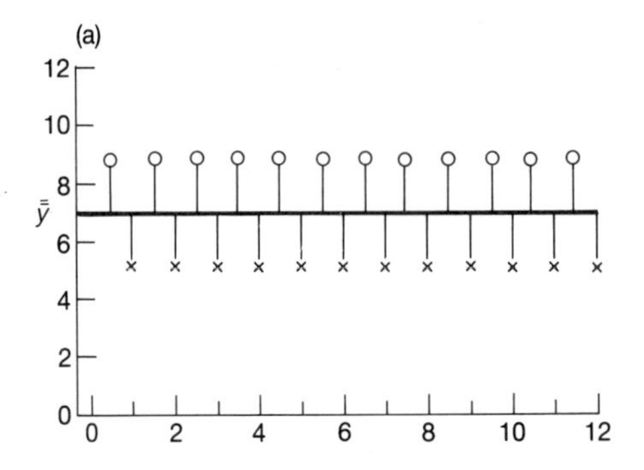

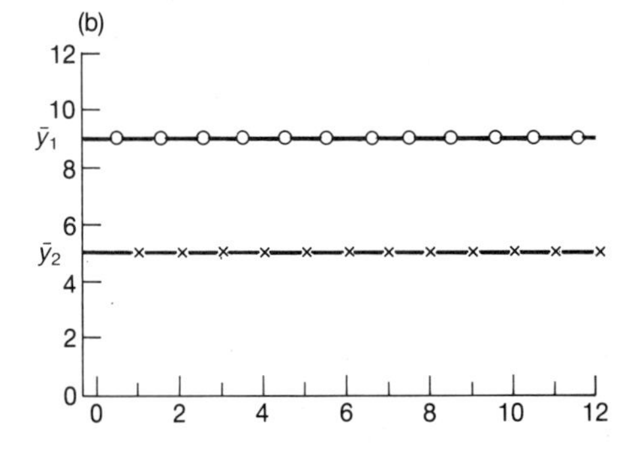

Fig. 8.2 Extreme cases of ANOVA sums of squares. When there is no sampling error, (a) SST is large but (b) SSE is zero.

by dividing by their degrees of freedom. In our example, there are two levels of the factor and so there is $2 - 1 = 1$ degree of freedom for SSA. In general, we might have k levels of any factor and $k - 1$ d.f. for treatments. Each level of irrigation was replicated 20 times, so there are $20 - 1 = 19$ degrees of freedom for error in each level, and hence $2 \times 19 = 38$ degrees of freedom for error in all. In general, there might be n replicates per level, and with k levels this would give $k(n-1)$ degrees of freedom for error. The total degrees of freedom is always one less than the total number of samples ($40 - 1 = 39$ or, in general, $kn - 1$). As a check in more complicated designs, it is useful to make sure that the individual component degrees of freedom add up to the correct total. In the present case we have one degree of freedom for irrigation plus 38 degrees of freedom for error, which gives the correct total of 39. The divisions for turning the sums of squares into variances are usually carried out in an ANOVA table (Table 8.1) so that each element in the sums of squares column is simply divided by the number in the adjacent degrees of freedom column to give the variances in the mean square column (headed MS; see the previous chapter for details). The significance of the difference between the means is then assessed using an F-test (a variance ratio test). The irrigation variance is divided by the error variance, and the value of

Table 8.1 ANOVA table for a one-way design

Source	SS	d.f.	MS	F-ratio	Probability
Irrigation	SSA	$k-1$	$MSA = \frac{SSA}{k-1}$	$\frac{MSA}{s^2}$	Tables
Error	SSE	$k(n-1)$	$S^2 = \frac{SSE}{k(n-1)}$		
Total	SST	$kn-1$			

this test statistic is compared with the value of F in tables (this has $k-1$ degrees of freedom in the numerator (looked up in the columns of the F table) and $k(n-1)$ degrees of freedom in the denominator (looked up in the rows of the table)).

Thus, the analysis of variance *compares means by looking at variances*. The test works because when the means are the same, SSE is equal to SST, so SSA is zero and the F-ratio is zero. When the means are significantly different, then SSE is less than SST, so SSA is relatively large, and the ratio of MSA to the error variance will be larger than the value of F in tables.

Another way of visualizing this is to think of the relative amounts of sampling variation between replicates receiving the same treatment (i.e. between individual samples in the same level), and between different treatments (i.e. between-level variation). When the variation between replicates within a treatment is large compared with the variation between treatments, we are likely to conclude that the difference between the treatment means is insignificant. Only if the variation between replicates within treatments is relatively small compared with the differences between treatments will we be justified in concluding that the treatment means are significantly different.

8.2 Calculations in ANOVA

The definitions of the various sums of squares can now be formalized, and ways found of calculating their values from samples. The total sum of squares, SST, is defined as:

$$SST = \Sigma(y - \bar{\bar{y}})^2 \qquad (8.1)$$

where the difference between each y-value and the overall mean value of y ($\bar{\bar{y}}$) is squared and then added up (the second summation sign and the subscripts referring to treatment, i, and replicate within treatment, j, have been omitted for simplicity). This is not the most convenient form for hand-calculation, because it requires that the overall mean value of y is calculated first, then subtracted from each y-value in turn, before the differences are squared and added up (note, however, that this form of the equation may be best from the point of computational precision, especially if the individual data values are large, because it does not suffer from rounding errors). A little algebra, following expansion of the squared bracketed term, leads to a more satisfactory formula for calculating SST as follows:

$$SST = \Sigma y^2 - \frac{(\Sigma y)^2}{kn} \tag{8.2}$$

The formula is the same as in regression analysis, but the total number of samples is written as kn (the number of levels times the number of replicates per level). As before, note the distinction between the sum of the squares Σy^2 and the square of the sum $(\Sigma y)^2$.

The error sum of squares is defined as the sum of the squares of the departures of the individual y-values from their respective treatment means (where we shall use $\bar{y}_i$ to represent the mean value of y in the ith treatment):

$$SSE = \sum_{i=1}^{k} \Sigma(y - \bar{y}_i)^2 \tag{8.3}$$

where the squared differences are added together over all k levels of the factor (i.e. over all treatments). As with SST, this defining formula leads to cumbersome calculation, and is not used in the subsequent arithmetic.

The treatment sum of squares, SSA, is defined as:

$$SSA = SST - SSE \tag{8.4}$$

but it is actually more straightforward to calculate SSA, and to obtain SSE by subtraction from SST. The short-cut formula for SSA is this:

$$SSA = \frac{\Sigma C^2}{n} - \frac{(\Sigma y)^2}{kn} \tag{8.5}$$

where the new term is C, the *treatment total*. This is the sum of all the n replicates within a given level. Each of the k different treatment totals is squared, added up, and then divided by n (the formula is slightly different if there is unequal replication in different treatments, as we shall see below). The meaning of C will become clear when we work through the example later on. Finally,

$$SSE = SST - SSA \tag{8.6}$$

to give all the elements required for completion of the ANOVA table.

8.3 Assumptions of ANOVA

You should be aware of the assumptions underlying the analysis of variance. They are all important, but some are more important than others:

1 random sampling;
2 equal variances;
3 independence of errors;
4 normal distribution of errors;
5 additivity of treatment effects.

These assumptions are well and good, but what happens if they do not apply to the data you propose to analyse? We consider each case in turn.

8.3.1 Random sampling

If samples are not collected at random, then the experiment is seriously flawed from the outset. The process of randomization is sometimes immensely tedious, but none the less important for that (Section 4.16). Failure to randomize properly often spoils what would otherwise be well-designed ecological experiments (see Hairston, 1989, for examples). Unlike the other assumptions of ANOVA, there is nothing we can do to rectify non-random sampling after the event. If the data are not collected randomly, they will probably be biased. If they are biased, their interpretation is bound to be equivocal.

8.3.2 Equal variances

It seems odd, at first, that a technique which works by comparing variances should be based on the assumption that variances are equal. What ANOVA actually assumes, however, is that the *sampling errors* do not differ significantly from one treatment to another. The comparative part of ANOVA works by comparing the variances that are due to differences between the treatment means with the variation between samples within treatments. The contributions towards *SSA* are allowed to vary between treatments, but the contributions towards *SSE* should not be significantly different.

An example should demonstrate the folly of attempting to compare treatment means when the variances of the samples are different. Consider the data in Table 8.2, which come from three commercial gardens producing lettuce. The numbers refer to daily concentrations of ozone in parts per hundred million (pphm) on 10 consecutive days in summer. We calculate the mean and variance for each garden separately.

There are three important points to be made from these calculations: (i) two treatments can have different means and the same variance (gardens A and B); (ii) two treatments can have the same mean but different

Table 8.2 Ozone levels (pphm) in three gardens

	Garden A	Garden B	Garden C
	3	5	3
	4	5	3
	4	6	2
	3	7	1
	2	4	10
	3	4	4
	1	3	3
	3	5	11
	5	6	3
	2	5	10
$\bar{y}$	3	5	5
s^2	12/9	12/9	128/9

variances (gardens B and C); (iii) when the variances are different, the fact that the means are identical does *not* mean that the treatments are ecologically identical.

This last point is the most important. Suppose that the threshold concentration of ozone that causes damage to this lettuce variety is 8 pphm, and then ask on what proportion of days did air pollution reach damaging levels in gardens B and C? Analysis of the *mean* concentration suggests that pollution levels in both gardens are the same and are safely below the damage threshold (5 pphm is substantially less than the damage threshold of 8 pphm). Inspection of the raw data, however, shows that while damaging levels were *never* observed in garden B, the ozone level was high enough to damage plants on 3 days out of 10 in garden C. This problem of interpretation arises because the variances from gardens B and C are significantly different (variance ratio = 128/12 = 10.6; F in 5% tables with 9 and 9 degrees of freedom = 3.18). Attempting to base our conclusions about the prevalence of air pollution damage upon a comparison of the means would lead to us missing the fact that, in garden C, ozone levels were high enough to cause damage on 30% of days.

We often encounter data with non-constant variance in ecological work, and GLIM is ideally suited to deal with this. With Poisson-distributed data, for example, the variance is equal to the mean, and with binomial data the variance (npq) increases to a maximum and then declines with the mean (np). Many ecological data sets exhibit the property that the coefficient of variation is roughly constant, which means that a plot of log(variance) against log(mean) increases with a slope of approximately 2 (this is known as Taylor's power law; see Section 4.19); gamma errors can deal with this. Alternatively, the response variable can be transformed (Section 3.4) and the analysis carried out on the new variable.

8.3.3 Independence of errors and pseudoreplication

In ecology it is very common to find that the errors are not independent from one experimental unit to another. We know, for example, that the response of an organism is likely to be influenced by its sex, age, body size, neighbours and their sizes, history of development, genotype and many other things. Differences between our samples in these other attributes would lead to non-independence of errors, and hence to bias in our estimation of the difference that was due to our experimental factor. For instance, if more of the irrigated plots happened to be on good soil, then increases in shoot weight might be attributed to the additional water when, in fact, they were due to the higher level of mineral nutrient availability on those plots. The ways to minimize the problems associated with non-independence of errors are:

1 use block designs, with each treatment applied in each block (i.e. repeat the whole experiment in several different places);

2 divide the experimental material up into homogeneous groups at the outset (e.g. large, medium and small individuals), then make these groups into statistical blocks, and apply each treatment within them;

3 have high replication; and

4 thorough randomization.

No matter how careful the design, however, there is always a serious risk of non-independence of errors in ecological work. A great deal of the skill in designing ecological experiments is in anticipating where such problems are likely to arise, and in finding ways of blocking the experimental material to maximize the power of the analysis of variance (see Hurlbert, 1984). This topic is considered in more detail in Section 4.14.

8.3.4 Normal distribution of errors

Ecological data often have non-normal error distributions (Fig. 8.3a). The commonest form of non-normality is skewness, in which the distribution has a long 'tail' at one side or the other. A great many collections of data on plant size, for example, are skewed to the right; most individuals in a population are small, but a few individuals are very large indeed (well out towards the right-hand end of the size axis).

The second kind of non-normality that is encountered in ecological data is kurtosis (Fig. 8.3c,d). This may arise because there are long tails on both sides of the mean so that the distribution is more 'pointed' than the bell-shaped normal distribution (so-called *leptokurtosis*), or conversely because the distribution is more 'flat-topped' than a normal one (so-called *platykurtosis*).

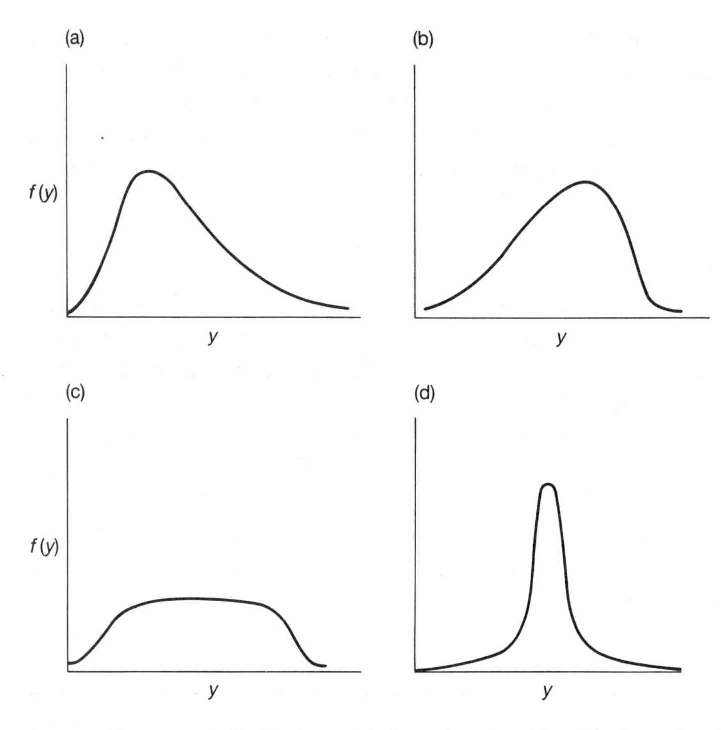

Fig. 8.3 Non-normal distributions: (a) skewed to the right; (b) skewed to the left; (c) platykurtotic (flat-topped); (d) leptokurtotic (pointed). The commonest causes of non-normality in ecological data are positive (i.e. right) skewness and leptokurtosis (e.g. size distributions in sessile organisms).

Finally, data may be bimodal or multi-modal, and look nothing at all like the normal curve. This kind of non-normality is most serious, because skewness and kurtosis are often cured by the same kinds of transformations that can be used to improve homogeneity of variance. If a set of data is strongly bimodal, then it is clear that any estimate of central tendency (mean or median) is likely to be *unrepresentative of most of the individuals* in the population.

8.3.5 Additivity of treatment effects in multifactor experiments

In two-way analysis there are two kinds of treatments (say irrigation and nitrogen fertilization), each with two or more levels. ANOVA is based on the assumption that the effects of the different treatments are additive. This means that if irrigation produces a yield increase of $0.4\,\mathrm{t\,ha^{-1}}$ at low nitrogen, then irrigation will produce the same yield increase at high nitrogen. Where this is not the case, and the response to one factor depends upon the level of another factor (i.e. where there is *statistical interaction*), we must use factorial experiments. Even here, however, the model is still assumed to be additive (see Section 8.6). When treatment effects are multiplicative (e.g. temperature and dose are multiplicative in some insect toxicity experiments) then it may be appropriate to transform the data by taking logarithms in order to make the treatment effects additive. In GLIM, we would specify a log link function in order to achieve additivity in a case like this.

8.4 A worked example of one-way ANOVA

To draw this background material together, we shall work through an example by hand. In so doing, it will become clear what GLIM is doing during its analysis of the same data. The example comes from an experiment on mineral nutrition in wild plants, in which the treatments are four different kinds of fertilizer (see Table 8.3). In this case we have four levels of a single factor (fertilizer) and five replicates within each level. Thus $k = 4$ and $n = 5$. The data have been rounded to whole numbers to simplify the arithmetic, and we assume that the errors are normal for the purpose of demonstration. It would not be good practice to analyse a set of small integer data with a normal likelihood (see Chapter 14).

Table 8.3 Plant yields from five replicates of four nutrient treatments

Control	Potassium	Phosphorus	Nitrogen
3	3	4	6
4	5	2	7
3	6	3	5
3	4	4	8
2	5	4	7

As ever, we begin by looking at the data (Table 8.4), considering the treatment means, inspecting the error variances, and checking for outliers (these may turn out to be nothing more than mistakes in data entry, but they may be real extreme values of y).

Table 8.4 Means and standard deviations (s) of plant yields

Treatment	Level	Mean	s
Control	1	3.0	0.707
Potassium	2	4.6	1.140
Phosphorus	3	3.4	0.894
Nitrogen	4	6.6	1.140
Overall		4.4	1.698

From this preliminary look at the data it does seem that there are differences between the mean responses (the largest mean, level 4 = nitrogen, is more than double the smallest mean, the control level 1). Also, the standard deviations within each level are sufficiently similar that the assumption of homogeneity of variance is upheld. Inspection of the raw data reveals no obvious outliers, so we appear to be justified in proceeding with the analysis of variance, in order to estimate the standard errors of the differences between the means.

First we calculate the grand total of the y's and the grand total of the y^2's; $\Sigma y = 88$ and $\Sigma y^2 = 442$. The *correction factor*, required in calculating both SST and SSA, is $(\Sigma y)^2/kn$ which is $88^2/20 = 387.2$ so that:

$$SST = 442 - 387.2 = 54.8$$

The next task is to compute SSA. This involves finding the treatment totals, i.e. the values of the C's (which stands for column totals) in equation (8.5). The first treatment, level 1, adds up to 15; this is C_1. The second to 23, and so on. Thus:

$$SSA = \frac{15^2 + 23^2 + 17^2 + 33^2}{5} - 387.2 = 39.2$$

We divided by 5 because we added together 5 numbers to get each treatment total. SSE is simply $54.8 - 39.2 = 15.6$.

There are four levels of fertilizer, and hence $4 - 1 = 3$ degrees of freedom for treatments. There are five replicates in each treatment and hence $5 - 1 = 4$ degrees of freedom for error in each treatment. There are four treatments, so there are $4 \times 4 = 16$ degrees of freedom for error overall. There are 20 samples in the compete experiment, so the total degrees of freedom is $20 - 1 = 19$. We check that the component degrees of freedom add up to give the correct total: $3 + 16 = 19$. The ANOVA table can now be completed, as shown in Table 8.5.

Table 8.5 One-way ANOVA table for plant yields

Source	SS	d.f.	MS	*F*-ratio	Probability
Treatment	39.2	3	13.067	13.40	0.0001
Error	15.6	16	0.975		
Total	54.8	19			

The error variance is 0.975, so we can test the null hypothesis that all the means are the same against the alternative hypothesis that at least one of the means is significantly different from the others, by computing the ratio of the treatment variance to the error variance. This gives a highly significant F-value of 13.4 (the value in 1% tables with three degrees of freedom in the numerator – column 3 of the table – and 16 degrees of freedom in the denominator – row 16 in the table – is 5.29). In fact, the probability of an F-ratio this large arising by chance alone is less than 0.0001. Rejecting this null hypothesis is often of little or no practical interest. Generally, we are much more concerned with individual contrasts (see below).

Next we compute the standard error of the difference between two means:

$$SE_{\text{diff}} = \sqrt{2\frac{s^2}{n}} = \sqrt{2\frac{0.975}{5}} = 0.6245$$

which, using Student's t, suggests that the least significant difference between a pair of means is:

$$LSD = t_{\text{tables},\nu=8,\alpha=0.05} SE_{\text{diff}} = 2.306 \times 0.6245 = 1.44$$

Thus, the nitrogen mean appears to be significantly higher than any of the other means (it is greater by 2.0 than the next highest mean) and the potassium mean is marginally greater than the control. The differences between the means can be computed in lots of different ways (see aliasing in Section 4.20). We shall take the mean of the first level as a reference point, and compare the other means with level 1. This makes good sense in the present example, because the first level is the control to which no extra fertilizer was added. The differences between the means of the control and levels 2, 3 and 4 are therefore as shown in Table 8.6.

Table 8.6 Differences between control and treatment means

Level	Difference	*SE*	*LSD*
2	1.6	0.6245	1.44
3	0.4	0.6245	1.44
4	3.6	0.6245	1.44

Before repeating the same one-way analysis of variance using GLIM, it is important to understand how the model underlying analysis of variance is handled by GLIM. In regression, the GLIM model was:

$$y = \beta_0 + \beta_1 x_1$$

which was easy to visualize, with β_0 representing the intercept and β_1 representing the slope of the graph. To see how analysis of variance can be formulated as a general linear model requires some thought.

The traditional model for one-way analysis of variance is this:

$$\hat{y}_i = \mu + \alpha_i \tag{8.7}$$

where $\hat{y}_i$ is the expected value of the ith treatment, μ is the overall mean, and α_i is the difference between the overall mean and the mean of the ith treatment. In this form, the model does not look at all like the generalized linear model for regression (e.g. there is no x-variable), and it is not obvious what are the equivalents of the intercept and the slope of that model. The trick is simple, but you will probably need to read the remaining part of this section twice to understand the principle. All the response data are entered in a single vector, just as in a regression. A different x vector is then set up *for each of the treatment levels* in the model (there are four in our worked example). Only a sub-set of the y-values is associated with any one treatment (only five out of the 20 y-values in our example, because the four treatments had equal replication). Next, all the x vectors are filled up with 0's, except for those rows that correspond to y-values that came from the treatment in question. In these rows, the x vector is filled with 1's. So, for example, in our treatment level 2 the x_2 vector looks like this:

y	x_2
3.0000	0
4.0000	0
3.0000	0
3.0000	0
2.0000	0
3.0000	1
5.0000	1
6.0000	1
4.0000	1
5.0000	1
4.0000	0
2.0000	0
3.0000	0
4.0000	0
4.0000	0
6.0000	0
7.0000	0
5.0000	0
8.0000	0
7.0000	0

Each of the four x vectors is constructed along the same lines, and the model is written:

$$\hat{y}_i = \beta_0 + \beta_1 x_1 + \beta_2 x_2 + \beta_3 x_3 + \beta_4 x_4 \tag{8.8}$$

which should look familiar as a linear model (in fact it is a multiple regression with four explanatory variables, four slopes and an intercept).

What it means in the context of ANOVA is that any expected value of y can be predicted by taking some overall mean β_0, and adding to it *just one* of the other β's. Only one of the other β's is added, because any y-value can belong only to a single treatment (1, 2, 3 or 4 in this example), so all but one of the x vectors must contain a zero. Adding up the product of lots of zeros gives zero. So, for instance, if we want to find the expected value of y in treatment 3, the x-terms for treatments 1, 2 and 4 are all zero, and we are left with:

$$\hat{y}_3 = \beta_0 + \beta_3 \tag{8.9}$$

and the term x_3 disappears because x_3 is 1.0. All the other terms disappear because their x's are 0.

This formula says that in order to calculate the expected value of y in treatment 3 we add β_3 to an 'intercept' β_0. So the 'slopes' in the generalized linear model for an analysis of variance are actually *differences between means*. In the example just given, the 'slope' β_3 is the difference between the 'intercept' β_0 and the mean value of y in treatment 3.

This approach takes some getting used to, and it may take several readings before the penny drops. However, it is worth persevering with the underlying model before going through the worked example, because if this part is understood, then what follows will be relatively easy. It is worth noting that the dummy variables for each of the factor levels (the x's in this example) are not set up explicitly in GLIM, so there is no wastage of space. Further discussion of these issues can be found in Section 4.20 which deals with aliasing.

8.5 ANOVA in GLIM

As in regression analysis, we begin with the preliminaries of telling GLIM the number of rows of data (20 = five replicates in each of four treatments), and the number of variables (two in this case, the value of y and the treatment number (the factor level) A, which lies in the range 1–4). Notice that we follow the convention of representing factors by capital letters and continuous variables by lower-case letters where they appear in the text. In GLIM the case of the letters you use is irrelevant, and we use lower case throughout.

```
$units 20 $

$data y a $

$dinput 6 $

File name? glex10.dat

$look y a $
```

```
Y        A
3.0000   1.0000
4.0000   1.0000
3.0000   1.0000
3.0000   1.0000
2.0000   1.0000
3.0000   2.0000
5.0000   2.0000
6.0000   2.0000
4.0000   2.0000
5.0000   2.0000
4.0000   3.0000
2.0000   3.0000
3.0000   3.0000
4.0000   3.0000
4.0000   3.0000
6.0000   4.0000
7.0000   4.0000
5.0000   4.0000
8.0000   4.0000
7.0000   4.0000
```

Note that the variable A contains the treatment number; the first five y-values belong to treatment number 1, the next five to treatment number 2, and so on.

We inform GLIM that the variable A is a factor with four levels, as follows:

```
$factor a 4 $
```

which distinguishes A from the continuous variable x we used in the regression example. We begin by inspecting the individual treatment means and variances using the **tab** directive:

```
$tab the y mean for a $

        1       2       3       4
[ ]  3.000   4.600   3.400   6.600
```

to give the means, and we can practise using the abbreviated form of the tabulate directive, **t**, to get the standard deviations (**d**):

```
$t t y d f a $

          1        2        3        4
[ ]  0.7071   1.1402   0.8944   1.1402
```

These look satisfactory, so we can proceed with the modelling.

The next stage is to declare which of the variables is the response variable. As before, we write:

```
$yvar y $
```

and the analysis of variance proceeds in a very similar manner to the regression. We begin by calculating *SST*. This is done by fitting the null model using the **fit** directive, and produces the following output:

```
$fit $

deviance = 54.800
      d.f. = 19
```

which means that $SST = 54.8$ with 19 degrees of freedom (which agrees with our long-hand calculation). Next, we can look at the overall mean by using the **disp e** directive:

```
$disp e $

   estimate  s.e.     parameter
1  4.400     0.3798   1
scale parameter taken as 2.884
```

which tells us that the grand mean (the average over all four treatments) is 4.4, the standard error of the overall mean is 0.3798 and the overall error variance is 2.884. Although we did not calculate the overall variance and standard error, you can readily check their values using *SST*, the overall degrees of freedom and the total sample size (recall that the standard error of a mean is $\sqrt{(s^2/n)}$. Now we fit the treatments to the model by adding the factor A, as follows:

```
$fit +a $

deviance = 15.600 (change = −39.20)
      d.f. = 16       (change = −3)
```

which shows that the error sum of squares is 15.6 with 16 degrees of freedom (four treatments each with $5 - 1 = 4$ degrees of freedom). The change in deviance gives *SSA*, the amount of *SST* that was explained by fitting treatment A to the model. Thus, out of a total *SST* of 54.8, differences between the means account for 39.2 (about 72%, a substantial part). There is a change of three in the number of degrees of freedom associated with this fit (4 treatments − 1).

The next stage involves a comparison of the different means. We begin by looking at the parameter estimates and their standard errors. As with regression, we do this with the **disp e** directive:

```
$disp e $

   estimate  s.e.     parameter
1  3.000     0.4416   1
2  1.600     0.6245   A(2)
3  0.400     0.6245   A(3)
4  3.600     0.6245   A(4)
scale parameter taken as 0.9750
```

This looks different from the equivalent regression table, because there are four rows rather than two. In general, GLIM will produce *as many rows in the* disp e *table as there are parameters in the model*. In this simple one-way ANOVA there are four treatment means, so there are four rows in the table. There were two rows in regression because we estimated two parameters — the slope and the intercept.

The scale parameter, as in regression, is the error variance ($s^2 = SSE/[k(n-1)]$ where there are k treatments, each with n replicates), taken from the ANOVA table we completed earlier.

One question arises immediately: what has become of the mean for treatment level 1? The second, third and fourth levels of A are listed in the *parameter* column and labelled by their *subscripts*, but A(1) is missing. This is one of the most difficult things to understand when GLIM is first encountered. What the output means is this. Rather than have an overall mean, with four separate departures from this mean (a total of five parameters in equation 8.8), GLIM makes do with only four. Since the experiment can be summarized by four treatment means, one of the five parameters is redundant (it is said to be *aliased*; see Section 4.20). GLIM eliminates the redundant parameter by making the 'intercept' of the linear model (*parameter 1*) the mean for treatment 1. You can confirm this by inspecting the individual y means (above).

Notice that the standard error in row 1 is different from the standard errors in the other three rows. The first row gives the standard error of a single mean (the mean of treatment one). You will recall that the formula for the standard error of a mean is $\sqrt{(s^2/n)}$ where $n = 5$ (the number of replicates from which an individual mean was calculated). Note that you divide by $n = 5$ and not by $kn = 20$ (this is a common mistake).

The second row of the **disp e** table refers to treatment number 2. As we have seen, the mean of treatment 2 is 4.6, but the estimate for treatment 2 is 1.6 and its standard error is larger than that of *parameter 1*. Both these points should become clear when it is realized that the parameter estimate in row 2 is *the difference between the means* of treatments 1 and 2, and the standard error is *the standard error of the difference between two means*. Thus, if the estimate in row 2 is added to the intercept we obtain the mean of treatment 2: $1.6 + 3.0 = 4.6$.

The standard error of the difference between two means is given by:

$$SE_{\text{diff}} = \sqrt{\frac{s_1^2}{n_1} + \frac{s_2^2}{n_2}} \qquad (8.10)$$

but in our analysis the two variances are assumed to be the same (we use the scale parameter which is the pooled estimate of variance), and the sample sizes are the same. So we can use the simplified formula:

$$SE_{\text{diff}} = \sqrt{2\frac{s^2}{n}} \tag{8.11}$$

You should check that the standard error of 0.6245 is indeed given by this formula. Note that all the remaining standard errors have the same value, because they, too, relate to differences between two means (the 'intercept' (the mean of treatment 1) and the treatment mean in question), and replication is the same in all four treatments. You should confirm that the means of treatments 3 and 4 are given by adding the parameter estimates in rows 3 and 4 to the mean of treatment 1.

8.5.1 Hypothesis testing

There are two ways of testing hypotheses in GLIM:

1 using the estimated standard errors to do *t*-tests;

2 by deletion, i.e. removing a given factor (or factor level) from the model and observing the increase in deviance that results.

In general, the second method is preferable because it is more robust. For the present, however, we shall use the first method because it is likely to be more familiar.

In comparing any two of the means we have $5 + 5 - 2 = 8$ degrees of freedom, so the appropriate two-tailed value of t from tables is 2.306 at 95%. We regard any estimates that are more than 2.306 times their standard errors as being significant at 95%. Thus, working down the list, we ask is the mean of treatment 2 significantly greater than the mean of treatment 1? The second parameter, 1.60, is the difference between the means of treatments 1 and 2, and our test consists of determining whether this difference is significantly different from zero. The ratio of the difference to its standard error is 2.56, just larger than the value in tables, and so we conclude that the difference between means 1 and 2 is marginally greater than zero. The ratio for treatment 3 is only 0.64, so there is obviously no significant difference between the means of treatments 3 and 1. The difference between the means of treatments 1 and 4, however, is much greater than zero and has a *t*-ratio of 5.77; this is significant at better than 5 in 1000.

We have compared treatment 1 with all the others. But what about other comparisons? For example, is the mean of treatment 2 significantly lower than the mean of treatment 4? To answer this question, we need to know the standard error of the difference between parameters 2 and 4. This is obtained by using the display standard errors directive, **disp s**:

```
$disp s $

S.E.s of differences of parameter estimates
```

```
1  0.000
2  0.9874  0.000
3  0.9874  0.6245  0.000
4  0.9874  0.6245  0.6245  0.000
      1       2       3       4
scale parameter taken as 0.9750
```

This triangular matrix contains the standard errors of the differences between each and every one of the parameters. We are interested to test the significance of the difference between the parameters in rows 2 and 4, so we look along the fourth row of the matrix until we get to column 2; the entry is 0.6245. We should not be surprised to find that this is the same standard error as was used for comparing any two differences between means, because the replication is uniform throughout.

The difference between the potassium and nitrogen means is $1.6 - 3.6 = -2.0$. The sign can be ignored, so we compute the t-ratio $2.0/0.6245 = 3.20$. This is substantially larger than the tabulated t-value of 2.306, so we are justified in concluding that these two means are significantly different at better than the 5% level.

To determine the F-ratio associated with fitting a given term to the ANOVA model, we *convert the change in deviance into a variance* by dividing by the change in degrees of freedom. Looking back, you will see that when we typed **fit** + A we obtained a reduction in deviance of 39.2 and a reduction in d.f. of 3, so the treatment mean square is 13.067 and the F-ratio is obtained as:

$$F = \frac{MSA}{s^2} = \frac{13.067}{0.975} = 13.40$$

You should check this against the earlier hand-calculations.

In later chapters we discuss a more thorough approach to hypothesis testing through model simplification; a factor is taken as being statistically significant *if its removal from the model causes a significant increase in the deviance*. The significance of the change in deviance is assessed by F-tests when we have normal or gamma errors, and by χ^2 tests for Poisson or binomial errors (see Section 11.4).

8.5.2 Using the scale directive in ANOVA

We used the **scale** directive in fitting regression models in order to obtain direct estimates of F-ratios for assessing the significance of the contribution of various terms to the overall explanatory power of the model. We can do the same in analysis of variance. The pooled error variance of 0.975 is given as the scale parameter following the **disp e** directive, above. If we now declare $scale 0.975 $ and refit the model, then, instead of getting sums of squares as output, the figures are multiples of the error variance. This is how it is done:

```
$scale 0.975 $

-- model changed

$fit $

scaled deviance = 56.205
            d.f. = 19

$fit +a $

scaled deviance = 16.000 (change = -40.21)
            d.f. = 16     (change = -3)

$disp e $

   estimate  s.e.     parameter
1  3.000     0.4416   1
2  1.600     0.6245   A(2)
3  0.400     0.6245   A(3)
4  3.600     0.6245   A(4)
scale parameter taken as 0.9750
```

The estimates and their standard errors are unaffected, but now instead of deviance ($SST = 54.8$), we obtain scaled deviance ($SST/0.976 = 56.205$). When the treatments are added to the model (**fit** + A), there is a reduction in scaled deviance of 40.21 with three degrees of freedom. The F-ratio is simply the change in scaled deviance divided by the degrees of freedom:

```
$calc 40.21/3 $

13.40
```

•

You need to remember which error structure is being used (normal, by default, in our case), in order to understand the consequences of using the **scale** directive. With normal errors we use it only as a quick way of finding the significance of terms added to or subtracted from a model. For Poisson or binomial errors it is used as a means of dealing with over-dispersion (see Section 13.7).

8.6 Factorial ANOVA: an example

Factorial experimental designs are used to investigate the *interaction between factors*. Interaction occurs when the way that the response variable behaves in relation to changes in the level of one factor depends upon the level of another factor (see Section 4.10). Thus, for example, in a desert ecosystem, the yield response to added nitrogen might be positive under an irrigation treatment, but zero in the unirrigated treatments where yield was water-limited. The crucial thing about factorial ANOVA is the addition of an *interaction term* to the model:

$$\hat{y}_{ij} = \mu + \alpha_i + \beta_j + \gamma_{ij} \tag{8.12}$$

where α and β are called the main effects of *treatments A* and *B* and γ is the interaction term. Note that even though the interaction term can vary from treatment to treatment, the model is still *additive*.

The worked example concerns a two-factor factorial experiment, in which the growth increments of insect larvae (increases in length; mm) were measured after one week of feeding on artificial diets under controlled temperature conditions. The first factor, dietary protein, had three levels (low, medium and high) and the second factor, alkaloid, had two (absent or present). Each of the $3 \times 2 = 6$ treatment combinations was replicated four times, so the whole experiment contained $6 \times 4 = 24$ larvae (note that this would probably be an inadequate sample size in a real insect experiment). We begin, as usual, by looking at the data, calculating treatment means, inspecting the variances and looking for outliers (Table 8.7).

It appears that the insects grow more quickly with alkaloid in the diet than without it (perhaps it is a feeding stimulant for this species), but that the effects of protein level are complicated. Growth is least on diets containing alkaloid that are high in protein, suggestive of an interaction between these two dietary components.

The calculations are as follows. As with one-way ANOVA, we begin by determining the variation around the overall mean (i.e. the sum of the squares of the deviations of each of the 24 datum points from the grand mean of 5.0):

Table 8.7 Data on insect growth (mm) from a factorial experiment

		Growth (mm)									
Factor	*B*	Alkaloid absent				Alkaloid present				Mean	Row total
A											
Protein	1	3	5	4	6	6	7	5	8	5.50	44
	2	3	2	4	5	7	8	6	7	5.25	42
	3	7	6	5	4	5	2	3	2	4.25	34
Mean		4.5				5.5				5.00	
Column total		54				66					120

$$SST = \Sigma y^2 - \frac{(\Sigma y)^2}{rcn} = 80$$

where r and c refer to the numbers of rows and columns in the data matrix, and n is the number of replicates in each treatment combination (i.e. in each cell of the matrix). The number of levels of factor A (protein level) is given by r (3), and the number of levels of factor B (alkaloid) by c (2). The *row totals*, therefore, reflect differences between the mean lengths of insects in the different protein treatments, and the *column totals* reflect differences between the alkaloid treatment means. We denote the row totals by R and the column totals by C. Now the treatment sum of squares for protein, SSA, is given by:

$$SSA = \frac{\Sigma R^2}{cn} - \frac{(\Sigma y)^2}{rcn} = 7$$

The point that always causes most difficulty in ANOVA is calculation of the term $\Sigma R^2/cn$. The row totals are found by adding up all of the replicates separately for each of the three levels of factor A (44, 42, 34). These totals are squared, then added up ($44^2 + 42^2 + 34^2 = 4856$). This total is then divided by *the number of numbers that were added together to get any one of the row totals*. There were eight numbers in every row in this case: the number of columns, $c = 2$, times the number of replicates, $n = 4$, in each cell of the data matrix; $cn = 8$. No matter how complex the analysis of variance becomes, the logic is the same. The sum of the squares of the sub-totals is divided by the number of numbers making up any one of the sub-totals.

Likewise, the treatment sum of squares for alkaloid, SSB, is:

$$SSB = \frac{\Sigma C^2}{rn} - \frac{(\Sigma y)^2}{rcn} = 6$$

where the sum of the squares of the column totals ($54^2 + 66^2 = 7272$) is divided by 12, the number of numbers in a column (namely the number of rows, r, times the number of replicates per cell, n: $3 \times 4 = 12$).

The calculations so far have been exactly the same as for one-way ANOVA except that we have two main effects rather than one. The novel aspect of factorial ANOVA comes in the calculation of the *interaction sum of squares*. To calculate this we need to write down a matrix containing the sub-totals of all the replicates in each of the treatment combinations (Table 8.8).

It is important to inspect this table carefully, because it exposes the nature of any interaction that might be present (it is often useful to plot these totals – or the individual cell means – against the level of factor A, drawing a separate graph for each of the levels of factor B; *if there is an*

Table 8.8 Interaction sub-totals: the Q matrix

Protein	Alkaloid $B(1)$	$B(2)$
$A(1)$	18	26
$A(2)$	14	28
$A(3)$	22	12

interaction, then the lines will not be parallel; see below). In the present example, it is clear that adding alkaloid to the diet increases larval length in all cases except the highest level of dietary protein, $A(3)$, when added alkaloid appears to cause a reduction in growth. We need to test the statistical significance of this interaction.

To do this, we calculate an *interaction sum of squares*, using the sub-totals in the matrix above. The individual sub-totals are denoted by Q. Each sub-total is squared, and the six squared sub-totals are added together ($18^2 + 14^2 + 22^2 + 26^2 + 28^2 + 12^2 = 2608$), then divided by the number of numbers that were added together to get each sub-total ($n = 4$). The term just calculated contains the interaction sum of squares that we require, but it also contains the sum of squares due to the main effects of both factors. In order to compute the interaction sum of squares, therefore, we need to subtract both of the main effect sums of squares as well as the correction term $CT = (\Sigma y)^2/rcn$. Try the calculation of the interaction sum of squares, $SSAB$, for yourself:

$$SSAB = \frac{\Sigma Q^2}{n} - \frac{\Sigma R^2}{cn} - \frac{\Sigma C^2}{rn} + \frac{(\Sigma y)^2}{rcn} \tag{8.12}$$

or, alternatively:

$$SSAB = \frac{\Sigma Q^2}{n} - SSA - SSB - CT$$

After this, it is simple to obtain the error sum of squares by subtraction:

$$SSE = SST - SSA - SSB - SSAB \tag{8.13}$$

The ANOVA table (Table 8.9) can now be drawn up. The only new element here concerns the degrees of freedom. There are r levels of treatment A, so there are $r - 1$ degrees of freedom for protein. There are c levels of treatment B, so there are $c - 1$ degrees of freedom for alkaloid. *Interaction degrees of freedom* are found by *multiplying together the relevant main effect degrees of freedom* to obtain $(r - 1)(c - 1)$. There are n replicates at each of the rc treatment combinations, and since there are $n - 1$ degrees of freedom for error in each cell of the matrix, there

Table 8.9 ANOVA table for two-factor factorial experiments

Source	SS	d.f.	MS	*F*-ratio	Probability
Factor *A*	*SSA*	$r-1$	*MSA*	MSA/s^2	Tables
Factor *B*	*SSB*	$c-1$	*MSB*	MSB/s^2	Tables
Interaction *A.B*	*SSAB*	$(r-1)(c-1)$	*MSAB*	$MSAB/s^2$	Tables
Error	*SSE*	$rc(n-1)$	$s^2 = \dfrac{SSE}{rc(n-1)}$		
Total	*SST*	$rcn-1$			

must be $rc(n-1)$ in all. There are rcn samples in the whole experiment, so there are $rcn-1$ degrees of freedom in total. As an algebraic exercise, you might like to check that $(r-1)+(c-1)+(r-1)(c-1)+rc(n-1)$ does indeed equal $rcn-1$.

For the present example, the ANOVA table can be completed as shown in Table 8.10. Check to make sure that you get the same sums of squares and degrees of freedom, then divide through to obtain the mean squares. The error variance is 1.556 (28/18) and we can compute three different *F*-ratios: one to test the main effect of protein, one the main effect of alkaloid, and one the interaction. It is always wise to begin the interpretation by looking at the significance of the interaction term first. This is because if the interaction *is* significant, then it is telling us that both factors are important in determining insect length. We do not gain a great deal from looking at the *F*-ratios for the main effects because we know that both of them are important already. The main-effect *F*-ratios can sometimes be positively misleading, as in the case of protein in this example. A naive interpretation of the analysis might be that protein is unimportant because it has a non-significant main effect ($F = 2.25$; $p = 0.1342$). But we know that protein *is* important, because it affects the way that insect length responds to alkaloid. The moral is simple. Always look at the interaction *F*-ratio first. If it is significant, then both factors are important. If the interaction is not significant, then inspection of the main-effect *F*-ratios will show which factors have important main effects and which do not.

Table 8.10 ANOVA table for factorial experiment on insect growth

Source	SS	d.f.	MS	*F*-ratio	Probability
Protein	7	2	3.5	2.25	0.1342
Alkaloid	6	1	6.0	3.86	0.0652
Interaction	39	2	19.5	12.54	0.0004
Error	28	18	$s^2 = 1.556$		
Total	80	23			

The standard error for any one of the cell means is $SE = \sqrt{(s^2/n)} = \sqrt{(1.556/4)} = 0.6236$ and the standard error of the difference between any two cell means is 0.8819 ($\sqrt{2(s^2/n)}$).

To conclude, it seems that in the absence of alkaloid, protein content has no significant effect on insect length. On the other hand, the presence of alkaloid leads to significant increases in length on low and medium protein diets, but to reduced growth on high protein diets. In practice, the next step would be to establish the mechanism by which this effect came about. In truth, however, the analysis is unconvincing with such small sample sizes, and further work would be necessary to substantiate the result and to demonstrate that this was a repeatable effect, before any effort was expended on trying to determine the mechanism.

8.6.1 Factorial ANOVA **in** *GLIM*

The procedure is a simple extension of one-way analysis except that now there are two kinds of treatments, and therefore two factors in the model (we shall call them A and B). In the present example, A has three levels and B has two levels. You should be able to follow the first part of the GLIM code:

```
$units 24 $

$data y a b $

$dinput 6 $

File name? glex11.dat

$factor a 3 b 2 $
```

For data inspection we use the **tab** directive to summarize the means and variances, including the interaction means, like this:

```
$tab the y mean for a $

       1      2      3
[ ]  5.500  5.250  4.250

$tab the y mean for b $

       1      2
[ ]  4.500  5.500

$tab the y mean for a;b $

      1      2
1  4.500  6.500
2  3.500  7.000
3  5.500  3.000
```

```
$tab the y var for a $

       1       2       3
[ ]  2.571   4.500   3.357

$tab the y var for b $

       1       2
[ ]  2.091   4.636

$tab the y var for a;b $

      1        2
1   1.6667   1.6667
2   1.6667   0.6667
3   1.6667   2.0000
```

A useful trick is to save the table of interaction means so that it can be plotted as a graph to look for interaction effects (i.e. for non-parallel lines). The six means are stored in a new vector called ym, then plotted like this:

```
$tab the y mean for a;b into ym by aa;bb $

$factor aa3 bb 2 $

$plot ym aa '*+' bb $
```

The **plot** directive says plot mean caterpillar growth increment on the *y*-axis against the three levels of factor A (i.e. dietary protein is the *x*-axis) using a different symbol for each level of B (* for alkaloid absent, + for alkaloid present). The graph will be easier to interpret if you join the dots together with straight lines between each symbol:

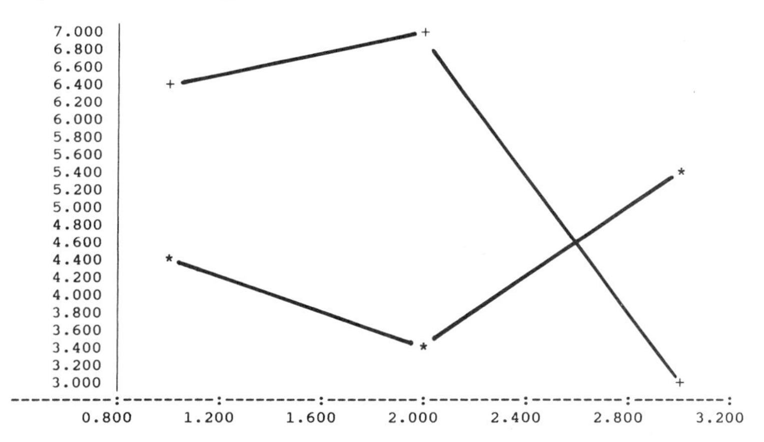

Far from being parallel, the two graphs actually cross over one another at the third level of A, which suggests a strong interaction effect. To begin the analysis we declare the response variable to be caterpillar growth increment, y, then estimate *SST* by fitting the null model:

```
$yvar y $

$fit $

deviance = 80.000
      d.f. = 23
```

The total sum of squares (*SST*) is 80.0 with 23 degrees of freedom. We now fit the *maximal model*. In a factorial we may be interested in the *main effect* of factor A, the main effect of factor B, and the *interaction* between factors A and B. There are several ways of fitting this model in GLIM. The long way is to write:

```
$fit + a + b + a.b $
```

which fits the main effects (as above) plus the interaction term (written a.b). A shorthand equivalent is:

```
$fit + a*b $
```

which fits the main effects and all the interaction terms in a single statement. The important thing to understand is the difference between '.' and '*':

a.b means the interaction between factor A and factor B;
a*b means the main effects of A and B plus the interaction a.b.

The GLIM output from the second part of the analysis is as follows:

```
deviance = 28.000 (change = −52.00)
      d.f. = 18        (change = −5)

$disp e $

     estimate  s.e.      parameter
1     4.500    0.6236    1
2    −1.000    0.8819    A(2)
3     1.000    0.8819    A(3)
4     2.000    0.8819    B(2)
5     1.500    1.247     A(2).B(2)
6    −4.500    1.247     A(3).B(2)
scale parameter taken as 1.556
```

The deviance shows that the error sum of squares is 28.0 with 18 degrees of freedom, so the error variance is $28/18 = 1.56$. You will notice that this is the value taken by the *scale parameter*, and you should check that this is the value obtained in the long-hand calculations earlier. The full model explains $100 \times 52/80 = 65\%$ of the total deviance; a reasonable, but not an excellent, description of the data.

The table obtained by the **disp e** directive looks somewhat different than in the case of one-way analysis. The rules for its construction, however, are consistent. As with one-way ANOVA, there are as many rows in the table as there are means to be estimated from the experiment. The three levels of A and the two levels of B allow six means overall, and that is why there are six rows in the table.

It is important that you understand what the parameter estimates mean, and where they come from. The general linear model for this particular factorial ANOVA is:

$$\hat{y} = \beta_0 + \beta_1 x_1 + \beta_2 x_2 + \beta_3 x_3 + \beta_4 x_4 + \beta_5 x_5 + \beta_6 x_6 \qquad (8.14)$$

where the x vectors define the treatment combinations, and there is one x vector for each combination of A and B (a total of six vectors in the present case). The x vectors contain 0's except where the y-values derive from the given combination of levels of A and B, in which case the x vector contains 1's. Thus, the vector x_1 represents the treatment combination A(1)B(1), and it has a value of 1 wherever the y-value came from this treatment combination, and 0 otherwise. As before, the β's are *differences between means*. As with one-way ANOVA, the first parameter β_0 is subsumed within β_1 because its value is aliased. There is no point in having seven parameters to estimate six means (see Section 4.20).

This section is important, and you should work through it slowly, checking the calculations as you go. Consider the table produced by **disp e**; as with the GLIM output from one-way ANOVA, the first parameter is labelled 1, and refers to the mean value of y when all factors are set to level 1. In our case this is the mean under the treatment combination A(1)B(1). Thus the 'intercept' is 4.5. Obtaining the mean for A(2)B(1) is easy; we just add the parameter (difference) A(2) on to the intercept: $4.5 + -1.0 = 3.5$. The mean for A(3)B(1) is obtained in exactly the same way: $4.5 + 1.0 = 5.5$. The A(1)B(2) mean is easy as well: $4.5 + 2.0 = 6.5$. Now comes the only tricky bit. When neither of the subscripts is 1, we must employ the interaction terms in order to calculate the means. Thus the mean for A(3)B(2) is obtained by adding the parameter for A(3) and the parameter for B(2) to the intercept, and then adding the interaction parameter for A(3)B(2), thus: $4.5 + 1.0 + 2.0 - 4.5 = 3.0$ (check this in the table, and work out the mean for A(2)B(2) for yourself).

In the standard error column, you will see that there are now three different values: one for the intercept (parameter 1), one for the other main effects (A(2), A(3) and B(2)), and one for the interaction terms (A(2)B(2) and A(2)B(3)). As with one-way ANOVA, the standard error of the intercept is given as the standard error of a single mean, $\sqrt{(s^2/n)} = \sqrt{(1.556/4)} = 0.6236$.The main effect terms are the differences between two means, and so the appropriate standard error is $\sqrt{2(s^2/n)} = \sqrt{2(1.556/4)} = 0.8819$. The interaction means were computed as the sum of four numbers: the intercept, plus the A main effect, plus the B main effect, plus the AB interaction effect. As you will recall, the standard error of a sum is the root of the sum of the individual variance components. Thus, the standard error of the interaction terms is $\sqrt{4(s^2/n)} = \sqrt{4(1.556/4)} = 1.2474$.

The next step is to compare the parameter estimates with their standard errors, and to determine which, if any, of the terms are statistically significant, and need to be retained in the model. There are two steps to this procedure. The first is to carry out simple t-tests on the parameters, by asking which of the parameters is more than 2 times its standard error (the value 2 is a reasonable rule of thumb for the value of Student's t; for degrees of freedom less than 6, it rises to about 2.5; see Appendix Table 1). The second step is to determine how the removal of a particular term from the model affects the amount of deviance explained. We will want to retain all parameters that are responsible for explaining a significant amount of the deviance, but the principle of parsimony requires that we remove terms from the model if they account for an insignificant amount of the deviance.

First, the t-tests. It is important to remember that if a parameter does not appear in the parameter list, its level is assumed to be 1. Thus, because neither A nor B is named in the parameter list for row 1, they are both at level 1. Starting at row 1 it is clear that the intercept is significantly greater than zero (this is the mean of the cell in the data table that has all its factor levels (subscripts) equal to 1; A(1)B(1)). Rows 2 and 3 show that none of the other mean differences for treatment A is significantly different from zero, so we conclude that A(1)B(1) = A(2)B(1) = A(3)B(1). The main-effect mean difference for treatment B(2) is just significant, so A(1)B(2) may be larger than A(1)B(1) ($t = 2.0/0.8819 = 2.250$; 5% tables have $t = 2.101$). The first interaction term is insignificant, but the second has a t-ratio of 3.609, a highly significant difference.

Second, we can try removing various terms from the model, in order to assess their contribution to the total deviance explained. In general, these *deletion tests* are to be preferred to t-tests as the means of hypothesis testing. Things are made a little easier if we use the scale directive to set the scaled error variance to unity, because then the change in scaled deviance resulting from removal of a given term from the model can be easily converted into an F-ratio. This makes the assessment of significance very simple. We set the **scale** parameter equal to the error mean square, then refit the full model:

```
$scale 1.556 $

-- model changed

$fit a*b $

scaled deviance = 17.995
           d.f. = 18
```

after which, the ratio of the scaled deviance to its degrees of freedom is equal to 1 (to a good approximation). Now we can ask whether the interaction term needs to be retained in the model, by removing A.B as follows:

```
$fit -a.b $

scaled deviance = 43.059 (change = +25.06)
           d.f. = 20     (change = +2)
```

which shows that removal of the interaction term causes an increase in scaled deviance of 25.06 with two degrees of freedom (recall that interaction degrees of freedom is the product of the main-effect degrees of freedom: $(3-1) \times (2-1) = 2$). We divide the change in scaled deviance by the change in degrees of freedom to find the F-ratio of 12.53; this is highly significant (F tables with 2 and 18 degrees of freedom give 6.01 at 1%). You should check back to the long-hand calculations and confirm that this is the same F-ratio as was obtained using familiar ANOVA methods (apart from a difference in the last decimal, resulting from GLIM's rounding of the scaled deviance to two places). Clearly, the interaction term is required in the model, so we add it back:

```
$fit +a.b $

scaled deviance = 17.995 (change = -25.06)
           d.f. = 18     (change = -2)
```

Testing significance by deletion of terms is generally preferred to the use of t-tests on parameter estimates, though in cases like the present example with normal errors and the identity link the procedures are identical. The process of hypothesis testing by step-wise deletion of explanatory variables is explained in more detail in Section 11.4.

It is useful at this stage to ask whether all the levels of each factor are required on the model, or whether one or more of them could sensibly be grouped together. For example, it is clear that the 'medium' protein treatment (the second level of treatment A) figures in neither significant main effects nor interaction terms, and we might consider combining the first and second levels of A (i.e. instead of having low, medium and high protein diets, we lump low and medium together, and just have high and low protein diets). We need to retain the third level of A because it enters into a significant interaction with B(2) (see row 6). These matters are dealt with fully in Chapter 12 on model simplification.

In conclusion, the GLIM analysis has demonstrated that there is a highly significant interaction between protein and alkaloid in their effects on insect growth increment. It does not appear to be necessary to distinguish between the low and medium protein diets, but high dietary protein is important because the impact of alkaloid is reversed with high protein diets. ●

8.7 Contrasts

The comparison of individual treatment means, or groups of treatment means, involves the technique known as *contrasts*. When considering comparisons between groups of means we have to be careful to distinguish groups:

1 where there were sensible ecological reasons for grouping together particular treatments *before* we saw the results of the experiment (these are known as *a priori* contrasts);

2 where we think we can see groups within the results after the analysis of the data (*a posteriori* contrasts).

In the past, tests like Duncan's multiple range test were used to carry out multiple comparisons between means. These days, the emphasis is on parameter estimation rather than hypothesis testing, and it is generally more informative to present a series of means and their standard errors than it is to carry out a range test. The problems of *a posteriori* tests are discussed by Day & Quinn (1989), and Rice (1989) outlines the problems of assessing tables of multiple statistical tests.

Contrasts are dealt with as follows. Suppose we have an experiment with five different treatments. The first step is to set up a vector of contrast coefficients c which reflects the *a priori* grouping to be tested. We work under the constraint that the sum of the contrast coefficients must be 0:

$$\Sigma c_i = 0$$

If we want to compare treatments 1 and 2 grouped together with treatments 3 and 5 grouped together (ignoring treatment 4), this would be:

$$c = 1\ 1\ -1\ 0\ -1$$

and if we wanted to contrast treatment 4 with all the others we would put:

$$c = 1\ 1\ 1\ -4\ 1$$

These contrast coefficients are then used to compute a contrast sum of squares, like this:

$$SSC = \frac{\left(\sum \frac{c_i T_i}{n_i}\right)^2}{\sum\left(\frac{c_i^2}{n_i}\right)}$$

and the significance of the single-degree-of-freedom contrast is assessed by calculating the F-ratio:

$$F = \frac{SSC}{s^2}$$

and comparing the result with F tables (with 1 d.f. in the numerator and the error d.f. in the denominator).

Two contrasts $\Sigma b_i \mu_i$ and $\Sigma c_i \mu_i$ are said to be *orthogonal* when $\Sigma b_i c_i / n_i = 0$ (there are n_i replicates in each treatment). When contrasts are orthogonal, their sums of squares are components of the total treatment sum of squares, SSA. Thus, the treatment sum of squares can be partitioned into at most $k-1$ independent, single-degree-of-freedom contrast sums of squares. You should check that the two contrasts presented above are orthogonal.

In GLIM, contrasts are dealt with very simply by **fitting** redefined factor levels. For example, in an experiment on plant growth we might have six replicates of five different clipping treatments: (1) 25% defoliation of neighbours; (2) 50% defoliation of neighbours; (3) root pruning to a depth of 5 cm; (4) controls with no disturbance of surrounding plants; and (5) root pruning to 10 cm depth. It was suspected in advance that treatments 1 and 2 might produce a similar response because they both dealt with defoliation alone, and that treatments 3 and 5 might be similar because they involved root-clipping. Treatment 4 stood out as being the only one which involved no disturbance of neighbouring plants (the control). The analysis proceeds as follows:

```
$units 30 $

$data y $

$dinput 6 $

File name? glex39.dat
```

The data are individual plant dry weights (mg) and the six replicates of each treatment T are in groups. Therefore:

```
$calc t=%gl(5,6) $

$factor t 5 $

$yvar y $
```

```
$fit:+t$disp e $

deviance = 209 377.
      d.f. = 29

deviance = 124 020. (change = -85 356.)
      d.f. = 25        (change = -4)

    estimate  s.e.   parameter
1   553.3     28.75  1
2    16.00    40.66  T(2)
3    57.17    40.66  T(3)
4   -88.17    40.66  T(4)
5    57.33    40.66  T(5)
scale parameter taken as 4961.
```

It looks as though the mean weight of plants from the control (treatment 4) is significantly lower than that following treatment 1 (and from all the other means as well, since they are bigger than in treatment 1), and that the means of treatments 3 and 5 are similar to one another (and probably not significantly greater than those of treatments 1 and 2). We test these ideas using contrasts as follows, beginning with the contrast between treatment 4 and all the others. We create a new factor T2 which has the value 1 for all treatments except 4, and a value of 2 for treatment 4:

```
$calc t2=1+(t==4) $

$factor t2 2 $
```

then fit the new contrast to the data:

```
$fit:+t2 $

deviance = 209 377.
      d.f. = 29

deviance = 139 342. (change = -70 035.)
      d.f. = 28        (change = -1)
```

This contrast has explained a major part of the total sum of squares attributable to treatments ($SSA = 85\,356$), and is highly significant ($F =$ 14.12). We conclude that pruning neighbours leads to a significant increase in plant dry weight. As a second contrast, we wish to test our original hypothesis that defoliation might differ from trenching. The means certainly show the right pattern, but are the differences significant? Because we want to ignore treatment 4 in this contrast, we must weight it out of the analysis using the **weight** directive. We create a new factor T3 that has a value of 1 for the defoliation treatments (1 or 2) and a value of 2 for the root-pruning treatments (3 or 5). It does not matter which of the values we give to treatment 4 since we intend to weight it out of the analysis (note, however, that it must be 1 or 2 otherwise the factor-out-of-range error will occur when we try to fit the factor). We use the **group** directive with specified *values* to calculate T3 as follows:

```
$assign values=1,1,2,1,2 $

$group t3=t v values $
```

then weight out treatment 4:

```
$calc wt=(t/=4) $

$weight wt $
```

and fit the new contrast:

```
$fit:+t3 $

deviance = 122749.
    d.f. = 23 from 24 observations

deviance = 108196. (change = -14553.)
    d.f. = 22       (change = -1) from 24 observations
```

This is not significant ($F = 2.93$) so we conclude that defoliation and root pruning were not significantly different in their effects on plant dry weight. Table 8.11 shows the full ANOVA for the contrasts.

Note that the two orthogonal contrasts account for over 99% of the total treatment sum of squares ($100 \times 84588/85356$). What other orthogonal contrasts account for the remainder? ●

Table 8.11 ANOVA table with contrasts

Source	SS	d.f.	MS	F
Clipping	85356	4	21339	4.30
(1,2) vs (3,5)	14553	1	14553	2.93
(1,2,3,5) vs (4)	70035	1	70035	14.12
Error	124021	25	$s^2 = 4961$	
Total	209377	29		

8.8 Nested analysis in GLIM

GLIM is built on the assumption that the linear model has a single error term. This means that nested experiments, like split-plot designs, where there is a different error variance for each different plot size, cannot be handled directly in GLIM. There are ways around this, as we shall see in Exercise 8.1, but, in general, a larger program like Genstat should be used for complex experimental designs that have several error terms (see Section 4.12).

GLIM is perfectly capable of handling the analysis of nested samples, however, so long as you know exactly what you are doing. In the **fit** directive we can use the slash symbol / to mean 'nested within'. Thus,

```
$fit a/b $
```

means 'fit the factors A and B with factor B nested within factor A'. We might have measured six trees at each of three sites. Tree number 1 at site 1 has nothing in common with tree 1 at sites 2 or 3, so the trees are nested within sites. Thus, it is sensible to estimate a term for differences between the sites, but not for differences between the trees numbered 1, 2 or 3. What we are really interested in is the degree to which trees within sites differ from one another. The terms we want to estimate are a main effect for site (A) and an interaction effect (A.B) for trees within sites. We could spell this out in long hand by writing:

```
$fit a + a.b $
```

which is synonymous with a/b.

We take an example from Sokal & Rohlf (1981). The experiment involved a simple one-factor ANOVA with three treatments given to six rats. The analysis was complicated by the fact that three preparations were taken from the liver of each rat, and two readings of glycogen content were taken from each preparation. This generated six pseudoreplicates per rat to give a total of 36 readings in all. Clearly, it would be a mistake to analyse these data as if they were a straightforward one-way ANOVA, because that would give us 33 d.f. for error. In fact, since there are only two rats in each treatment, we have only 1 d.f. per treatment, giving a total of 3 d.f. for error.

The variance is likely to be different at each level of this nested analysis because:

1 the readings differ because of variation in the glycogen detection method within each liver sample;

2 the pieces of liver may differ because of heterogeneity in the distribution of glycogen within the liver of a single rat;

3 the rats will differ from one another in their glycogen levels because of sex, age, size, genotype, etc.

If we want to test whether the experimental treatments have affected the glycogen levels, then we are not interested in (1) or (2). We could add all the pseudoreplicates together, and analyse the six averages. This would have the virtue of showing what a tiny experiment this really was (see

Section 8.9, below). But to analyse the full data set, we proceed as follows:

```
$units 36 $

$data y $

$dinput 6 $

File name? glex38.dat
```

The only trick is to ensure that the factor levels are set up properly. There were three treatments, so we make a treatment factor T with three levels. While there were six rats in total, there were only two in each treatment, so we declare rats as a factor R with two levels (not six). There were 18 bits of liver in all, but only three per rat, so we declare liver-bits as a factor L with three levels (not 18).

```
$calc t=%gl(3,12) $

$calc r=%gl(2,6) $

$calc l=%gl(3,2) $

$factor t 3 r 2 l 3 $
```

Initial data inspection using **tab for** will give us the treatment means for each rat:

```
$t t y m f t;r $

      1       2
1  132.5   148.5
2  149.7   152.3
3  134.3   136.0
```

Perhaps treatment 2 had somewhat greater glycogen (it gave the two largest means, 149.7 and 152.3) but the rats within a given treatment differed from one another by a greater degree (e.g. in treatment 1 the two rat means ranged from 132.5 to 148.5). We obtain *SST* in the usual way:

```
$yvar y $

$fit $

deviance = 3330.2
    d.f. = 35
```

and obtain the treatment sum of squares as normal:

```
$fit +t $
```

```
deviance = 1772.7 (change = −1558.)
    d.f. = 33     (change = −2)
```

The difference in a nested design is that we estimate a term for rats-within-treatments that will be used as the error variance for comparing the treatments:

```
$fit+t/r $

deviance = 975.00 (change = −797.7)
    d.f. = 30     (change = −3)
```

Next we compute liver bits within rats; note the double nesting:

```
$fit +t/r/l $

deviance = 381.00 (change = −594.0)
    d.f. = 18     (change = −12)
```

We now have all the components necessary for drawing up the nested ANOVA table. The term for readings-within-liver-bits is simply the residual deviance with 18 d.f. In nested designs, the *F*-ratios are computed by *dividing each variance by the variance immediately below it* in the table (not by the error variance as in a factorial design) – see Table 8.12.

Table 8.12 Nested ANOVA for rat data

Source	SS	d.f.	MS	*F*	Significance
Treatment	1557.55	2	778.78	2.93	n.s.
Rats within treatments	797.67	3	265.89	5.37	$P<0.05$
Liver bits within rats	594.00	12	49.50	2.34	$P=0.05$
Readings within liver bits	381.00	18	21.17		
Total	3330.22	35			

The *F*-value in tables with 2 d.f. in the numerator and 3 d.f. in the denominator is 9.55, so the calculated value of 2.93 falls well short of significance. There is no evidence that treatment affected the glycogen levels in the liver of these rats. Note, however, that the rats differed significantly from one another in mean glycogen level (5.37 is greater than the value in tables).

You would make a serious mistake if you were to try to interpret significance by looking at GLIM's parameter estimates in a nested design:

```
$disp e $
```

```
         estimate   s.e.    parameter
    1    130.5      3.253   1
    2     20.50     4.601   T(2)
    3     -1.000    4.601   T(3)
    4     18.50     4.601   T(1).R(2)
    5      2.000    4.601   T(2).R(2)
    6      9.500    4.601   T(3).R(2)
    7     -2.500    4.601   T(1).R(1).L(2)
    8      8.500    4.601   T(1).R(1).L(3)
    9     -7.500    4.601   T(1).R(2).L(2)
   10      6.000    4.601   T(1).R(2).L(3)
   11     -3.000    4.601   T(2).R(1).L(2)
   12     -1.000    4.601   T(2).R(1).L(3)
   13     -6.000    4.601   T(2).R(2).L(2)
   14      4.000    4.601   T(2).R(2).L(3)
   15      8.500    4.601   T(3).R(1).L(2)
   16      6.000    4.601   T(3).R(1).L(3)
   17     -0.500    4.601   T(3).R(2).L(2)
   18     -8.500    4.601   T(3).R(2).L(3)
   scale parameter taken as 21.17
```

This suggests that treatment 2 produced significantly higher glycogen levels ($t = 20.5/4.601 \gg T$ from tables). This has happened because GLIM takes the readings within liver bits as the error term (compare the scale parameter of 21.17 with our ANOVA table). With nested designs you can use GLIM to estimate the changes in deviance, but you will need to draw up the ANOVA table yourself in order to test the significance correctly. *Don't do t-tests on the parameter estimates in a nested design.*

8.9 Handling pseudoreplication in GLIM

In a pseudoreplicated study, like the rat's liver example in the previous section, it is both simpler and better to analyse the means rather than the raw data. By averaging over the pseudoreplicates, we reduce the amount of data but obtain a substantial benefit because the resulting GLIM analysis will be correct. The danger of misinterpretation which we saw in the previous analysis will thus be eliminated.

Continuing where we left off, the first step is to reduce the data and eliminate the pseudoreplication. This is extremely straightforward in GLIM because we can save summary tables for subsequent analysis using the **tab into** directive. Create a new vector called ym which contains the mean glycogen level for each rat:

```
$tab the y mean for t;r into ym by tt;rr $
```

Next, we change the **units** directive to reflect the reduced size of the new data set, and then declare the vector of means ym as the response variable:

```
$units 6 $
```

```
$factor tt 3 $

$yvar ym $
```

To test the effect of treatments we use the shortened vector of treatment codes, TT, that was generated in the **tab-into-by** expression, above:

```
$fit:+tt$disp e $

deviance = 392.54
      d.f. = 5

deviance = 132.94 (change = −259.6)
      d.f. = 3          (change =−2)
```

This gives *SST* for the averages, *SSA* and *SSE*, as shown in Table 8.13.

Table 8.13 Non-pseudoreplicated ANOVA for rat data

Source	SS	d.f.	MS	*F*	Significance
Treatment	259.60	2	129.80	2.93	n.s.
Rats within treatments	132.94	3	44.31		
Total	392.54	5			

Obviously, the numerical values of the sums of squares are different than they were in the raw data, but the *F*-ratio for the comparison of the treatment means is identical. The big advantage of this approach is that when we use **disp e** we get the correct standard errors for the differences between treatment means:

```
$disp e $

   estimate  s.e.    parameter
1  140.5     4.707   1
2   10.50    6.657   TT(2)
3  −5.333    6.657   TT(3)
scale parameter taken as 44.31
```

In contrast to the pseudoreplicated case, the standard errors now demonstrate that there is no significant difference between the average glycogen levels in the three different treatments.

The moral is that nested and other pseudoreplicated designs need to be analysed with care. For large, complex experiments, it is better to use Genstat because it can handle multiple error terms. For simpler designs with pseudoreplication it is best to use GLIM to create a table of the averages, and then to analyse these rather than the raw data. If you do this, the error degrees of freedom will be correct, and the **disp e** directive will give you the proper standard errors for the differences between treatment means. ●

CHAPTER 9

Analysis of covariance

Analysis of covariance combines elements from regression and analysis of variance. Thus, experiments with one or more factors may also have one or more explanatory variables associated with each measurement of the response variable. For example, in an experiment on crop yields with three different kinds of soil nutrients, the initial concentrations of nitrogen, phosphorus and potassium in the soil might serve as covariates. Again, in a study of seed production in plants under different levels of herbivory, the initial size of the plants would be a useful covariate.

The aim is to take account of measurable variability that we anticipate might influence the results. Of course, if we were in a position to block the experimental design to take account of this variability, then this might be preferable to the use of covariates (see Section 4.14). We can never measure everything that we would like to, and Murphy's law of covariates says that the things we choose to measure will turn out to be unimportant, while the things we do not (or cannot) measure will turn out to be vitally significant.

The calculations are as follows. We compute the sums of squares for treatments as in straightforward ANOVA. We also calculate the sums of products for the regression of the response variable on the covariate, keeping accounts separately for each of the treatments. The idea is to fit a common regression slope through each of the treatments (unless we can demonstrate that the slopes are significantly different in each treatment; see below). A comparison of the treatments is then based on a comparison of the intercepts of the various parallel lines (see Fig. 9.1). If the regression slopes are not parallel within the different treatments, then it makes no sense to compare their intercepts, unless some point on the x-scale acts as a natural reference point, and this is set as the origin.

9.1 ANCOVA: a worked example

The worked example concerns an experiment on the impact of grazing on the fecundity (seed production) of a biennial plant. Forty plants were allocated at random to two treatments, grazed and ungrazed, and the grazed plants were exposed to rabbits during the first two weeks of stem elongation. They were then protected from subsequent grazing by the erection of a fence and allowed to regrow. Because initial plant size was thought likely to influence fruit production, the diameter of the top of the rootstock was measured before each plant was potted up. At the end of

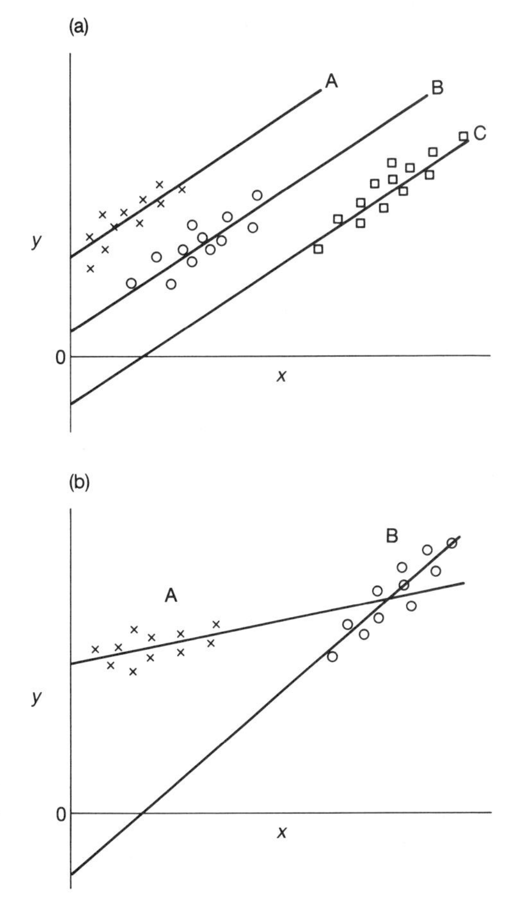

Fig. 9.1 Analysis of covariance. (a) Simple ANCOVA fits lines of common slope through data from different levels of a factor. (b) When the best-fit lines are not parallel, then tests based on comparisons of the intercepts will usually be uninformative.

the growing season, the fruit production (dry wt, mg) was recorded on each of the 40 plants, and this forms the response variable in the following analysis. (Table 9.1 records the data.)

Now work out the sums, sums of squares and sums of products for each treatment separately (20 pairs of numbers) and for the whole data set combined (40 pairs of numbers). You should get the totals shown in Table 9.2.

We start by performing a straightforward one-way ANOVA on the total data set, calculating *SST*, *SSA* and *SSE* in the usual way (see Section 8.2):

$$SST = 164\,928 - \frac{2376.42^2}{40} = 23\,743.7$$

$$SSA = \frac{1017.61^2 + 1358.81^2}{20} - \frac{2376.42^2}{40} = 2910.4$$

Table 9.1 Grazing and fecundity: fruit (mg dry weight); root size (mm)

Ungrazed plants		Grazed plants	
Fruit	Roots	Fruit	Roots
59.77	6.225	80.31	8.988
60.98	6.487	82.35	8.975
14.73	4.919	105.1	9.844
19.28	5.130	73.79	8.508
34.25	5.417	50.08	7.354
35.53	5.359	78.28	8.643
87.73	7.614	41.48	7.916
63.21	6.352	98.47	9.351
24.25	4.975	40.15	7.066
64.34	6.930	116.1	10.25
52.92	6.248	38.94	6.958
32.35	5.451	60.77	8.001
53.61	6.013	84.37	9.039
54.86	5.928	70.11	8.910
64.81	6.264	14.95	6.106
73.24	7.181	70.70	7.691
80.64	7.001	71.01	8.515
18.89	4.426	83.03	8.530
75.49	7.302	52.26	8.158
46.73	5.836	46.64	7.382

$$SSE = SST - SSA = 23\,743.7 - 2910.4 = 20\,833.3$$

Now we work out the regression slopes in each of the two treatments separately (see Section 7.5 for details). For the ungrazed plants, we have:

$$SSXY_1 = 6509.5 - \frac{121.06 \times 1017.61}{20} = 349.91$$

$$SSX_1 = 747.34 - \frac{121.06^2}{20} = 14.56$$

$$b_1 = \frac{SSXY}{SSX} = \frac{349.91}{14.56} = 24.03$$

Table 9.2 Sums of squares and products

	Ungrazed	Grazed	Total
ΣX	121.06	166.19	287.25
ΣY	1017.61	1358.81	2376.42
ΣX^2	747.34	1400.83	2148.17
ΣY^2	60772	104156	164928
ΣXY	6509.5	11753.6	18263.2

and for the grazed plants, we obtain:

$$SSXY_2 = 11\,753.6 - \frac{166.19 \times 1358.81}{20} = 462.57$$

$$SSX_2 = 1400.83 - \frac{166.19^2}{20} = 19.87$$

$$b_2 = \frac{SSXY}{SSX} = \frac{462.57}{19.87} = 23.28$$

Table 9.3 Corrected sums of squares and products

	Ungrazed	Grazed	Total
SSX	14.56	19.87	85.36
SSY	8995.49	11 837.77	23 743.70
$SSXY$	349.91	462.57	1 197.53

The two slopes are clearly not significantly different, so we use ANCOVA to determine a single, common slope for the two grazing treatments. This involves a new technique, in that the common slope is estimated from the error line in the ANOVA table.

There is only one new term to be computed, namely the *between-treatments corrected sum of products*. This is calculated from the subtotals of x and y within each treatment; in the present case, each of these is the sum of 20 replicates:

$$SSXY_{\text{between}} = \Sigma\left(\frac{\Sigma x\ \Sigma y}{n}\right) - \frac{\Sigma\Sigma x \Sigma\Sigma y}{kn}$$

so, for our example, this is:

$$\frac{121.06 \times 1017.61 + 166.19 \times 1358.81}{20} - \frac{287.25 \times 2376.42}{40} = 384.96$$

Now the within-treatments sum of products is obtained by subtraction:

$$SSXY_{\text{within}} = SSXY_{\text{total}} - SSXY_{\text{between}} = 1197.53 - 384.96 = 812.57$$

The within-treatments sum of squares for x is just the sum of the two separate terms we have computed already:

$$SSX_{\text{within}} = SSX_1 + SSX_2 = 14.56 + 19.87 = 34.43$$

Now the common slope is given by:

$$b = \frac{SSXY_{\text{within}}}{SSX_{\text{within}}} = \frac{812.57}{34.50} = 23.60$$

and the regression sum of squares:

$$SSR = b.SSXY_{\text{within}} = \frac{SSXY^2_{\text{within}}}{SSX_{\text{within}}} = \frac{812.57^2}{34.50} = 19\,177.17$$

The analysis of covariance table can now be completed, as shown in Table 9.4 where the error sum of squares (deviations from the regression, *SSE*) is obtained by subtraction.

Table 9.4 ANCOVA table for regrowth data

Source	d.f.	*SSX*	*SSXY*	*SSY*
Total	39	85.36	1197.53	23 743.70
Between grazing treatments	1	50.93	384.96	2910.44
Within grazing treatments	38	34.43	812.57	20 833.26
Regression	1			19 177.17
Error	37			1656.1

The next step is to calculate the intercepts for the two parallel regression lines. This is done exactly as before, by rearranging the equation of the straight line to obtain $a = y - bx$. For each line we can use the mean values of x and y, with the common slope in each case. Thus:

$$a_1 = \bar{Y}_1 - b.\bar{X}_1 = 50.88 - 23.60 \times 6.053 = -91.729$$

$$a_2 = \bar{Y}_2 - b.\bar{X}_2 = 67.94 - 23.60 \times 8.309 = -127.82$$

This demonstrates that the grazed plants produce, on average, 36.09 fruits fewer than the ungrazed plants $(127.82 - 91.73)$.

The next step is to calculate the standard errors for the common regression slope and for the difference in mean fecundity between the treatments. First we calculate the error variance. This is obtained by dividing the error sum of squares (above) by the error degrees of freedom. We have estimated three parameters in order to calculate SSE_r (i.e. the common slope and the two treatment means), so the error degrees of freedom is $40 - 3 = 37$ d.f. The error variance, s^2, is therefore:

$$s^2 = \frac{1656.1}{37} = 44.76$$

The standard errors are obtained as follows. The standard error of the common slope is found in the usual way (Section 7.4.1):

$$SE_b = \sqrt{\frac{s^2}{SSX}} = \sqrt{\frac{44.76}{14.56 + 19.87}} = 1.140$$

The standard error of the intercept of the regression for treatment number 1 (ungrazed) is also found in the usual way (see Section 7.4.1):

$$SE_a = \sqrt{s^2\left[\frac{1}{n_1} + \frac{(0 - \bar{X}_1)^2}{SSX}\right]} = \sqrt{44.76\left[\frac{1}{20} + \frac{6.053^2}{34.43}\right]} = 7.062$$

It is clear that the intercept of −91.73 is very significantly less than zero, suggesting that there is a threshold rootstock size before reproduction can begin.

Finally, the standard error of the difference between the elevations of the two lines (the grazing effect) is given by:

$$SE_{\hat{y}_1 - \hat{y}_2} = \sqrt{s^2\left[\frac{2}{n} + \frac{(\bar{x}_1 - \bar{x}_2)^2}{SSX}\right]}$$

which, substituting the values for the error variance and the mean rootstock sizes of the plants in the two treatments, becomes:

$$SE_{\hat{y}_1 - \hat{y}_2} = \sqrt{44.76\left[\frac{2}{20} + \frac{(6.053 - 8.3095)^2}{34.43}\right]} = 3.331$$

This suggests that any lines differing in elevation by more than 6.66 mg dry wt would be regarded as significantly different. Thus, the present difference of 36.09 represents a highly significant reduction in fecundity caused by grazing ($t = 10.83$).

9.3 ANCOVA in GLIM

We now repeat the analysis using GLIM. The response variable is fecundity, and there is one experimental factor (grazing) with two levels (ungrazed and grazed) and one covariate (initial root stock diameter). There are 40 values for each of these variables. We can read the fruit production (y) and initial root size (x) from a data file called glex3.dat and create the grazing factor levels (G) using %gl.

```
$units 40 $

$data x y $

$dinput 6 $

File name? glex3.dat

$calc g=%gl(2,20) $

$factor g 2 $
```

In order to see the effect of grazing on plant fecundity we simply tabulate the means for each level of G:

```
$tab the y mean for g $
```

and GLIM responds with:

```
        1      2
[ ]   50.88  67.94
```

This analysis gives the puzzling result that the grazed plants appear to have produced more fruits than the ungrazed plants. In a case like this, the first thing to do is check that the factor levels are the right way round, and that level 2 really does represent the grazed treatment. It does, and so the result appears to be sound; the shoots that regrow after grazing produce more fruits than the ungrazed shoots. The next step is to see whether this increase is statistically significant. We declare fruits to be the response variable:

```
$yvar y $

$fit $
```

and GLIM prints the total sum of squares:

```
deviance = 23 744.
    d.f. = 39
```

Note that because of rounding errors, the numerical values of the GLIM results are slightly different than those in the example worked by hand. How much of the variation in fruit production represented by *SST* is explained by differences due to the grazing treatment? To see this, we simply fit the factor G to the model:

```
$fit+g $
```

and GLIM calculates that grazing accounts for 2910 (about 12%) of the variation in fecundity:

```
deviance = 20 833. (change = -2910.)
    d.f. = 38      (change =     1)
```

There are two ways we can test the significance of this change in deviance (both are equivalent in this simple example with normal errors and an identity link; see Section 11.4). We can do a *t*-test on the parameter estimates for grazing:

```
$disp e $

   estimate  s.e.    parameter
1  50.88     5.236   1
2  17.06     7.404   G(2)
scale parameter taken as 548.2
```

The mean fecundity of the grazed plants, G(2), is higher than the ungrazed by 17.06 and this difference is statistically significant; a *t*-test gives a value of 17.06/7.404 = 2.30 (higher than the value in tables with 38 d.f.). In general, it is better to carry out significance tests on the change in deviance that results when a factor is removed from the maximal model. Here, removal of the grazing factor would cause an increase in deviance of 2910 with 1 d.f. We test the significance of this using an *F*-test. The variance in the numerator is 2910/1 = 2910. For the denominator, we use the residual variance 20 833 divided by the residual degrees of freedom 38 to give 548.24 (note that this is the scale factor, given after the **disp e** directive). Thus, the *F*-ratio can be calculated as:

```
$calc 2910/548.2 $
```

giving a value of 5.308. This is larger than the value in tables with 1 and 38 d.f., and so the increase in fecundity following grazing is statistically significant.

Now we might leave it at this, were it not for the fact that the result is so odd; examples of overcompensation for grazing are very rare, and the experiments often turn out to be flawed in one way or another. We have extra information on the initial size of the plants that has not been used so far. It is plausible, for example, that the randomization led to a greater number of large plants being allocated to the grazed treatment. We begin by assessing the strength of the relationship between fecundity and initial root size:

```
$plot y x $
```

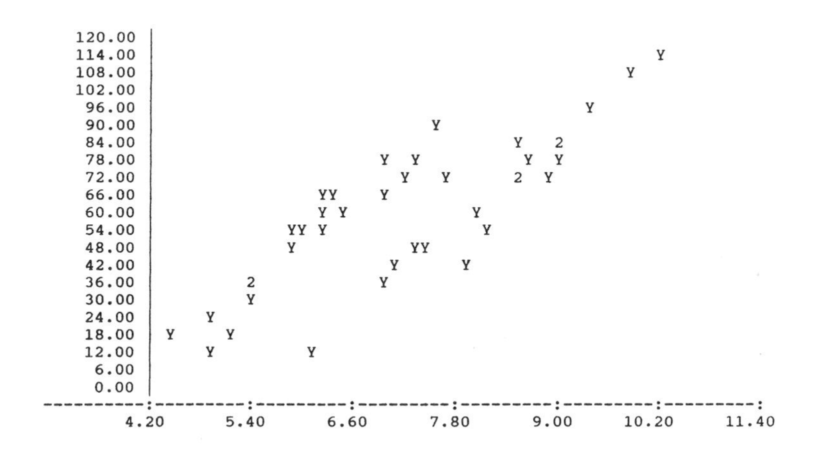

There is a clear positive correlation, but it would be informative to see which of the plants were grazed and which were ungrazed. To do this, we use different plotting symbols for the two levels of G: ⋆ to represent ungrazed and o to represent grazed:

$plot y x '⋆o' g $

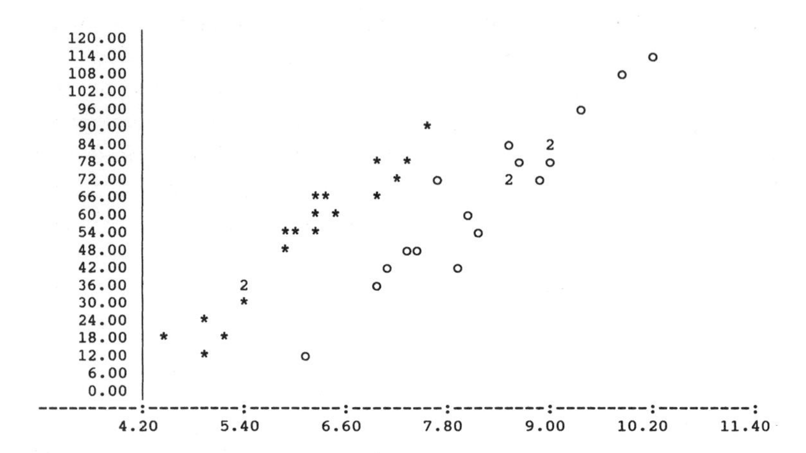

This does indeed suggest that the largest plants were allocated to the grazed (o) treatments. It also suggests that for a given rootstock diameter (say 6.60 mm) the grazed plants produced *fewer* fruits than the ungrazed plants. We test the significance of this by adding root diameter to the model as a covariate:

$fit +x $

to which GLIM responds with:

```
deviance = 1684.5 (change = −19149.)
    d.f. = 37     (change = −1)
```

This has produced an enormous reduction in deviance of 19 150 (81% of the total deviance in fruit production), and it is evident that initial plant size has a highly significant effect on fruit production of both grazed and ungrazed plants. To see the effect of adding the covariate on the other terms in the model we display the parameter estimates:

$disp e $

	estimate	s.e.	parameter
1	−91.73	7.115	1
2	−36.10	3.357	G(2)
3	23.56	1.149	X

scale parameter taken as 45.53

The first thing you see is the highly significant impact of initial root diameter (x) on fecundity ($t = 20.5$). The important point, however, is that by adding the covariate to the model, we have *reversed* the impact of the grazing treatment, G(2), on predicted fecundity. Without the covariate, the model predicted that grazing led to an increase in mean fecundity of 17.06 mg dry wt of fruit. With the covariate, on the other hand, the model predicts that grazing leads to a *reduction* in fecundity of 36.1 mg.

There is an extremely important general message in this example for experimental design. No matter how carefully we randomize at the outset, our experimental groups are likely to be heterogeneous. Sometimes, as in this case, we may have made initial measurements that we can use as covariates later on, but this will not always be the case. There are bound to be important factors that we did not measure. If we had not measured initial root size in this example, we would have come to entirely the wrong conclusion about the impact of grazing on plant performance.

A far better design for this experiment would be to measure the rootstock diameters of all the plants at the beginning of the experiment (as was done here), but then to place the plants in matched pairs with similarly sized rootstocks, and allocate one of the pair to the grazing treatment by tossing a coin. Then the size ranges of the two treatments would have overlapped and the analysis of covariance would have been unnecessary.

The model appears to be minimal adequate, and all the parameters are significant (i.e. they are significantly different from zero). You should check to ensure that the data do not support the need for different regression slopes for each grazing treatment by fitting the interaction term (**fit** + G.x, which gives a reduction in deviance of 4.81). It simply remains to test the adequacy of the error distribution by inspecting the residuals:

```
$calc resid=y−%fv $

$plot resid %fv '+' $

$plot resid x '*' $
```

The plots of residuals are satisfactory, showing no systematic patterns. We can get an impression of the overall fit of the model by plotting the observed and fitted values on the same axes. We shall plot + for both sets of fitted values with * for the ungrazed and o for the grazed plants as before.

```
$plot %fv y x '++*o' g $
```

Note carefully the order of the plotting symbols '++*o'. All the symbols for the first response variable (%fv) come first (++), then both the symbols for the measured y values (*o). The order of these is important, the first being associated with G=1 (⋆ = ungrazed) and the second with G=2 (o = grazed).

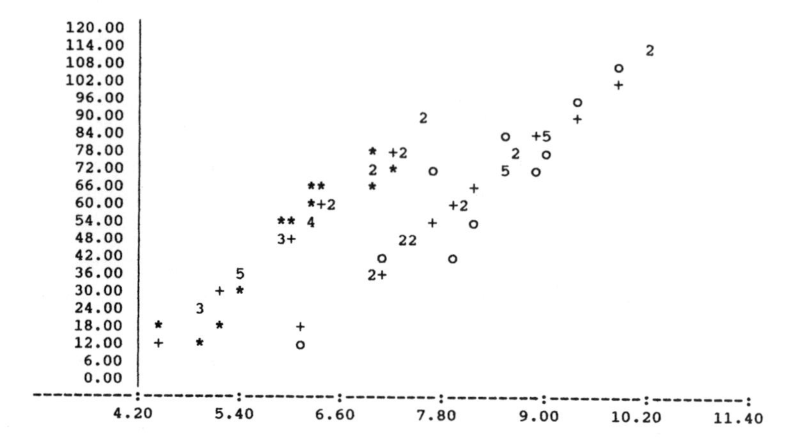

The fit of the model to the data is good, and we conclude that there is a reasonably linear relationship between initial plant size and fecundity and that grazing reduces fecundity by about 36 mg per plant.

More examples of analysis of covariance are to be found in Sections 12.5.4 and 15.7 and in Exercise 9.1. ●

CHAPTER 10

Generalized linear models

One of the common misconceptions about GLIM is that linear models involve a straight-line relationship between the response variable and the explanatory variables. This is not the case, as you can see from Figs 10.1 and 17.2, which show a variety of useful linear models. The definition of a linear model is an equation that contains mathematical variables, parameters and random variables that is *linear in the parameters and in the random variables*.

What this means is that if a, b and c are parameters, then obviously

$$y = a + bx \tag{10.1}$$

is a linear model, but so is

$$y = a + bx + cx^2 \tag{10.2}$$

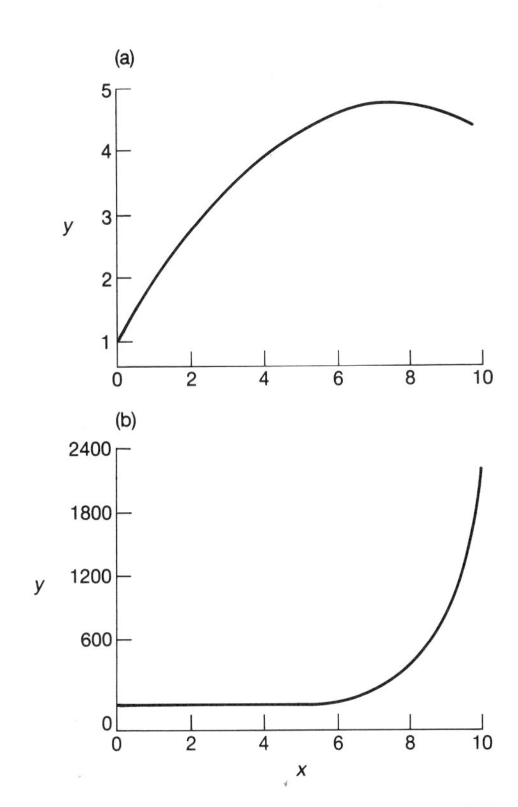

Fig. 10.1 Linear models are not necessarily straight-line models: (a) polynomial ($y = 1 + x - x^2/15$); (b) exponential ($y = 3 + 0.1e^x$).

because x^2 can be replaced by z, which gives

$$y = a + bx + cz \tag{10.3}$$

and so is

$$y = a + be^x \tag{10.4}$$

because we can create a new variable $z = \exp(x)$, so that

$$y = a + bz \tag{10.5}$$

Some models are non-linear but can be readily linearized by transformation. For example:

$$y = \exp(a + bx) \tag{10.6}$$

on taking logs of both sides, becomes

$$\ln y = a + bx$$

GLIM can deal with this model by specifying the log link (see Section 10.4.1). Again, the Holling 'disk equation' (also known as the Michaelis–Menten or Briggs–Haldane equation) is a much-used asymptotic relationship, given by:

$$y = \frac{ax}{1 + bx} \tag{10.7}$$

This is non-linear in the parameter b, but it is readily linearized by taking reciprocals. Thus

$$\frac{1}{y} = \alpha + \beta\frac{1}{x}$$

GLIM handles this family of equations by a transformation of the explanatory variable ($z = 1/x$) and using the reciprocal link (often, for data like this, associated with gamma rather than normal errors; see Section 17.2).

Other models are *intrinsically non-linear* because there is no transformation that can linearize them in all the parameters. Some important ecological examples include the hyperbolic function

$$y = a + \frac{b}{c + x} \tag{10.8}$$

and the asymptotic exponential

$$y = a(1 - be^{-cx}) \tag{10.9}$$

where both models are non-linear unless the parameter c is known in advance.

In cases like this, GLIM is unable to estimate the full set of parameters using the **fit** directive, but all is not lost. It is straightforward to write a macro that will estimate the value of c by trial and error. The model is fitted to the data many times with different values of c each time, and the value of c that gives the smallest deviance when the remaining parameters are estimated by maximum likelihood is selected. Examples of this technique are given in Chapter 19 and Exercise 19.3.

The only difficulty with fitting these intrinsically non-linear models with GLIM is that we need: (i) to compensate for the loss of an extra degree of freedom for every parameter that we estimate from the data in advance of fitting the linearized model; and (ii) to interpret the results with care because of the extra uncertainty introduced by adding externally determined parameters to the model (e.g. an unknown degree of underestimation of the standard errors).

10.1 Generalized linear models

A generalized linear model has three important properties:

1 the *error structure*;

2 the *linear predictor*;

3 the *link function*.

These are all likely to be unfamiliar concepts. The ideas behind them are straightforward, however, and it is worth learning what each of the concepts involves.

10.2 The error structure

Up to this point, we have dealt with the statistical analysis of data with normal errors. Many kinds of ecological data, however, have non-normal errors. For example:

1 errors that are strongly skewed;

2 errors that are kurtotic;

3 errors that are strictly bounded (as in proportions);

4 errors that cannot lead to negative fitted values (as in counts).

In the past, the only tools available to deal with these problems were transformation of the response variable or the adoption of non-parametric methods. GLIM allows the specification of a variety of different error distributions:

1 Poisson errors, useful with count data;

2 binomial errors, useful with data on proportions;

3 gamma errors, useful with data showing a constant coefficient of variation;

4 exponential errors, useful with data on time to death (survival analysis).

The error structure is defined by means of the **error** directive, used like this:

```
$yvar z $

$error p $
```

which means that the response variable z has Poisson (P) errors. The only exception to this format involves binomial errors (B). Here we need an extra variable in the **error** directive which contains the sample size (n) from which the binomial sample (x) was drawn (this is called the *binomial denominator* and is accessible via the system pointer called %bd). For binomial errors and a fixed sample size of 50, we would write:

```
$yvar x $

$calc n=50 $

$error b n $
```

If the sample size varies from case to case, it will be necessary to read in each different value of n with its matching x-value. It is important to ensure that none of the values in the n vector is zero or negative, and that none of the x's is bigger than its matching n.

10.3 The linear predictor

The structure of the model relates each observed y-value to a predicted value. The predicted value is obtained *by transformation of the value emerging from the linear predictor*. The linear predictor, η, pronounced 'eta', is a linear sum of the effects of one or more explanatory variables, x_j:

$$\eta_i = \sum_{j=1}^{p} x_{ij}\beta_j \tag{10.10}$$

where the x's are the values of the p different explanatory variables, and the β's are the (usually) unknown parameters to be estimated from the data. The right-hand side of the equation is called the *linear structure*.

There are as many terms in the linear predictor as there are parameters to be estimated from the data. Thus, with a simple regression, the linear predictor is the sum of two terms: the intercept and the slope. With a one-way ANOVA with four treatments, the linear predictor is the sum of four terms: the mean for treatment 1 and the three differences of the other treatment means when compared with treatment 1. If there are covariates in the model, they add one term each to the linear predictor. Each interaction term in a factorial ANOVA adds one more term to the linear predictor.

To determine the fit of a given model, GLIM evaluates the linear predictor for each value of the response variable, then compares the predicted value with a *transformed* value of y. The transformation to be employed is specified in the link function (see below). The fitted value, %fv, is computed by applying the inverse of the link function, in order to get back to the original scale of measurement of the y-variable. Thus, with a log link, the fitted value is the antilog of the linear predictor, and with the reciprocal link it is the reciprocal of the linear predictor.

10.4 The link function

One of the difficult things to grasp about GLIM is the relationship between the values of the response variable (as measured in the data and predicted by the model in fitted values) and the linear predictor. The thing to remember is that the *link function relates the mean value of* y *to its linear predictor* as given in the **disp e** directive. In symbols, this means that:

$$\eta = g(\mu) \tag{10.11}$$

which is simple, but needs thinking about. The linear predictor, η, emerges from the linear model as a sum of the terms for each of the p parameters. *This is not a value of* y (except in the special case of the *identity link* $\eta = \mu$). The value of η is obtained by transforming the value of y by the link function, and the predicted value of y is obtained by applying the inverse link function to η.

The most frequently used link functions are shown in Table 10.1. An important criterion in the choice of link function is to ensure that the fitted values stay within reasonable bounds. We would want to ensure, for example, that counts were all greater than or equal to zero (negative count data would be nonsense). Similarly, if the response variable was the proportion of animals that died, then the fitted values would have to lie between zero and one (fitted values greater than 1 or less than 0 would be meaningless). In the first case, a log link is appropriate because the fitted values are antilogs of the linear predictor, and all antilogs are greater than or equal to zero. In the second case, the logit link is appropriate because the estimated proportions are calculated from the antilogs of the log-odds ratio, $\log(p/q)$, and must lie between 0 and 1.

By using different link functions in GLIM, the performance of a variety of models can be compared directly. The total deviance is the same in each case and we can investigate the consequences of altering our assumptions about precisely how a given change in the linear predictor brings about a response in the fitted value of y. The most appropriate link function is the one that produces the minimum residual deviance (see the examples in Section 13.5.1).

10.4.1 The log link

The log link has many uses, but the most frequent are:

1 for count data, where negative fitted values are prohibited;

Table 10.1 The link functions used by GLIM. The canonical link function for normal errors is the identity link, for Poisson errors the log link, for binomial errors the logit link and for gamma errors the reciprocal link. These canonical link functions are defined by default when the error structure is declared

Symbol	Link function	Formula	Use
I	Identity	$\eta = \mu$	Regression or ANOVA with normal errors
L	Log	$\eta = \log \mu$	Count data with Poisson errors
G	Logit	$\eta = \log\left(\frac{\mu}{n - \mu}\right)$	Proportion data with binomial errors
R	Reciprocal	$\eta = \frac{1}{\mu}$	Continuous data with gamma errors
P	Probit	$\eta = \Phi^{-1}(\mu/n)$	Proportion data in bioassays
C	Complementary log-log	$\eta = \log[-\log(1 - \mu/n)]$	Proportion data in dilution assay
S	Square root	$\eta = \sqrt{\mu}$	Count data
E	Exponent	$\eta = \mu$***number*	Power functions

2 for explanatory variables that have multiplicative effects, where the link introduces additivity.
The model parameters inspected with **disp e** are in natural logarithms, and the fitted values are the natural antilogs (%exp) of the linear predictor.

10.4.2 The logit link

This is the link used for proportion data, and the logit link is generally preferred to the more old fashioned *probit* link (see Chapter 15). If a fraction p of the insects in an experiment died, then a fraction $q = (1 - p)$ must have survived out of the original cohort of n animals. The logit link is:

$$\text{logit} = \ln\left(\frac{p}{q}\right) \tag{10.12}$$

This is beautifully simple, and it ensures that the fitted values are bounded both above and below (the predicted proportions may not be greater than 1 or less than 0). The details of how the logit link linearizes proportion data are explained in Section 15.2.

It does make calculations of p from the parameter estimates a little tedious. Suppose the predicted logit at $z = 10$ was -0.328. To get the value of p we find the antilog of the fitted value x = %exp(−0.328), then evaluate:

$$p = \frac{1}{1 + \frac{1}{x}} \tag{10.13}$$

which gives $p = 0.42$. Note that confidence intervals on p will be asymmetrical when back-transformed, and it is good practice to draw histograms and error bars on the logit scale rather than the proportion (or percentage) scale to avoid this problem.

10.4.3 Other link functions

Two other commonly used link functions are the *probit* and the *complementary log-log* links. They are used in bioassay and in dilution analysis respectively, and examples of their use are to be found in Chapter 15.

Use of probits for bioassay is largely traditional, because probit paper used to be available for converting percentage mortality to a linear scale against log dose. Since computers have become widely available the need for the probit transformation

$$\frac{y_i}{n_i} = \Phi(\eta_i) + \varepsilon_i \tag{10.14}$$

has declined. The proportion responding (y/n) is linked to the linear predictor by $\Phi(.)$, the unit normal probability integral. Because the logit is so much simpler to interpret, and because the results of modelling with the two transformations are almost always identical, the logit link function is nowadays recommended for bioassay work, even though probits are based on a reasonable distributional argument for the tolerance levels of individuals.

The complementary log-log link:

$$\theta = \ln[-\ln(1 - p)] \tag{10.15}$$

is not symmetrical about $p = 0.5$ and is often used in simple dilution assay. If the proportion of tubes containing bacteria p is related to dilution x like this:

$$p = 1 - e^{(-\lambda x)}$$

then the complementary log-log transformation gives:

$$\eta = \ln[-\ln(1 - p)] = \ln\lambda + \ln x$$

which means that the linear predictor has a slope of 1 when plotted against $\ln(x)$. We fit the model, therefore, with $\ln(x)$ as an offset, and GLIM estimates the maximum likelihood value of $\ln(\lambda)$ (see Section 12.3).

The complementary log-log link should be assessed during model criticism for binary data and for data on parasitism rates and other proportional responses (see Sections 13.4 and 15.7.2). It will sometimes lead to a lower residual deviance than the symmetrical logit link.

10.4.4 Canonical link functions

The canonical link functions are the default options employed when a particular error structure is specified in the **error** directive. Omission of a **link** directive means that the settings shown in Table 10.2 are used. You should try to memorize these canonical links and to understand why each is appropriate to its associated error distribution.

Table 10.2 Canonical link functions used by GLIM

Error	Symbol	Link	Symbol
Normal	N	Identity	I
Poisson	P	Log	L
Binomial	B	Logit	G
Gamma	G	Reciprocal	R

10.5 Ecological examples of generalized linear models

The following chapters contain many worked examples of problems requiring different kinds of linear models. Table 10.3 shows some of the combinations of response variable, explanatory variable, link function and error distribution that are used frequently in ecological applications.

Table 10.3 Examples of some error distributions and link functions

Type of analysis	Response variable	Explanatory variable	Link function	Error distribution
Regression	Continuous	Continuous	Identity	Normal
ANOVA	Continuous	Factor	Identity	Normal
ANCOVA	Continuous	Both continuous and factor	Log	Gamma
Regression	Continuous	Continuous	Reciprocal	Gamma
Contingency table	Count	Factor	Log	Poisson
Proportions	Proportion	Continuous	Logit	Binomial
Probit	Proportion	Continuous (dose)	Probit	Binomial
Survival	Binary (alive or dead)	Factor	Complementary log-log	Binomial
Survival	Time to death	Continuous	Reciprocal	Exponential

The generality and flexibility of GLIM means that you can devise combinations of link function and error distribution to suit the specific problem in hand.

10.6 The likelihood function

The concept of maximum likelihood is unfamiliar to most non-statisticians. Fortunately, the methods that ecologists have encountered in linear regression and traditional ANOVA (i.e. least squares) are the maximum likelihood estimators when the data have normal errors and the model has an identity link. For other kinds of error structure and different link functions, however, the methods of least squares do not give unbiased parameter estimates, and maximum likelihood methods are preferred. It is easiest to see what maximum likelihood involves by working through two simple examples based on the binomial and Poisson distributions.

10.6.1 The binomial distribution

Suppose we have carried out a single trial, and have found $r = 5$ parasitized animals out of a sample of $n = 9$ insects. Our intuitive estimate of the proportion parasitized is $5/9 = r/n = 0.555$. What is the maximum likelihood estimate of the proportion parasitized? With $n = 9$ and $r = 5$ the formula for the binomial looks like this:

$$P(5) = \left(\frac{9!}{5!(9-5)!}\right)\theta^5(1-\theta)^{(9-5)}$$

Now the likelihood L does not depend upon the combinatorial part of the formula, because θ, the parameter we are trying to estimate, does not appear there. This simplifies the problem, because all we need to do now is to find the value of θ that maximizes the likelihood:

$$L(\theta) = \theta^5(1-\theta)^{(9-5)}$$

To do this we might plot $L(\theta)$ against θ as in Fig. 10.2, from which it is

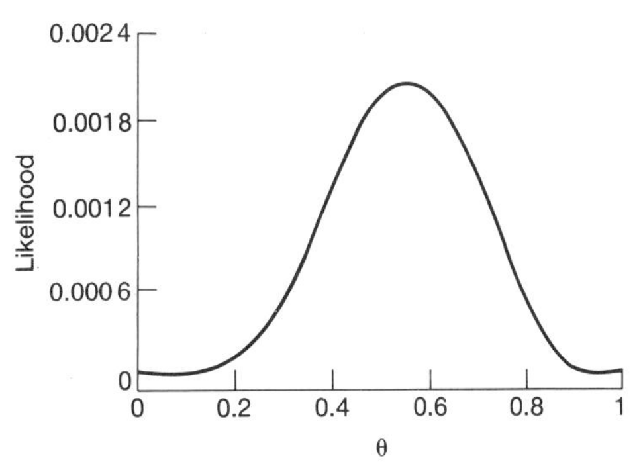

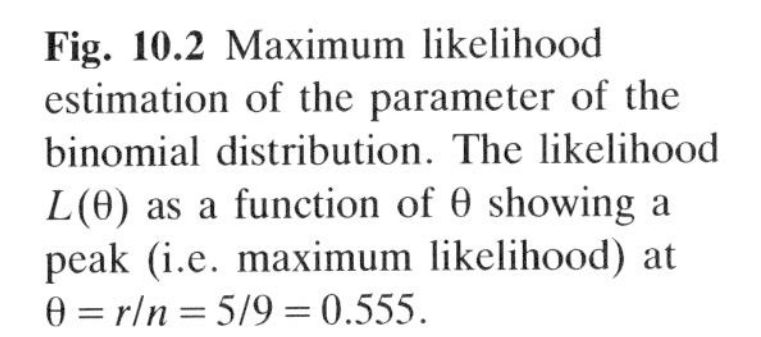
Fig. 10.2 Maximum likelihood estimation of the parameter of the binomial distribution. The likelihood $L(\theta)$ as a function of θ showing a peak (i.e. maximum likelihood) at $\theta = r/n = 5/9 = 0.555$.

clear that the maximum likelihood occurs at $\theta = r/n$. It is reassuring that our intuitive estimate of the proportion parasitized is r/n as well.

A more general way to find the maximum likelihood estimate of θ is to use calculus. We need to find the derivative of the likelihood with respect to θ, then set this to zero, and solve for θ. In the present case it is easier to work with the log of the likelihood. Obviously, the maximum likelihood and the maximum log likelihood will occur at the same value of θ.

$$L(\theta) = r \ln(\theta) + (n - r)\ln(1 - \theta)$$

so the derivative of the log likelihood with respect to θ is:

$$\frac{dL(\theta)}{d\theta} = \frac{r}{\theta} - \frac{n - r}{1 - \theta}$$

remembering that the derivative of $\ln\theta$ is $1/\theta$ and of $\ln(1-\theta)$ is $-1/(1-\theta)$. We set this to zero, rearrange, then take reciprocals to find θ:

$$\frac{r}{\theta} = \frac{n - r}{1 - \theta} \qquad \text{so } \theta = \frac{r}{n}$$

The maximum likelihood estimate of the binomial parameter is the same as our intuitive estimate.

10.6.2 The Poisson distribution

As a second example, we take the problem of finding the maximum likelihood estimate of μ for a Poisson process in which we observed, say, r leaf mines per leaf in n leaves, giving a total of Σr leaf mines in all. The probability density function for the number of mines per leaf is:

$$P(r) = \frac{e^{-\mu}\mu^r}{r!}$$

so the initial likelihood is the density function multiplied by itself as many times as there are individual leaves:

$$L(\mu) = \prod_1^n \frac{e^{-\mu}\mu^r}{r!} = e^{-n\mu}\mu^{\Sigma r}$$

because the constant $r!$ can be ignored. Note that nr is replaced by Σr, the observed total number of mines. Now it is straightforward to obtain the log likelihood:

$$L(\mu) = -n\mu + \Sigma r \ln \mu$$

The next step is to find the derivative of the log likelihood with respect to μ:

$$\frac{dL}{d\mu} = -n + \frac{\Sigma r}{\mu}$$

We set this to zero, and rearrange to obtain:

$$\mu = \frac{\Sigma r}{n}$$

Again, the maximum likelihood estimator for the single parameter of the Poisson distribution conforms with intuition; it is the mean (in this case, the mean number of mines per leaf).

10.6.3 Maximum likelihood estimation

The object is to determine the values for the parameters of the model that lead to the best fit to the data. It is in the definition of what constitutes 'best' that maximum likelihood methods can differ from the more familiar least squares estimates. There are three components to understand about the notion of maximum likelihood:

1 the data that we have gathered on the response variable, y;

2 the model, as presently specified;

3 the estimates of the parameters in that model, determined by more or less laborious calculation, using both the data and the model.

The data are sacrosanct, and they tell us what actually happened under a given set of circumstances. It is a common mistake to say 'the data were fitted to the model' as if the data were something flexible, and we had a clear picture of the structure of the model. On the contrary, what we are looking for is the minimal adequate model to describe the data. The model is fit to data, not the other way around. The best model is the model that produces the minimal residual deviance, subject to the constraint that all the parameters in the model should be statistically significant.

You have to specify the model. It embodies your best hypothesis about the factors involved, and the way they are related to the response variable. We want the model to be minimal because of the principle of parsimony, and adequate because there is no point in retaining an inadequate model that does not describe a significant fraction of the variation in the data. It is very important to understand that *there is not one model*; this is one of the common implicit errors involved in traditional regression and ANOVA, where the same models are used, often uncritically, over and over again. In most circumstances, there will be a large number of different, more or less plausible, models that might be fit to any given set of data. Part of the job of data analysis is to determine which, if any, of the possible models are adequate, and then, out of the set of adequate models, which is the minimal adequate model. In some cases there may be no single best model and a set of different models may all describe the data equally well.

Armed with the data, and having selected a particular model, the task is to estimate numerical values for the model's parameters. In a simple linear regression, for example, we estimated the values for the slope and the intercept using least squares techniques. For non-normal errors and with different kinds of link function, GLIM uses more general techniques, and the parameters are estimated using maximum likelihood methods. The way to picture the procedure is as follows:

1 given that we have the data;

2 and given that we have a specific model to describe the data;

3 what values for the parameters of the model make the data most likely?

If the model is appropriate and the parameter estimates are good, then it is very likely that the data could have arisen by the mechanisms proposed. On the other hand, if the model is good, but the parameter estimates are poor, then it is much less likely that the observed data would have been gathered. The object, then, is to *find the values for the parameters which maximize the likelihood of the data being observed*. You will see that this procedure depends not just on the data, but also on the structure of the model.

10.7 Parameter estimation in generalized linear models

The method of parameter estimation is *iterative, weighted least squares*. You know about least squares methods from Chapters 7–9. GLIM is different in that the regression is not carried out on the response variable, y, but on *a linearized version of the link function applied to* y. The *weights* are functions of the fitted values, %fv. The procedure is *iterative* because both the adjusted response variable and the weight depend upon the fitted values.

This is how it works (the technical details are on pp. 31–4 in McCullagh & Nelder, 1983). Take the data themselves as starting values for estimates of %fv. Use this to derive the linear predictor, the derivative of the linear predictor ($d\eta/d\mu$) and the variance function. Then re-estimate the adjusted response variable z and the weight W, as follows:

$$z_0 = \eta_0 + (y - \mu_0)\left(\frac{d\eta}{d\mu}\right)_0$$

where the derivative of the link function is evaluated at μ_0 and

$$W_0^{-1} = \left(\frac{d\eta}{d\mu}\right)_0^2 V_0$$

where V_0 is the variance function of y (see Table 19.2). Keep repeating the cycle until the changes in the parameter estimates are sufficiently small. It is the difference $(y - \mu_0)$ between the data y and the fitted values μ_0 that lies at the heart of the procedure. The maximum likelihood parameter estimates are given by:

$$\Sigma W(y - \mu)\frac{d\eta}{d\mu}x_i = 0 \qquad (10.16)$$

for each explanatory variable x_i (summation is over **units**; the rows of the data matrix). For more detail, see McCullagh & Nelder (1989) and Aitkin *et al.* (1989); a good general introduction to the methods of maximum likelihood is to be found in Edwards (1972).

10.8 Deviance: measuring the goodness of fit

The fitted values produced by the model are most unlikely to match the values of the data perfectly. The size of the discrepancy between the model and the data is a measure of the inadequacy of the model; a small discrepancy may be tolerable but a large one will not be. The measure of discrepancy used by GLIM to assess the goodness of fit of the model to the data is called the *deviance*: it is the logarithm of the ratio of two likelihoods.

In particular, the discrepancy of the fit is proportional to twice the difference between the maximum log likelihood achievable and that attained using the particular model under investigation. The formulas for calculating the deviance associated with the error distributions used by GLIM are shown in Table 10.4; exercises involving hand calculation of deviance are given in Exercises 14.1 and 15.1.

Table 10.4 The formulas used by GLIM in calculating deviance, where y is the data and μ is the fitted value under the model in question (the grand mean in the simplest case); note that, for the grand mean, the term $\Sigma(y - \mu) = 0$ in the Poisson deviance, and so this reduces to $2\Sigma y \ln(y/\mu)$; in the binomial deviance, n is the sample size (the binomial denominator), out of which y successes were obtained

Error structure	Deviance
Normal	$\Sigma(y - \mu)^2$
Poisson	$2\Sigma[y \ln(y/\mu) - (y - \mu)]$
Binomial	$2\Sigma\{y \ln(y/\mu) + (n - y)\ln[(n - y)/(n - \mu)]\}$
Gamma	$2\Sigma[-\ln(y/\mu) + (y - \mu)/\mu]$
Inverse Gaussian	$\Sigma(y - \mu)^2/(\mu^2 y)$

CHAPTER 11

Modelling in GLIM

11.1 Fitting models to data

Model-fitting is the central function of GLIM. The process is essentially one of exploration; there are no fixed rules and no absolutes. The object is to determine a minimal adequate model from the large set of potential models that might be used to describe the given set of data. In this book we discuss five types of model, as shown in Table 11.1.

Table 11.1 Glossary of model terms used in GLIM

Model	Description
Null model	Only the grand mean is fitted; deviance = SST
Full model	Zero deviance; a parameter for each observation
Maximal model	Contains all factors, interaction terms and covariates
Current model	The model specified by the last **fit** directive; details displayed by **disp m**. Probably simpler than the maximal model, but perhaps more complex than the minimal adequate model
Minimal adequate model	The object of the exercise; the model with the minimal number of terms, in which all the parameters are significantly different from zero, and no important terms have been excluded

The step-wise progression from the full model (or the maximal model), through a series of simplifications to the minimal adequate model is made on this basis of *deletion tests*; F-tests or χ^2 tests that assess the significance of the increase in deviance that results when a given term is removed from the current model.

11.2 Fitting models in GLIM

Models are fit using the **fit** directive. The simple command:

```
$fit $
```

causes the null model to be fit. This works out the grand mean (the overall average) of all the data and works out the total deviance (the total sum of squares, SST, in models with normal errors and the identity link). In some cases, this may be the minimal adequate model; it is possible that none of the explanatory variables we have measured contributes anything significant to our understanding of the variation in the response variable (e.g. Section 18.4). This is normally what you do not want to happen at the end of your three-year research project.

To add terms to the model we use + and to remove them, −. Thus

```
$fit +x $
```

adds x to the null model and

```
$fit −x $
```

takes it back out again. If you fit without a + or a −, then a new model is initiated with only the specified terms in it. So

```
$fit x+z $
```

replaces whatever model was previously in use. It is better to use + and − when fitting terms, because then GLIM prints the *change in deviance* which accompanies that particular fit. If you do not use + or −, then GLIM will print only the residual deviance. A useful trick is to use a colon : (which means repeat the last directive) to get both *SST* and the change in deviance, like this:

```
$fit : +x $
```

This performs two fits: it fits the null model first, printing the total deviance, then adds the explanatory variable x to the model, printing the change in deviance and the residual deviance.

The maximal model is usually fitted first. The simplest way to do this is to use * between all the factors and covariates you want to use, because this fits all the main effects and all their interactions. Thus

```
$fit : + a * b * c $
```

will fit the main effects of factors A, B and C, the three two-way interactions A.B, A.C and B.C and the single three-way interaction A.B.C. You cannot fit interactions between continuous variables using '.' in older versions of GLIM like 3.77. If you really want to look at this kind of interaction, you need to calculate a new vector as the product of the two continuous variables in question (see p. 292). For example, to look at the interaction between x and z you would type:

```
$calc xz = x * z $

$fit + xz $
```

Model simplification now begins. This involves removing terms from the current model using **fit** −. A significant term is one that causes a significant increase in deviance (as judged by an *F*-test in the case of normal or gamma errors, or a χ^2 test in the case of Poisson or binomial errors) when removed from the current model. Significant terms should be added back into the model using **fit** +.

For nested designs, you can use /. For example,

```
$fit + a/b $
```

means fit B nested within A. It is exactly equivalent to writing:

```
$fit + a + a.b $
```

because in nested analysis we are not interested in the main effect of B, but only in the interaction term A.B (see Section 8.8).

Model-fitting features of GLIM 4 not found in 3.77 include the ability to specify orthogonal polynomials of factors to a given order (say cubic):

```
$fit + a <3> $
```

contrasts defined by a given matrix (say C) for a factor:

```
$fit + C(a) $
```

and an exponentiation facility, giving easy specification of *n*-way interactions, so

```
$fit + (a+b+c)**2 $
```

is equivalent to

```
$fit + a + b + c + a.b + a.c + b.c $
```

See the GLIM 4 Manual for details.

11.3 System vectors and scalars

After a **fit** has been carried out, a range of vectors and scalars is available to the user to carry out further calculations, plots and tables. The most important of these are shown in Table 11.2.

11.4 Hypothesis testing

The tests you are likely to be familiar with (like Student's t and Fisher's F) give *exact probabilities*. Thus $t = 1.96$ has a probability of exactly 5% if the means of two treatments are identical and we have 30 degrees of freedom (see Appendix Table 1). In GLIM we need to get used to a new convention. Except in the case of normal errors and the identity link, we cannot assign exact probabilities to our tests of hypotheses. For large samples ($n > 30$) the probabilities are asymptotically correct, which means that the larger the sample, the closer to the probability will be to the values printed in tables. For small samples, exact probabilities cannot be calculated. This means that the smaller the sample, the more careful we need to be in assigning probability values to tests. If we have more

Table 11.2 System vectors and system scalars produced by the **fit** directive

Symbol	Meaning	Use
Vectors		
%fv	Fitted value	For plotting on graphs along with the raw data, and for calculating residuals
%lp	Linear predictor	Contains the current values of the linear predictor (these are the same as the fitted value %fv only with normal errors and identity link)
%wt	Weight	The relative importance given to a particular value of the y-variable
%wv	Working value	The same as the y-variable %yv except in iterative fits, when it has the current iterative values
%re	Restrict	User-defined vector, created in a **calc** directive, for restricting the y-values displayed in **plot** commands; if %re is zero, the point will not be plotted; see also **disp w**
Scalars		
%dv	Deviance	Assessing the contribution of an explanatory variable
%df	Degrees of freedom	
%x2	Pearson's χ^2	Testing for over-dispersion
%pl	Number of parameters	The number of non-aliased parameters in the current model; stands for parameter list
%nu	Number of units	Rows of data
%sc	Scale parameter	Error variance with normal or gamma errors; overdispersion adjustment with Poisson or binomial errors

than about 30 degrees of freedom for error, however, this should present no real problem. The magic number 30 is a useful rule of thumb in ecological work for separating big samples ($n > 30$) from small samples ($n < 30$).

We can test hypotheses in GLIM in two ways: (i) by carrying out t-tests on parameter estimates; and (ii) by deletion (i.e. carrying out likelihood ratio tests on the increase in deviance that results when terms are removed from the model). With normal errors and the identity link, the t-tests are exact and no problems arise. For other combinations of error distribution and link function, however, the results are only asymptotically correct, and there is no firm theory on the precision of the probability estimates for the significance tests. This is not a major problem, so long as you do not want to provide 'exact P values' for your tests. There are some good rules of thumb about t-tests in GLIM. Usually, a parameter that is less than 1 standard error away from zero will not be significant. Similarly, a parameter that is greater than 3 standard errors away from zero will normally be significant. The grey area stretches from about 1.5 to 2.5 standard errors. Murphy's law of hypothesis testing in GLIM states that most of your results will fall in this equivocal range.

A further problem for t-tests arises because of correlations of parameters. When correlations are high, the t-test can be highly misleading. Because of this ambiguity, it is better to test marginal cases by deletion rather than by t-tests. The change in deviance following removal of the factor from a maximal model is generally a more robust test of the significance of that factor, and the probability level is closer to the exact value than with the t-test. If you are in doubt, however, you might as well do both tests, and see whether they differ in their assessment of the significance. If they do not, then you can be reasonably confident in the result. If they do differ, then believe the deletion test rather than the t-test.

11.5 Likelihood ratios

By comparing the likelihood of the current model (l_c) to the likelihood of the full model (l_f) using the data in hand, we obtain a measure of the acceptability of the current model relative to that of the full model. The scaled deviance is:

$$S(c,f) = -2\ln\left(\frac{l_c}{l_f}\right)$$

Thus, large values of S reflect low likelihood of the current model (i.e. increasing lack of fit); this is why S is called *deviance*.

The difference in the explanatory power of two models can be assessed as follows. If l_1 is the likelihood for model 1 and l_2 the likelihood for model 2, and model 2 is a simplified version of model 1 (so that all of model 2's attributes are included in model 1), then the deviance $-2\ln(l_2/l_1)$ is asymptotically distributed as χ^2 with $t_2 - t_1$ d.f. This result is exact for normal errors and the identity link, and is approximate for other error and link combinations. Thus, removing a component from a model causes a significant increase in deviance when the scaled deviance is greater than the value of χ^2 in tables with $t_2 - t_1$ d.f.

An analysis of deviance table can be drawn up by tabulating the scaled deviances that result when each component is removed from the maximal model. If the factors are orthogonal, then the order in which they are removed will be irrelevant. But if the factors are not orthogonal (or the model is non-linear), then the deviance will depend upon the order in which the factors are deleted from the model. This is why it is a good idea to calculate the analysis of deviance table by subtracting each factor in turn from the maximal model, then restoring the maximal model before subtracting the next factor. Adding factors to the null model is a bad idea because this will make factors that are fitted early in the sequence appear more significant than they really are.

11.6 The disp directive

The **disp** directive has a wide range of options that allow access to different aspects of the linear model. **Disp** can be used only after the **fit** directive, and its function can best be understood by working through an example. Suppose we have a graph with 30 points, generated by adding normally distributed noise with a mean of zero and standard deviation of 1 to the equation $y = 11 - 0.2\ x$:

```
$units 30 $

$calc x=%cu(1) $

$calc y=11-0.2*x+%nd(%sr(0)) $

$plot y x $
```

```
   11.600 |
   11.200 | YY        Y
   10.800 |
   10.400 |
   10.000 |                Y
    9.600 |       Y    Y
    9.200 |    Y Y  Y       Y                Y
    8.800 |                          YY
    8.400 |                     Y  Y      Y
    8.000 |              Y    Y  Y
    7.600 |
    7.200 |                                       Y
    6.800 |                                Y        Y       Y
    6.400 |                                     Y        Y
    6.000 |                             Y                 Y
    5.600 |                                    Y
    5.200 |                                            Y
    4.800 |
    4.400 |
    4.000 |                                          Y
    3.600 |
----------:---------:---------:---------:---------:---------:---------:
        0.00      6.00     12.00     18.00     24.00     30.00     36.00
```

Now we fit a simple linear regression, declaring y to be the response variable and x to be the sole explanatory variable:

```
$yvar y $

$fit x $

deviance = 28.671
      d.f. = 28
```

so that the error sum of squares is 28.671 on 28 degrees of freedom. To see the parameter estimates, we use the now familiar **disp e**:

```
$disp e $
```

```
     estimate   s.e.      parameter
1    10.74      0.3789    1
2    -0.1708    0.02134   X
scale parameter taken as 1.024
```

the intercept is 10.74 with a standard error of 0.3789 while the slope is −0.1708 with a standard error of 0.02134. This is the equation of the linear predictor.

It is often of interest to know the degree to which different parameter estimates are correlated with one another. This is seen with the **disp c** directive:

```
$disp c $
```

```
Correlations of parameter estimates
1     1.0000
2    -0.8731   1.0000
        1        2
```

which indicates that there is a high negative correlation between the estimates of the slope and the intercept. This means that making the slope steeper (i.e. more negative) would substantially increase the intercept.

In order to be reminded of the current deviance (the error sum of squares in the present example) we use **disp d**:

```
$disp d $
```

```
Deviance is 28.67 on 28 d.f. from 30 observations
n.b. current and previous models are not necessarily nested
```

This gives the deviance (or the scaled deviance as appropriate) and the residual degrees of freedom. In the present example this is two less than the sample size because two parameters have been estimated from the data (the mean value of y and the regression slope).

In a long modelling session, it is often useful to be reminded of the current structure of the linear predictor. We obtain this by typing **disp l**:

```
$disp l $
```

```
Linear predictor:
terms = 1 + X
```

This shows the composition of the linear predictor as a sum of simple terms. In this case the intercept (1) and a single explanatory variable (x).

For a full description of the current model, **disp m** is used:

```
$disp m $
```

```
Current model:
```

```
number of units is 30

y-variate Y
weight    *
offset    *

probability distribution is NORMAL
link function            is IDENTITY
scale parameter          is to be estimated by the mean deviance
    terms = 1 + X
```

This is a more detailed version than disp l, and gives **units**, **yvar**, **weight** and **offset** (with asterisks for the last two if, as in this case, they are not in operation). It also prints the current specification for the probability distribution and link function, and shows how the scale parameter (the error variance in the present case) is to be calculated. As we shall see later (Sections 14.9 and 15.6) the default scale parameter will be 1.0 if we have declared Poisson or binomial errors.

To see the values of the (scaled) residuals, we use the **disp r** directive:

$disp r $

unit	observed	fitted	residual
1	11.025	10.570	0.455
2	11.087	10.399	0.688
3	9.319	10.229	−0.910
4	9.330	10.058	−0.728
5	9.417	9.887	−0.470
6	9.159	9.716	−0.557
7	11.214	9.546	1.668
8	9.752	9.375	0.377
9	8.175	9.204	−1.029
10	9.930	9.033	0.897
11	9.048	8.863	0.185
12	8.051	8.692	−0.641
13	8.413	8.521	−0.108
14	8.128	8.350	−0.222
15	8.264	8.180	0.085
16	8.981	8.009	0.972
17	8.601	7.838	0.763
18	5.826	7.667	−1.841
19	8.502	7.496	1.005
20	6.836	7.326	−0.490
21	9.053	7.155	1.898
22	5.558	6.984	−1.427
23	6.401	6.813	−0.413
24	7.239	6.643	0.597
25	6.910	6.472	0.438
26	3.906	6.301	−2.395
27	5.291	6.130	−0.840
28	6.388	5.960	0.429
29	6.175	5.789	0.386
30	6.844	5.618	1.226

We obtain parallel listings of the response variable, the fitted values and the generalized residuals (in this case, with normal errors and the identity link, these are identical to the raw residuals %yv−%fv; see Section 5.11). The binomial denominator is also printed if binomial errors have been declared. The print-out shows that the y-value of 3.906 at unit 26 is substantially smaller than expected (6.301), having a residual of −2.395. It would be worth checking back to make sure that this was not a typing error on data entry. If the value is correct, it would be worth checking the influence of this point on the parameter estimates (see Section 5.14).

An important option in ANOVA and ANCOVA is **disp s** which prints a triangular matrix showing the standard errors of the differences between parameter estimates:

```
$disp s $

S.E.s of differences of parameter estimates
1  0.000
2  0.3977  0.000
      1       2
scale parameter taken as 1.024
```

Examples of the use of **disp s** in analysis of variance are given in Section 8.6 and in analysis of covariance in Section 9.3. In regression analysis, we are not much interested in the standard error of the difference between the intercept and the slope. This table is very important, however, for comparing the differences between individual interaction terms in factorial experiments (see Section 8.6.1).

A more technical application involves the **disp t** directive:

```
$disp t $

Working matrix

1  −0.1402
2    0.006897  −0.0004449
3   10.74       −0.1708       28.67
      1           2             3
```

The *working matrix* is the generalized inverse of the *SSP* matrix (*SSP* stands for sums of squares and products; this is the *information matrix*). The rows are ordered as in the **disp e** directive with signs reversed. Row 1 contains the asymptotic variance of the intercept without the scale parameter. The second row has the asymptotic covariance of the slope and intercept in the first column, and the asymptotic variance of the slope in the second. In this linear example, the variances and covariances are exact rather than asymptotic. The final row for the *y*-variate has the correct sign, and contains in columns 1−3 the intercept, slope and Pearson χ^2 statistic (in the present case this is identical to the residual deviance). See **disp v** for calculations. The Pearson χ^2 statistic is defined as:

$$\sum \frac{(\%yv - \%fv)^2}{V(\%fv)}$$

where the variance function $V(.)$ depends upon the error distribution (1 for normal, μ for Poisson, npq for binomial; see Table 19.2).

The *SSP* matrix does not normally contain rows for intrinsically aliased parameters (see Section 4.20). If there is aliasing, the diagonal element is zero, and the rows contain the coefficient of the linear dependency which produced the aliasing.

Perhaps the most important option after **disp e** is **disp v**, which prints the variance and covariance matrix:

```
$disp v $

(Co)variances of parameter estimates
1    0.1436
2  -0.007062  0.0004556
     1          2
scale parameter taken as 1.024
```

The principal use of this information is in calculating standard errors of differences. Remember that the standard error of a difference is:

$$SE = \sqrt{V_1 + V_2 - 2COV_{1,2}}$$

The covariance matrix of the parameter estimates is listed following **disp v**. The variances and covariances are divided by the scale parameter, and the signs reversed, to obtain the values displayed in the first two rows of the **disp t** option. The variances (on the diagonal) are the squares of the standard errors obtained in **disp e** and whose calculation is explained in Section 7.4.1.

The final option is **disp w**, which is identical to **disp r** except that residuals are printed only for %re > 0 (i.e. points that you have excluded from plotting by setting %re = 0 are not printed). This is very useful if you want to print, say, only the largest residuals from a big set of data. If only those residuals bigger than 2.0 were wanted, you could write:

```
$calc resid = %yv−%fv $

$calc %re = (resid>2)?(resid<–2) $

$disp w $

unit  observed  fitted   residual
26    3.906     6.301   −2.395
```

Modelling techniques are introduced throughout the following chapters, and examples of the various directives can be practised by working through the exercises. ●

CHAPTER 12

Model simplification

The principle of parsimony requires that the model should be as simple as possible. This means that the model should contain no redundant parameters or factor levels, and we achieve this by one or more of the following steps:

1 remove non-significant explanatory variables;
2 remove non-significant factors and interaction terms;
3 group together factor levels that do not differ from one another;
4 amalgamate explanatory variables that have similar parameter values;
5 set non-significant slopes to zero within ANCOVA;

subject, of course, to the caveats that the simplifications make good biological sense, and do not lead to significant increases in residual deviance (see Table 12.1). We deal with each of these techniques in the context of a series of worked examples, but first we need to master three basic principles of model simplification: *deletion*, *aggregation* and *offsets*.

Table 12.1 The aim of the exercise is to determine the minimal adequate model in which all the parameters are significantly different from zero. This is achieved by a step-wise process of model simplification, beginning with the full model, then proceeding by the elimination of non-significant terms, and the retention of significant terms

Full model	One parameter for each and every data point Fit: perfect Degrees of freedom: none Explanatory power of model: none
Maximal model	Contains all (p) factors, interactions and covariates that might be of interest Many of the model's terms may be insignificant Degrees of freedom: $n - p - 1$
Minimal adequate model	A simplified model with $0 \leqslant p' \leqslant$ p terms All the terms are significant (if there are no significant terms, then MAM = null model) Degrees of freedom: $n - p' - 1$ Explanatory power: $r^2 = SSR/SST$
Null model	A single parameter, the grand mean Fit: none Degrees of freedom: $n - 1$ Explanatory power of model: none

12.1 Deletion

Deletion uses the **fit** − directive to remove terms from the model. The simplest deletion is to remove the intercept from a regression study using the **fit −1** directive. This rotates the regression line about the point ($\bar{x}$, $\bar{y}$) until it passes through the origin. Because the regression line is rotated away from its maximum likelihood position, this will inevitably cause an increase in the residual deviance. If the increase in deviance is significant, as judged by a likelihood ratio test, then the simplification is unwarranted, and the intercept should be added back to the model. Forcing the regression to pass through the origin may also cause problems with non-constancy of variance, and is generally not to be recommended unless we have confidence in the linearity of the relationship all the way from the origin up to the right-hand end of the axis of the explanatory variable(s).

The standard practice with simplification is to remove high-order interaction terms first, then low-order interactions and then main effects. The justification for deletion is made with the maximal model *at the level in question* (see Table 12.2). Thus, in considering whether or not to retain a given second-order interaction, it is deleted with all other second-order interactions (plus any significant higher-order terms) fitted to the model. If its deletion leads to a significant increase in deviance, it must be

Table 12.2 The steps involved in model simplification. There are no hard and fast rules, and this is only a guide to one sensible way of approaching the problem of model simplification

Step	Procedure	Explanation
1	Fit the maximal model	Fit all the factors, interactions and covariates of interest. Note the residual deviance. Check for overdispersion (Poisson or binomial errors), and rescale if necessary
2	Begin model simplification	Inspect the parameter estimates **disp e** Remove the least significant terms first using **fit** −, starting with the highest order interactions
3	If the deletion causes an insignificant increase in deviance	Leave the term out of the model Inspect the parameter estimates Remove the least significant term remaining in the model
4	If the deletion causes a significant increase in deviance	Put the term back into the model using **fit** + These are the statistically significant terms as assessed by deletion from the maximal model
5	Keep removing terms from the model	Repeat steps 3 or 4 until the model contains nothing but significant terms The resulting model is the minimal adequate model If none of the parameters is significant, then the null model is the minimal adequate model

retained. In considering a given first-order interaction, all first-order interactions plus any significant higher-order interactions are included at each deletion. And so on.

Main effects that figure in significant interactions should not be deleted. If you do try to delete them there will be no change in deviance and no change in degrees of freedom, because the factor is aliased (see Section 4.20); the deviance and degrees of freedom will simply be transferred to an interaction term.

It is a moot point whether block effects should be removed during model simplification. In the unlikely event that block effects were *not* significant (bearing in mind that, in ecology, everything varies, and so insignificant block effects are the exception rather than the rule), then you should compare your conclusions with and without the block terms in the model. If the conclusions differ, then you need to think very carefully about why, precisely, this has happened. The balance of opinion is that block effects are best left unaltered in the minimal adequate model (but see Section 4.18 on Model II ANOVA and random effects).

12.2 Aggregation

By aggregation, I mean lumping together factor levels or explanatory variables that are not significantly different from one another in their parameter estimates. The variables in question are required in the model, because deletion tests have shown them to cause a significant change in deviance. What is at issue is whether the identity of all the factor levels needs to be maintained separately.

A frequent outcome during ANOVA is that, say, the 'low' and 'medium' levels of a treatment are not significantly different from one another, but both differ from the 'high' level treatment. Aggregation means calculating a new factor that has the same value for 'low' and 'medium' levels, then observing the change in deviance when this simplified model is fit to the data. If the change is not significant, then the simplification is justified. If a significant change in deviance occurs, then the original three levels must be restored.

In multiple regression, two explanatory variables may have similar parameter values, and it may make sense to create a single variable that is the sum of the original two variables. For example, if yield increases by 2 kg for every unit of Maxigrow and 2 kg for every unit of Yieldmore we might try calculating a single variable (Fertilizer = Maxigrow + Yieldmore) and fitting this to the model instead.

12.3 Offsets

When the parameter estimates are inspected at the end of the deletion and aggregation stages of model simplification, it may be possible to add a final flourish to the modelling by simplifying the numerical values of the parameters. It is much more straightforward, for example, to say that yield increases by 2 kg per hectare for every extra unit of fertilizer, than to say that it increases by 1.947 kg (assuming, of course, that removing the extra decimal places does not result in a significant decrease in the model's explanatory power). Similarly, it is preferable to be able to say that the odds of infection increase 10-fold under a given treatment, than it is to say that the logits increase by 2.321 (without model simplification this is equivalent to saying that there is a 10.186-fold increase in the odds).

The upshot of these kinds of modifications is that we would like to refit the model, but with one or more of the parameters specified by us, rather than estimated by GLIM. We do this by using **offsets**.

Sometimes we may wish to constrain the values of one or more of the parameters in a model because of some *a priori* theoretical expectation. In a study of plant reproduction, for example, we might wish to constrain the graph to pass through the origin (e.g. plants with zero mass cannot have positive fecundity) or to have a negative intercept, if there was a threshold size below which no seeds were produced. Similarly, we might want to know whether the parameterized allometric model $0.487x^{1.845}$ could be simplified to $x^2/2$ without loss of descriptive power (i.e. are the data capable of distinguishing between these two models). In a study of density-dependent mortality in plant populations, we might wish to test whether the self-thinning rule applied to a particular set of data relating mean plant weight (y) to mean plant density (x). The relationship is alleged to follow the allometric relationship $y = ax^{-3/2}$ (Yoda's rule), and we might wish to test whether fixing the exponent at $-3/2$ and estimating only the intercept of the log/log plot gives a significantly higher deviance, or predicts a significantly different intercept, than when the full regression model with two estimated parameters is fit to our data.

12.4 Caveats

Model simplification is an important process but it should not be taken to extremes. For example, the interpretation of deviances and standard errors produced with fixed parameters that have been estimated from the data should be undertaken with caution. Again, the search for 'nice numbers' should not be pursued uncritically. Sometimes there are good scientific reasons for using a particular number (e.g. a power of 0.66 in an allometric relationship between respiration and body mass), but it would be absurd to fix on an estimate of 6 rather than 6.1 just because 6 is a whole number.

12.5 Examples of model simplification

The following sections provide worked examples of deletion, aggregation and offsets in the context of different kinds of experiments.

12.5.1 Regression

The questions that arise most frequently in the context of model simplification in linear regression are: (i) is the intercept necessary, or would a graph passing through the origin serve equally well; and (ii) is there any evidence for significant curvilinearity?

We can assess the need for the intercept by means of the **fit −1** directive. If removing parameter 1 does not cause a significant increase in the residual deviance, then we might decide to leave it out of the model, and use the revised estimate of the slope in subsequent deliberations. Removing the intercept can cause problems due to non-constancy of variance (see above).

A simple test for curvilinearity is to calculate the square of the explanatory variable (see Exercise 12.2):

```
$calc x2 = x * x $

$fit + x2 $
```

then add this quadratic term to the model. If the addition causes a significant reduction in deviance, then there is evidence of curvilinearity.

12.5.2 Multiple regression

The usual question in simplifying multiple regression models is which subset of the explanatory variables is required in the minimal adequate model. There is considerable debate about the best practice for model simplification, but a good rule of thumb is to retain only those explanatory variables that cause a significant increase in residual deviance *when removed from the maximal model*. After all the variables have been assessed, then all the non-significant variables can be deleted and the reduced model can be assessed. Here again, all the terms should be deleted in turn, and only those leading to a significant increase in residual deviance should be retained. Finally, the need for the intercept could be assessed by **fit −1** (bearing in mind the caveats about linearity and non-constancy of variance; see above).

It is important to note that two variables may both appear to be insignificant when judged on the basis of a *t*-test on their parameter estimates, and yet both of them may make a significant contribution to explaining the deviance on their own. This usually occurs when the two explanatory variables are highly correlated with one another (like daily maximum and daily minimum temperature, for instance). Either one of them, fitted to the model on its own, may be highly significant when judged by a *t*-test, but not when they appear together in the model. *Backward elimination* does not lead to the mistake of concluding that neither variable is important, and leads to the acceptance of the variable that explains the greater deviance when removed from the maximal model. As we have seen elsewhere, deletion is a more robust test of significance than *t*-tests of parameter estimates.

Suppose that we are attempting to understand variation in body weight of adult rodents (y) in terms of the mean annual rainfall experienced in their habitat (x1), predator abundance (x2), vegetation cover (x3) and annual seed production (x4), all in arbitrary units. We can read the data from file glex23.dat:

```
$units 16 $

$data y x1 x2 x3 x4 $

$dinput 6 $

File name? glex23.dat
```

You should begin by plotting body mass (y) against each of the explanatory variables in turn. From these graphs it appears that there is a strong positive association between body size and rainfall and body size and seed production, but no hint of a relationship with predator abundance or vegetation cover. We declare body size to be the *y*-variable, and fit the null model:

```
$yvar y $

$fit $

deviance = 389.75
      d.f. = 15
```

We shall be fitting the maximal model x1 + x2 + x3 + x4 several times, so it is worth writing a text substitution macro to save typing effort:

```
$macro full x1 + x2 + x3 + x4 $endmac $
```

Don't forget to leave a space between x4 and the $ sign in front of **endmac**. Now fit the maximal model, and inspect the parameter estimates and their standard errors:

```
$fit #full $disp e $

deviance = 42.988
      d.f. = 11

     estimate     s.e.     parameter
1   -4.298        2.521    1
2    1.365        0.3752   X1
3   -0.003237     0.3105   X2
4    0.2333       0.2957   X3
5    0.7796       0.5030   X4
scale parameter taken as 3.908
```

Only rainfall, x1, looks to be significant on the basis of t-tests, but we shall check this by deleting each explanatory variable in turn from the maximal model.

```
$fit -x4 $

deviance = 52.375 (change = +9.386)
    d.f. = 12     (change = +1)

$fit #full:-x3 $

deviance = 42.988
    d.f. = 11

deviance = 45.419 (change = +2.431)
    d.f. = 12     (change = +1)

$fit #full:-x2 $

deviance = 42.988
    d.f. = 11

deviance = 42.989 (change = +0.0004234)
    d.f. = 12     (change = +1)

$fit #full:-x1 $

deviance = 42.988
    d.f. = 11

deviance = 94.714 (change = +51.73)
    d.f. = 12     (change = +1)
```

Although the change in deviance on removing seed production, x4, from the maximal model is quite large (9.386), it is not significant when judged against the error variance of 3.908. Only rainfall, x1, causes a significant increase in deviance when removed from the maximal model ($F = 51.73/3.908 = 13.24$; tables with 1,11 d.f. = 4.84).

It is informative to see what we would have concluded if we had not had the rainfall data. The model without rainfall looks like this:

```
$disp e $

   estimate  s.e.    parameter
1  -7.210    3.397   1
2   0.6402   0.3626  X2
3   0.1646   0.4194  X3
4   2.315    0.3888  X4
scale parameter taken as 7.893
```

```
$fit−x4 $

deviance = 374.62 (change = +279.9)
    d.f. = 13      (change = +1)
```

Notice that seed production is highly significant when removed from a model that does not contain rainfall ($F = 279.9/7.839 = 35.46$). This is because the two explanatory variables are so highly correlated with one another ($r = 0.8885$).

The removal of each of the factors from the maximal model allows us to draw up an analysis of deviance table, as shown in Table 12.3, from which it is clear that only rainfall has a significant effect on rodent body size. Presumably, rainfall affects rodent ecology in ways over and above its effect on the seed supply. The minimal adequate model for these data contains just two parameters: the intercept and the slope of the graph of body size against rainfall.

Table 12.3 Analysis of deviance (ANODEV) table for multiple regression with all deviances assessed by removal from the maximal model

Explanatory variable	Symbol	Deviance	Significance
Rainfall	x1	51.73	$F = 13.24$
Predators	x2	0.0004234	n.s.
Cover	x3	2.431	n.s.
Seed production	x4	9.386	n.s.

An alternative way to proceed in model simplification is to remove terms from the model in a sequence as judged by their t-values. The least significant parameters are removed first (i.e. those with the smallest t-values), and if the deletion causes an insignificant increase in deviance, then *the term is left out*. The parameter estimates of the remaining terms are then inspected, and the least significant term is deleted. And so on. For the present example, the procedure looks like this:

```
$fit #full $

deviance = 42.988
    d.f. = 11
```

Having fit the maximal model, we remove each explanatory variable, in turn, from the maximal model, starting with the least significant. To find the least significant terms we inspect the **disp e** table (note that we are *not* looking for the smallest parameter, but for the term with the smallest t-ratio):

```
$disp e $
```

```
      estimate    s.e.     parameter
 1  -4.298       2.521    1
 2    1.365      0.3752   X1
 3  -0.003237    0.3105   X2
 4    0.2333     0.2957   X3
 5    0.7796     0.5030   X4
 scale parameter taken as 3.908
```

We start, therefore, by taking out x2:

```
$fit-x2$disp e $

deviance = 42.989 (change = +0.0004234)
    d.f. = 12     (change = +1)

      estimate  s.e.     parameter
 1  -4.314    1.853    1
 2    1.363   0.2953   X1
 3    0.2340  0.2736   X3
 4    0.7826  0.3970   X4
 scale parameter taken as 3.582
```

The F-tests are calculated on the basis of *the variance of the model from which the term was removed* (i.e. $s^2 = 3.908$ on 11 d.f. in the present case). The removal caused an insignificant increase in deviance, so we leave out x2 and remove the next least significant explanatory variable, x3. Note that sometimes, as here, the increase in error degrees of freedom more than compensates for the increase in deviance, so that the error variance actually goes down when an insignificant term is removed.

```
$fit-x3$disp e $

deviance = 45.610 (change = +2.621)
    d.f. = 13     (change = +1)

      estimate  s.e.     parameter
 1  -3.376    1.478    1
 2    1.296   0.2818   X1
 3    0.8152  0.3910   X4
 scale parameter taken as 3.508
```

Both the remaining terms look as if they might be significant (x4 has a t-ratio >2)

```
$fit-x4 $

deviance = 60.858 (change = +15.25)
    d.f. = 14     (change = +1)
```

In fact, it is not significant ($F = 4.35$, and with 1 and 13 d.f. the value in tables is 4.67). Finally, we remove the last term, rainfall (x1):

```
$fit-x1 $

deviance = 389.75 (change = +328.9)
    d.f. = 15      (change = +1)
```

•

Now we can draw up the analysis of deviance table, as shown in Table 12.4. Notice that the numerical values in this table are different for the two procedures, but the interpretation is the same: only rainfall needs to be retained in the minimal adequate model. Can you see why the deviance values are different for the two procedures?

Table 12.4 ANODEV table for multiple regression with step-wise omission of non-significant terms, beginning from the maximal model

Explanatory variable	Symbol	Deviance	Significance
Rainfall	x1	328.9	$F = 75.66$
Predators	x2	0.00042	n.s.
Cover	x3	2.621	n.s.
Seed production	x4	15.25	n.s.

12.5.3 Analysis of variance

In ANOVA, questions of simplification include:

1 which, if any, of the interaction terms need to retained in the model;
2 can factor levels be combined into sensible sub-sets of the original classifications;
3 can ranked factor levels be replaced by regression variables?

We shall continue the example introduced in Section 8.6.

```
$units 24 $

$data y a b $

$dinput 6 $

File name? glex11.dat

$factor a 3 b 2 $

$yvar y $

$fit:+a*b $

$disp e $
```

It appeared from this initial analysis that the low and medium protein contents did not need to be distinguished, and that a new factor with only two levels ('lowish' and high) would suffice. In order to compute a new factor that lumps together different categories of an existing variable, we use a simple programming trick. The new factor, let us call it C, will have level 1 where factor A had either level 1 *or* level 2 (low or medium protein), and have level 2 where factor A had level 3 (high protein diet). We use the logical function:

```
(a > 2)
```

which is true (i.e. numerically equal to 1) when A is 3, and false (numerically zero) otherwise (i.e. when A is less than or equal to 2). We write:

```
$calc c = 1 + (a > 2) $
```

which says that the new vector C contains $1+0=1$ if the value in a given row of A is less than 3 (i.e. a 1 or a 2), and $1+1=2$ if it contains a value greater than 2 (i.e. a 3). If logical operators are new to you, you should practise Exercise 3.1 and read Section 3.8. While logical functions are difficult to grasp at first, they are immensely powerful in programming and model simplification, and you should persevere with them.

We now declare C to be a factor with two levels, calculate the new averages and refit the model.

```
$factor c 2 $

$tab the y means for c;b $

     1      2
1  4.000  6.750
2  5.500  3.000

$fit c*b $

deviance = 30.500
    d.f. = 20

$disp e $

     estimate  s.e.     parameter
1     4.000   0.4366   1
2     1.500   0.7562   C(2)
3     2.750   0.6175   B(2)
4    -5.250   1.069    C(2).B(2)
scale parameter taken as 1.525
```

The simplification of combining the lowest two levels of A has led to an increase in deviance (a reduction in explanatory power) of only $30.5-28.0=2.5$ with two degrees of freedom. This is clearly not statistically significant ($F=2.5/(2\times 1.5556)=0.804$), so we conclude that, so long as it was biologically sensible to combine the first two levels of factor A, the principle of parsimony requires the simple model with only two levels.

Note that there is no need to carry out a similar simplification on factor B. It has only two levels, and it is clear that we need to retain both, because level 2 appears in a highly significant interaction term (see row 4 of the **disp e** table, above). There is nothing to be gained in this example from attempting to find convenient whole number approximations to the parameter values, so we conclude that the new model is minimal adequate. ●

12.5.4 Analysis of covariance

In this example we investigate the effect of three different hormone treatments on the duration of the first instar of a caterpillar (y, measured in decimal days). There is one factor (hormone treatment) with three levels (control, hormone *A* and hormone *B*) and 15 replicates per level (a total of 45 measurements of the response variable). The covariate is initial caterpillar live weight (x, in mg). The preamble is:

```
$units 45 $

$data y x $

$dinput 6 $

File name: glex4.dat
```

We begin by plotting instar duration against initial weight:

```
$plot y x $
```

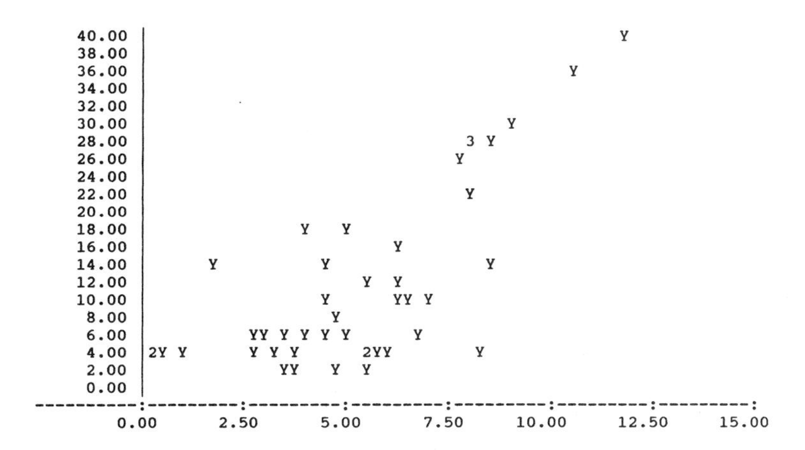

There is a lot of scatter, but it does look as though heavier caterpillars take substantially longer to pass through the first instar. We calculate the factor levels for the three hormone treatments (T) using %gl, then declare the response variable and determine the total sum of squares as usual:

```
$calc t=%gl(3,15) $

$factor t 3 $

$yvar y $

$fit $

deviance = 4497.1
     d.f. = 44
```

The main effect of hormone treatment is found simply:

```
$fit +t $

deviance = 1780.7 (change = −2716.)
     d.f. = 42      (change = −2)

$disp e $

   estimate  s.e.    parameter
1    4.088   1.681   1
2   18.32    2.378   T(2)
3    4.679   2.378   T(3)
scale parameter taken as 42.40
```

Hormone A (T(2)) clearly has a profound effect on the duration of the first instar, while hormone B has a lesser, and more marginally significant, impact. To see the effect of initial caterpillar weight, we add the covariate to the model:

```
$fit +x $

deviance = 447.96 (change = −1333.)
     d.f. = 41      (change = −1)

$disp e $

   estimate  s.e.     parameter
1   −6.633   1.293    1
2   15.15    1.241    T(2)
3    5.083   1.208    T(3)
4    2.184   0.1978   X
scale parameter taken as 10.93
```

This has a dramatic effect in reducing the deviance, so initial size is clearly of great importance in determining larval duration under the different hormone treatments. We can investigate the fit by plotting the fitted values for each treatment separately:

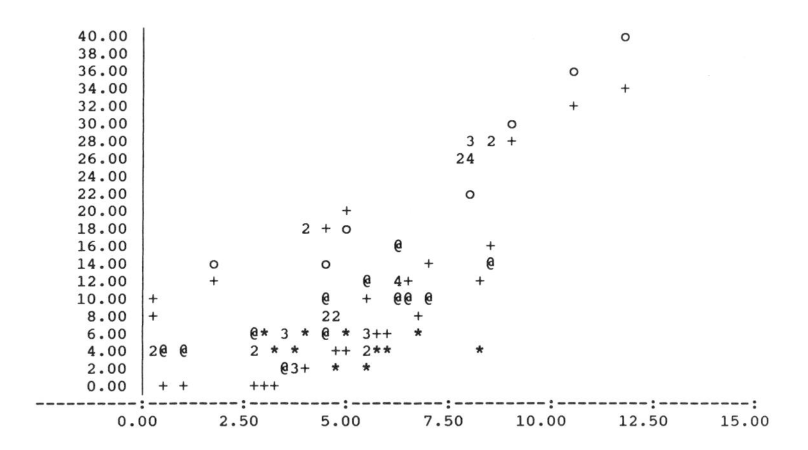

$plot %fv y x '+++⋆o@' t $

Now look carefully at the data and the fitted values. It is clear that hormone *A* (o) is different from the other two treatments (mean instar duration is much longer), but the fit of the model is rather poor. The residuals for the controls (⋆) are large, and all the residuals for hormone 1 are positive for high initial body mass. Perhaps the fit would be improved if we allowed that the slope of the graph could vary from one treatment to another? At the moment the three regression lines are parallel.

To fit non-parallel lines for the covariate in the different treatments, we use the interaction term T.x. We remove the overall covariate regression (−x) and add the interaction term (+T.x):

$fit −x+t.x $

```
deviance = 159.34 (change = −288.6)
    d.f. = 39     (change = −2)
```

The reduction in deviance of 288.6 with 2 d.f. is highly significant ($F = 144.3/4.08 = 35.4$). The reason for the significant interaction can be seen by inspecting the parameter estimates:

$disp e $

	estimate	s.e.	parameter
1	4.464	1.828	1
2	−0.4111	2.133	T(2)
3	−2.563	2.215	T(3)
4	−0.07659	0.3570	T(1).X
5	2.886	0.1521	T(2).X
6	1.454	0.2406	T(3).X

scale parameter taken as 4.085

The three treatments T(1), T(2) and T(3) all have different slopes (−0.07659, 2.886 and 1.454 respectively). The earlier parallel slope of 2.184 was much too steep for the control insects T(1), and too shallow for the caterpillars treated with hormone *A* (T(2)). Note that because we removed the original slope of the covariate (**fit**−**x**), the values of T(2).X and T(3).X are *slopes* (not differences between slopes as they would normally be, if x had been retained in the model). We can inspect the fit of the new model:

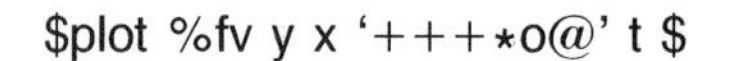

```
$plot %fv y x '+++*o@' t $
```

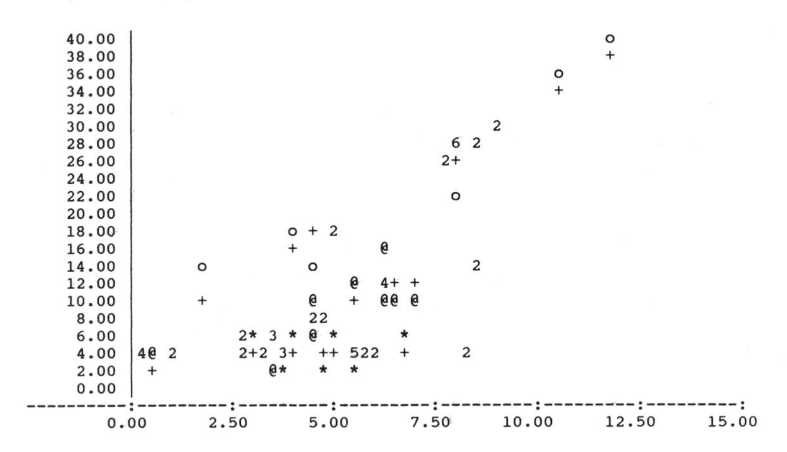

which is clearly much improved. It remains only to check that the error structure is satisfactory:

```
$calc resid=y−%fv $
$plot resid %fv '+' $
```

There is no pattern in the residuals, and the error structure can be regarded as satisfactory.

This is not the minimal adequate model, however, and a number of questions now arise under model simplification.

1 Does the control T(1).X need a non-zero slope?

2 Do the three regression lines require different intercepts?

3 Can we simplify the numerical values of the parameters?

In order to see whether a significant reduction in explanatory power occurs when we give the regression a zero slope for the controls, we need to compute a new vector of initial body weights that contains zeros for all the control animals. This causes the parameter to be aliased (see Section 4.20) so that a zero slope is assumed in calculating the fitted values for individuals in the control group. We use another simple computing trick to calculate the new vector (xx) from the old (x). This uses the logical function (T > 1) which is true for treatments 2 and 3 but false for treatment 1. The logical function therefore has the value 1 for T(2) and T(3) but 0 for T(1) (see Section 3.8).

```
$calc xx=x*(t>1) $
```

This means that the new vector xx has the same values as x except for the numbers relating to control animals where xx = 0. We now fit the model again, using xx instead of x:

```
$fit t+t.xx $

deviance = 159.52
      d.f. = 40
```

This causes only a tiny increase in deviance of $159.52 - 159.34 = 0.18$ on 1 d.f., which is nowhere near significant ($F = 0.18/3.99 = 0.045$). The simplification of replacing the slope of -0.07657 by a horizontal straight line for the control animals appears fully justified. Caterpillars in the control group take the same length of time, on average, to pass through the first instar, irrespective of their initial weight. The modification may have had a slight effect on the remaining parameter values:

```
$disp e $

     estimate    s.e.      parameter
1    4.088       0.5156    1
2   -0.03519     1.202     T(2)
3   -2.187       1.339     T(3)
4    0.000       aliased   T(1).XX
5    2.886       0.1502    T(2).XX
6    1.454       0.2377    T(3).XX
scale parameter taken as 3.988
```

Note that the slope within T(1) is zero and that the word 'aliased' appears in the standard error column (see above).

The second question concerns the intercepts. Would the fit of the model be significantly worse if all three graphs had a common origin? This is tested very simply. Instead of fitting the model T + T.xx we omit the main effect of hormone treatment and fit T.xx on its own:

```
$fit t.xx $

deviance = 170.40
      d.f. = 42
```

This simplification has caused a rather larger increase in deviance ($170.40 - 159.52 = 10.89$), and note that the degrees of freedom have increased by two because we are fitting only one origin instead of three.

The increase in deviance, however, is not significant ($F = 5.44/3.988 = 1.36$), so the simplification appears to be justified. The simplification may have affected the parameter values slightly:

```
$disp e $

   estimate  s.e.      parameter
1  3.810     0.4396    1
2  0.000     aliased   T(1).XX
3  2.916     0.08971   T(2).XX
4  1.120     0.1262    T(3).XX
scale parameter taken as 4.057
```

All the parameters in the model are now significantly different from zero (and significantly different from one another), and the model appears to be minimal adequate. Initial body weight does appear to influence the response of the caterpillars to the two hormones; initially heavier animals take longer to develop than initially smaller individuals. The magnitude of the response to initial body weight (the slope of the regression) is the only thing to differ between the two hormones (their intercepts are the same as the controls at approximately 4 days).

It only remains to check that we have not distorted the errors during the process of simplification:

```
$calc resid=%yv−%fv $

$plot resid %fv $

$plot resid xx $
```

The residual plots show no pattern, and we conclude that the error structure and link function of the model are satisfactory. ●

12.5.5 An example of offsets

In this example, we have gathered data on the basal diameters of 10 herbaceous plants (x, cm) and measured the dry mass of seeds produced by the same plant later in the season (y, g).

```
$units 10 $

$data x y $

$dinput 6 $

File name? glex30.dat

$plot y x $
```

After plotting the raw data, it looks as though we should carry out the analysis using log-log transformed data. The analysis proceeds as follows:

$calc ly=%log(y) $

$calc lx=%log(x) $

$plot ly lx $

```
  2.400 |
  2.280 |
  2.160 |                                                 Y
  2.040 |
  1.920 |
  1.800 |
  1.680 |
  1.560 |
  1.440 |                                            Y
  1.320 |
  1.200 |
  1.080 |                                       Y
  0.960 |
  0.840 |
  0.720 |                                  Y
  0.600 |
  0.480 |                             Y
  0.360 |
  0.240 |                        Y
  0.120 |                   Y
  0.000 |    Y    Y    Y
---------:---------:---------:---------:---------:---------:---------:
      0.000     0.200     0.400     0.600     0.800     1.000     1.200
```

```
  1.200 |
  0.900 |
  0.600 |                                                 L
  0.300 |                                               L
  0.000 |                                             L
 -0.300 |                                          L
 -0.600 |
 -0.900 |                                       L
 -1.200 |
 -1.500 |                                   L
 -1.800 |
 -2.100 |                               L
 -2.400 |
 -2.700 |
 -3.000 |                         L
 -3.300 |
 -3.600 |
 -3.900 |                 L
 -4.200 |
 -4.500 |    L
 -4.800 |
---------:---------:---------:---------:---------:---------:---------:
     -2.500    -2.000    -1.500    -1.000    -0.500     0.000     0.500
```

This looks like an excellent linear fit, apart from the smallest plant which appears to have a large positive residual. We shall carry out the regression with and without this point, to determine how influential it is in determining parameter estimates:

$yvar ly $

$fit lx $

deviance = 1.0377
d.f. = 8

We can inspect the residuals by using **disp r**:

```
$disp r $

unit  observed   fitted     residual
1     -4.60520   -5.24775   0.643
⋮        ⋮           ⋮          ⋮
```

This confirms our guess about the large positive residual of the smallest plant. But how influential is the point?

```
$disp e $

   estimate  s.e.     parameter
1  0.4580    0.1726   1
2  2.478     0.1638   LX
scale parameter taken as 0.1297
```

The slope of the log-log plot is 2.478 with the smallest plant included. We can see what difference it makes (if any) by taking it out of the analysis with a weight vector, w, which is 1 for all plants except the smallest (unit 1) for which the weight is set to 0:

```
$calc w=1 $

$calc w(1)=0 $

$weight w $

--model changed
```

We refit the model and inspect the residuals and the parameter estimates:

```
$fit lx $

deviance = 0.073584
      d.f. = 7 from 9 observations

$disp e $

   estimate  s.e.      parameter
1  0.6792    0.05430   1
2  2.947     0.06759   LX
scale parameter taken as 0.01051
```

The point turns out to be extremely influential, and tends to rotate the regression clockwise, reducing its slope from 2.947 to 2.478 (this change is highly significant, giving a t-test value of $(2.947 - 2.478)/0.1638 = 2.863$). Note how the residual deviance dropped from 1.0377 to 0.0736 after we removed the first point. This means that the remaining points follow a linear model very closely on a log-log scale. We consider whether removal of the point is justified later on, but here we continue the model simplification with the reduced data set.

Inspection of the parameter estimates now suggests a simplification that requires the use of offsets. The slope looks to be sufficiently close to 3.0 that constraining it to be 3 would not cause a significant change in deviance. This would make reasonable biological sense, because x is measured in cm and seed production is proportional to total shoot mass and this, in turn, is proportional to shoot volume (i.e. to cm^3). This is how we proceed. We calculate a new vector containing three times the log of x:

```
$calc slope = 3*lx $
```

and we declare that this new vector, called slope, is to be used as an offset:

```
$offset slope $

-- model changed
```

Technically, this means that GLIM will now use a new linear predictor:

$$\eta = \text{offset} + \Sigma\beta_i x_i$$

and estimate the remaining parameters by maximum likelihood. We need estimate only the intercept now, because we have specified the rest of the model in the offset. So we simply type:

```
$fit $

deviance = 0.080136
      d.f. = 8 from 9 observations
```

The deviance has increased slightly from 0.073584 to 0.080136, but the change is not significant ($F = (0.08 - 0.0735)/0.01 = 0.65$), and so the simplification was justified. The new estimate for the intercept is:

```
$disp e $

   estimate  s.e.       parameter
1  0.7126    0.03336  1
scale parameter taken as 0.01002
```

We need to know the antilog of this:

```
$calc %exp(.7125) $

2.039
```

which is very close to 2.0. We therefore try a further simplification, this time specifying the entire model in a new offset called model:

```
$calc model=%log(2)+3*lx $

$offset model $

-- model changed
```

There is just one snag. GLIM will not fit a model with no parameters. To get around this, we create a variable called zero containing only 0's:

```
$calc zero = 0 $
```

and then proceed to fit zero to the data with no intercept (i.e. using **fit −1**):

```
$fit zero−1 $

deviance = 0.083526
      d.f. = 9 from 9 observations
```

Again, there is an inevitable increase in deviance, but the change is very small (from 0.080136 to 0.083526) and this is not significant. You might wonder what the parameter estimates look like after such a curious procedure:

```
$disp e $

   estimate  s.e.      parameter
1  0.000     aliased   ZERO
scale parameter taken as 0.009281
```

The error variance is correct, and is marginally smaller than it was before we used the offset, because the increase in the degrees of freedom has more than compensated for the increase in deviance. The parameter zero is aliased (as we intended it should be) and adds 0.000 to the specified offset of $\ln(y) = 0.6931 + 3\ \ln(x)$.

From this we conclude that the data are just as well described by the model:

$$y = 2x^3$$

as by the initial regression model using the maximum likelihood estimates of the parameters:

$$y = 1.9723x^{2.947}$$

Let us give some more thought to the data point that we omitted. We saw that the seed production from the smallest plant (which had a y-value of 0.01) was highly influential. On closer inspection of the data, a possible reason for its influence emerges. Notice that all the data on seed weights are given to only two decimal places. Since the predicted value of y at $x =$ 0.1 would be substantially below 0.01 g, the outlier may have arisen from a measurement precision error or a data entry error. In this case, it turns out from inspection of the lab notebook that the y-value had been rounded up to the smallest value possible (e.g. from 0.002 to 0.01) at the data entry stage, since to round it down to zero would suggest that the plant produced no seed at all, and the computer operator assumed that the number had to fit into a field with space for only two decimal places.

On a log scale this change to the data makes an enormous difference. Let us compare the overall fit of the two models with and without the data point in question, and plot observed and fitted seed weights on the original axes. First we need to find the antilog of our fully specified model:

```
$calc em=%exp(model) $
```

Then we can plot this along with the observed seed weights against root diameter, using * to represent the fitted values and + the raw data:

```
$plot em y x '*+' $
```

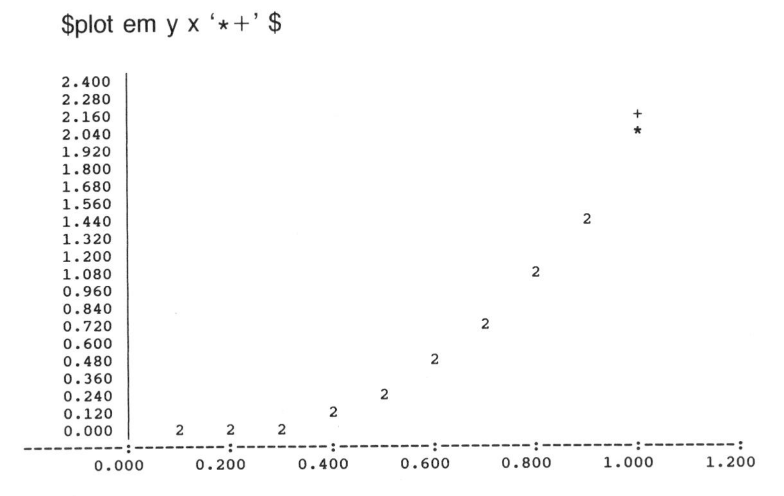

The fit of the fully specified model is excellent; the fitted values are indistinguishable from the data at this resolution, except for the largest plant, where the fitted value slightly underestimates the fecundity. Omission of the smallest plant makes little practical difference on a linear scale. This point did, however, have a substantial influence of the parameter estimates of the log-transformed model (it reduced the slope from 2.947 to 2.478). You can see just how influential this point was by using the parameter estimates from the original regression including the smallest plant, to calculate a new set of expected values, then transforming these back to the original axes:

```
$weight $

$fit lx $

$calc em2 = %exp(%fv) $

$plot em2 y x '*+' $
```

Inclusion of the seed mass from the smallest plant causes a dramatic reduction in the goodness of fit, such that observed and expected seed mass are distinguishable at five out of 10 positions (compared with only

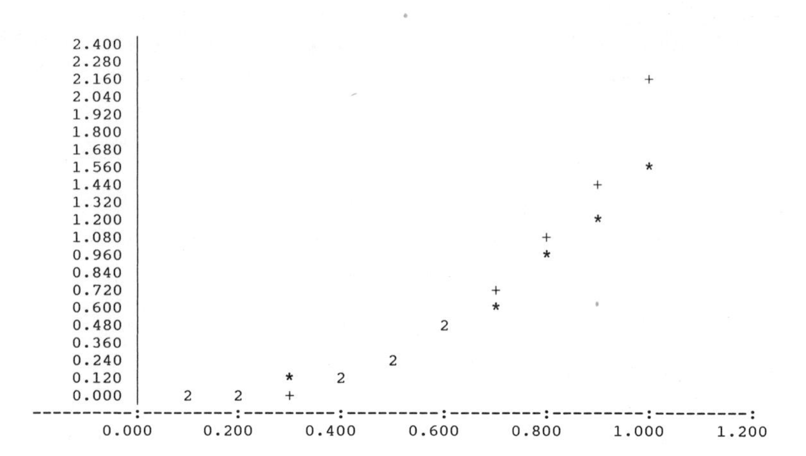

one out of 10 when the smallest plant was omitted). The moral is clear. Make sure that the decimal precision is correct when working with log-transformed data, and always test the influence of outlying points.

12.6 Summary

The fact that we have laboured long and hard to include a particular experimental treatment does not justify the retention of that factor in the model if the analysis shows it to have no explanatory power. ANOVA tables are often published in ecological studies and these typically contain a mixture of significant and non-significant effects. This is not a problem in orthogonal designs, because sums of squares can be unequivocally attributed to each factor and interaction term. But as soon as there are missing values or unequal weights, then it is impossible to tell how the parameter estimates and standard errors of the significant terms would have been altered if the non-significant terms had been deleted. The best practice is:

1 to say whether your data are orthogonal or not;
2 to present a minimal adequate model;
3 to give a list of the non-significant terms that were omitted, and the deviance changes that resulted from their deletion.

The reader can then judge the relative magnitude of the non-significant factors, and the importance of correlations between the explanatory variables.

The temptation to retain terms in the model that are 'close to significance' should be resisted. The best way to proceed is this: if a result would have been *important* if it had been statistically significant, then it is worth repeating the experiment with higher replication and/or more efficient blocking, in order to demonstrate the importance of the factor in a convincing and statistically acceptable way.

CHAPTER 13

Model criticism

There is a temptation to become personally attached to a particular model. Statisticians call this 'falling in love with your model'. It is as well to remember the following home truths about models:

1 all models are wrong;
2 some models are better than others;
3 the correct model can never be known with certainty;
4 the simpler the model, the better it is.

Just as there is no perfect model, so there may be no optimal scale of measurement for a model. Suppose, for example, we had a process that had Poisson errors and multiplicative effects amongst the explanatory variables. Then, one must chose between three different scales, each of which optimizes one of three different properties:

1 the scale of $\sqrt{y}$ would give constancy of variance;
2 the scale of $y^{2/3}$ would give approximately normal errors;
3 the scale of $\ln(y)$ would give additivity.

Thus, any measurement scale is a compromise, and is chosen to give the best overall performance of the model.

13.1 Residuals

After fitting a model to data we should investigate how well the model describes the data. In particular, we should look to see whether there are any systematic trends in the goodness of fit. For example, does the goodness of fit increase with the observation number, or is it a function of one or more of the explanatory variables? We can calculate the raw residuals:

```
$calc r = %yv - %fv $
```

or we can use the standardized residuals calculated within GLIM, by using **disp r**:

```
$disp r $
```

These do not generally give the same values as the raw residuals, because **disp r** uses standardized residuals, which are calculated as follows:

$$\text{residual} = (\textit{observed} - \textit{fitted}) \times \sqrt{\frac{\textit{prior weight}}{\textit{scale parameter} \times \textit{variance function}}}$$

With normal errors, the identity link, equal weights and the default scale factor, the raw and standardized residuals are identical. For Poisson errors, the standardized residuals are (%yv−%fv)/%sqrt(%fv), for the binomial they are (%yv−%fv)/%sqrt(%fv*(1−%fv/%bd)) and for gamma errors they are (%yv−%fv)/%fv (see Table 19.1). GLIM 4 holds the standardized residuals in a system vector called %rs.

We should routinely plot the residuals against:

1 the fitted values (%fv);
2 the explanatory variables;
3 the sequence of data collection;
4 standard normal deviates;

as illustrated in Section 5.8. A good model must also account for the variance mean relationship adequately and produce additive effects on the appropriate scale (as defined by the link function).

A plot of standardized residuals against fitted values should look like the sky at night, with no trend in the size or degree of scatter of the residuals. A common problem in ecological data is that the variance increases with the mean, so that we obtain an expanding, fan-shaped pattern of residuals (Fig. 5.1c).

One of the commonest reasons for a lack of fit is through the existence of outliers in the data. It is important to understand, however, that a point may *appear* to be an outlier because of mis-specification of the model, and not because there is anything wrong with the data. Given that the data are sound, there are five things we might do to improve the fit:

1 change the structure of the model;
2 change the scale of the axes;
3 alter the weights;
4 employ a different link function;
5 try a different error structure.

13.2 Mis-specified model

The model may have the wrong terms in it, or the terms may be included in the model in the wrong way. We have dealt with the selection of terms for inclusion in the minimal adequate model in Chapter 12. Here we simply note that *transformation of the explanatory variables* often produces improvements in model performance. The most frequently used transformations are logs and powers. Again, in testing for non-linearity in the relationship between y and x we might add a term in x^2 to the model; a significant parameter in the x^2 term indicates curvilinearity in the relationship between y and x (see Section 12.5.1).

A further element of mis-specification can occur because of *structural non-linearity*. Suppose, for example, that we were fitting a model of the form:

$$y = a + \frac{b}{x}$$

but the underlying process was really of the form:

$$y = a + \frac{b}{c + x}$$

GLIM cannot deal with non-linear modelling directly, but all is not lost. If the underlying structure is suspected, then we can create a new explanatory variable $z = 1/(c+x)$, then try fitting z to the data with a series of different guesses at the value of c. This is carried out most conveniently by writing a macro that cycles through different values of c (see Section 19.9); we select the value for c that gives the minimum deviance. In the present example with a roughly hyperbolic relationship, we would probably use gamma errors with a reciprocal link (see Section 17.2). Most general-purpose statistical packages provide facilities for non-linear modelling, and specialist programs like MLP (Maximum Likelihood Program) are also available.

13.3 Mis-specified error structure

A common problem in ecological data is that the variance increases with the mean. The assumption so far has been of normal errors with constant variance at all values of the response variable. For continuous measurement data with non-constant errors we can specify *gamma errors*. These are discussed in Chapter 17 along with worked examples, and we need only note at this stage that they assume a *constant coefficient of variation* (see Taylor's power law in Section 4.19).

With count data, we often assume Poisson errors, but the data may exhibit overdispersion (see below) so that the variance is actually greater than the mean (rather than equal to it, as assumed by the Poisson distribution). An important distribution for describing aggregated data in ecology is the *negative binomial*. While GLIM has no direct facility for specifying negative binomial errors, we can write our **own** model (see Section 19.10) or negative binomial errors can be approximated by assuming that the variance is a constant multiple of the mean (see Section 13.7 for details on using the **scale** directive).

13.4 Mis-specified link function

Although each error structure has a canonical link function associated with it (see Section 10.4.4), it is quite possible that a different link function would give a better fit for a particular model specification. For example, with normal errors we might try a log link or a reciprocal link to improve the fit. Similarly, with binomial errors we might try a complementary log-log link instead of the default logit link function.

An alternative to changing the link function is to transform the values of the response variable. The important point to remember here is that changing the scale of y will alter the error structure (see Section 3.4). Thus, if you take logs of y and carry out regression with normal errors, then you will be assuming that the errors in y were log-normally distributed. This may well be a sound assumption, but a bias will have been introduced if the errors really were additive on the original scale of measurement. If, for example, theory suggests that there is an exponential relationship between y and x:

$$y = ae^{bx}$$

then it would be reasonable to suppose that the log of y would be linearly related to x:

$$\ln y = \ln a + bx$$

Now suppose that the errors, ε, in y are multiplicative with a mean of zero and constant variance, like this:

$$y = ae^{bx} (1 + \varepsilon)$$

then they will also have a mean of zero in the transformed model. But if the errors are additive, like this:

$$y = ae^{bx} + \varepsilon$$

then the error variance in the transformed model will depend upon the expected value of y. In a case like this, it is much better to analyse the untransformed response variable and to employ the log link function, because this retains the assumption of additive errors.

When both the error distribution and functional form of the relationship are unknown, there is no single specific rationale for choosing any given transformation in preference to another. The aim is pragmatic, namely to find a transformation that gives:

1 constant error variance;
2 approximately normal errors;
3 a linear relationship between the response variables and the explanatory variables;
4 straightforward scientific interpretation.

The choice is bound to be a compromise and, as such, is best resolved by quantitative comparison of the deviance produced under different model forms. To demonstrate the procedure, we shall analyse the same set of data on timber volumes using a wide range of link functions and transformations.

13.5 Timber data: an example of model criticism in GLIM

Consider the question of estimating the volume of usable timber in tree trunks based on measurements of trunk height and girth at breast height.

We begin by thinking about the shape of tree trunks. Because they taper with height, we would expect that their volume would be less than the volume of a cylinder of uniform girth. Similarly, they do not taper to a sharp point (at least the usable timber does not), and so we might expect the volume to be somewhat larger than a cone with basal girth as measured at breast height (there is often not much taper between the breast height and the height at which the trunk is topped). Let us work out the models that might be appropriate for cones and cylinders, by expressing volume V in terms of girth G and height H. For the cone we have:

$$V = \frac{\pi}{3}\left(\frac{G}{2\pi}\right)^2 H = \frac{1}{12\pi}G^2H$$

and for the cylinder:

$$V = \pi\left(\frac{G}{2\pi}\right)^2 H = \frac{1}{4\pi}G^2H$$

Now with a multiplicative model like this (volume is proportional to girth squared *times* height), it is natural to take logs. For the cylinder, we get:

$$\ln V = \ln\left(\frac{1}{4\pi}\right) + 2\ln G + \ln H$$

and, for the cone, the log of the volume is given by:

$$\ln V = \ln\left(\frac{1}{12\pi}\right) + 2\ln G + \ln H$$

Thus log volume is the sum of linear functions of log girth (with slope 2) and log height (with slope 1), while the value of the intercept reflects the degree to which the tree trunk is cone-like or cylindrical. We can compute the expected values of the intercept in these two cases, using GLIM as a calculator:

```
$calc %log(1/(12*3.14159)) $

-3.630

$calc %log(1/(4*3.14159)) $

-2.531
```

So if the intercept is close to −3.6 the trees are more cone-like, whereas if it is more like −2.5 they are more cylindrical. Let us see what emerges from a study of 31 black cherry trees in the Allegheny National Forest of Pennsylvania (Ryan *et al.*, 1976). We read in the data on volume V (m^3), girth G (cm) and height H (m):

```
$units 31 $

$data v g h $

$dinput 6 $

File name? glex16.dat
```

Now we need to convert the girths into metres so that the units of height and girth are the same. Next, calculate the logs of volume lv, girth lg and height lh:

```
$calc g=g/100 $

$calc lg=%log(g) $

$calc lv=%log(v) $

$calc lh=%log(h) $
```

Let us see how well our model describes the data by fitting lg and lh to the response variable lv:

```
$yvar lv $

$fit:+lg+lh $

deviance = 8.3089
      d.f. = 30

deviance = 0.18555 (change = −8.123)
      d.f. = 28        (change = −2)
```

The model accounts for a very high proportion of the deviance ($r^2 = 8.123/8.3089 = 0.978$). We can see how closely the parameters match our expectations by inspecting the parameter values:

```
$disp e $

    estimate  s.e.      parameter
1   −2.899    0.6377    1
2    1.983    0.07503   LG
3    1.117    0.2045    LH
scale parameter taken as 0.006627
```

This shows a very good agreement: the slope for girth is 1.983 against a predicted value of 2, and the slope for height is 1.117 against an expected value of 1. The intercept of −2.9 indicates that the trunks are more cylinder-like than cone-like (see above).

If there is a good scientific reason for it, then it may be sensible to simplify the numerical values of the parameters (see Section 12.3). In the present case, we might try fitting a model with the two slopes fixed at 2 and 1, and have GLIM estimate the maximum likelihood value of the intercept. This procedure is carried out using the **offset** directive (see Section 12.5.5). We begin by calculating a new variable, m, whose value is:

$$m = 2\ln(g) + \ln(h)$$

and then define this new variable to be the offset. The model is then refitted:

```
$calc m=2*lg+lh $

$offset m $

-- model changed

$fit $

deviance = 0.18777
      d.f. = 30
```

The increase in deviance compared with the model in which the two slopes were estimated by GLIM is only 0.18769 – 0.18546 = 0.00223. This comes nowhere close to significance, and we can accept the simplified model with slopes 2 and 1.

How does fitting the offset change the value of the intercept?

```
$disp e $

    estimate  s.e.       parameter
1   -2.534    0.01421    1
scale parameter taken as 0.006259
```

The value of −2.534 is now very close to the value expected of a cylinder (−2.531), and we can conclude that the taper of these trees is not pronounced (at least up to the height at which usable timber is measured). We can finish with a flourish by fitting a completely specified model for a cylinder, using the theoretical intercept −2.531. We calculate a new offset:

```
$calc m2=-2.531+2*lg+lh $

$calc zero=0 $

$offset m2 $

$fit zero-1 $

deviance = 0.18806
      d.f. = 31
```

This is just as good a fit to the data as the model with parameters estimated by GLIM, and for practical purposes we can treat these particular tree trunks as if they were cylindrical. You should check the fit and look to see whether there are any unusually large residuals (see Exercise 13.2):

```
$calc resid=%yv-%fv $

$plot resid %fv $

$plot %yv %fv g '*+' $

$disp r $
```

13.5.1 Alternative model structures using link

If we want to modify the model structure, there are two alternatives:

1 transform the response variable;

2 change the link function.

The choice between these two options will usually be made on purely pragmatic grounds; other things being equal, the case that gives the lowest deviance will be preferred. Sometimes there may be *a priori* reasons for preferring one option over the other (e.g. the desire to have a dimensionally homogeneous equation), but there are no hard and fast rules. The important thing to understand is that transforming the response variable will change its error structure and alter the way that interaction terms will need to be interpreted (see Section 3.4.1).

We can compare different ways of transforming the response variable and different choices for the link function by reanalysing the tree data. Instead of taking log volume as the response variable, we could take other transformations like reciprocal or cube root. Again, if we took the untransformed volume as the response variable, then instead of the identity link, we could try the log, reciprocal or exponent links. The best model structure could then be decided on the basis of the minimal residual deviance.

For the purpose of demonstration, we work through the case of a cube root transformation for the response variable. This has the advantage that it is likely to produce a linear relationship between the response variables and the explanatory variables but it will have altered the error structure of the response variable (changing the link function would not do this; see above).

```
$offset $

$calc v3=v**(0.3333) $

$yvar v3 $

[w] -- model changed

$fit:+g+h $

deviance = 1.4930
      d.f. = 30
```

Notice how the value of *SST* is changed because we have transformed the response variable.

```
deviance = 0.033380 (change = −1.460)
    d.f. = 28       (change = −2)
```

```
$calc 1.460/1.4930 $
```

```
0.9779
```

Thus we find an r^2 of 0.9779 for the cube root transformation and we enter this value in the table below. Making the equation *dimensionally homogeneous* (so that all the terms have units of length, by using the square root of girth rather than girth as an explanatory variable) makes no great difference to the result in this case, though, in general, it is good practice to work with dimensionally homogeneous equations wherever possible.

To investigate the effects of changing the link function, we return to the original scale of measurement by specifying volume as the response variable:

```
$yvar v $
```

Now, to change the link from the default identity link, all we do is alter the **link** directive. For the log link, for instance, we write:

```
$link log $
```

then refit the model:

```
$fit : +g+h $
```

and note the reduction in deviance. To change the link to reciprocal, we just type:

```
$link r $
```

and so on. You should carry out the other links and transformations to complete Table 13.1 (a full transcript of this is in Exercise 13.2).

Table 13.1 Comparison of various model structures for tree data

Response variable	Link function	Deviance	r^2	x-axes
$y^{1/3}$	i	1.4933	0.9779	H, G
$y^{1/3}$	i	1.4933	0.9752	H, $\sqrt{G}$
$\ln y$	i	8.3089	0.9776	$\ln H$, $\ln G$
y	e 0.333	42.504	0.9773	H, G
y	log	42.504	0.9663	H, G
y	i	42.504	0.9479	H, G
y	r	42.504	0.8751	H, G

In summary, we have three equally plausible models for the timber volume data:

1 a regression of $(\text{volume})^{1/3}$ on height and girth; or
2 a regression of log volume on log height and log girth; or
3 a regression of volume on height and girth using a cube root link.

The data do not allow us to distinguish between these competing models. If you were writing up work like this, it would be a good idea to report all three models, and to explain that they described the data equally well. In the present example, the biological assumptions involved in the three models are the same, but if this were not the case, then a critical evaluation of the assumptions would be necessary.

13.6 The Box–Cox transformation

When no obvious transformation is suggested from theoretical considerations or from experience, then a useful method is to try different powers of y as the transformation, and to select the value of the power λ that gives the best fit of the model to the data. Specific values of λ produce some familiar transformations (see Table 13.2).

Table 13.2 Power transformations

λ	Transformation
−1	Reciprocal (or inverse)
0.5	Square root
0.333	Cube root

We would not want to use the transformation $\lambda = 0$ because this would turn our entire vector of y-values into 1's! The standard trick is to define $\lambda = 0$ as being the log transformation.

The best known family of transformations for positive-valued response variables is named after the famous statisticians G. E. P. Box and D. R. Cox (1964, 1982). The Box–Cox transformation look like this:

$$y^{(\lambda)} = \frac{y^{\lambda} - 1}{\lambda} \qquad \lambda \neq 0$$

$$y^{(\lambda)} = \ln(y) \qquad \lambda = 0$$

Other transformations that extend this power family include the *folded-power transformations* and the *modulus transformations*. These and other transformations are discussed by Atkinson (1985).

The GLIM library contains a macro that carries out transformations of the response variable for a range of different values of λ, fits the current model, then plots the residual deviance against the value of λ. This gives a U-shaped curve, and the value of λ giving the minimum deviance can be read from the graph. This gives the most appropriate transformation, which can then be specified as the numerical value in the **link e** directive. If changes in λ do not cause big changes in deviance, then it does not matter very much which transformation is chosen. The most sensible procedure is then to pick the value of λ that gives the simplest interpretation. Thus, if the minimum deviance occurs at $\lambda = -0.835$, it is sensible to round this down to -1 and to use the reciprocal transformation. Note that power transformations do not guarantee normality, so this should be checked independently, and that the choice of model influences the optimal λ. Note, also, that data with 0's cannot be Box–Cox-transformed.

Because λ is estimated from the data, the true variances of parameter estimates may be somewhat larger than the estimates that follow directly from model-fitting on the transformed data. Most workers assume that variance inflation due to estimating λ is neither severe nor important in practice.

We return to the timber volume data, and ask which power transformation gives the minimal deviance on a model of girth and height (g + h). The preamble is given above, and all we need to do is define the two very simple macros called yvar and model that must precede our use of boxcox. The first of these defines the name of the response variable, v:

```
$macro yvar v $endmac $
```

(don't forget the space after v and before $) and the second defines the model we wish to fit to the data on this particular run of boxcox (namely g + h):

```
$macro model g+h $endmac $
```

Next we read in the boxcox macro from GLIM's library file. The channel number is %plc (this stands for primary library channel) and the subfile is called boxcox:

```
$input %plc boxcox $
```

Finally, the macro is executed by typing:

```
$use boxcox $
```

The macro prints a heading containing the structure of the present model, then asks you to specify three values: the maximum value of λ, the increment between successive values of λ, and the minimum value of λ. If we have no real idea about the value of λ then a useful band of values is obtained by going down from 1 in steps of 0.1 to -1 like this:

```
--- Model is g+h

--- Y-Variate is v
```

```
Max. value of lambda?

$DIN? 1

Increment?

$DIN? .1

Min. value of lambda?

$DIN? −1
```

The computer then works out the deviance of the specified model at each of the incremental values of λ. If you specify too fine an increment, this stage can take a long time. Eventually, GLIM prints the graph (see p. 223), showing a minimum deviance close to $\lambda = 0.4$. The macro also tabulates the values of the deviance and of λ like this:

	DEV_	LMB_
1	56.804	−1.0000
2	51.471	−0.9000
3	46.006	−0.8000
4	40.396	−0.7000
5	34.630	−0.6000
6	28.705	−0.5000
7	22.634	−0.4000
8	16.457	−0.3000
9	10.264	−0.2000
10	4.227	−0.1000
11	−1.364	0.0000
12	−6.069	0.1000
13	−9.328	0.2000
14	−10.601	0.3000
15	−9.601	0.4000
16	−6.451	0.5000
17	−1.613	0.6000
18	4.338	0.7000
19	10.905	0.8000
20	17.737	0.9000
21	24.618	1.0000

which shows rather more clearly that the minimum lies closer to 0.3 than 0.4. To get a finer resolution, we can re-run the macro with a smaller increment (say 0.01):

```
$use boxcox $

--- Model is g+h

--- Y-Variate is v

Max. value of lambda?
```

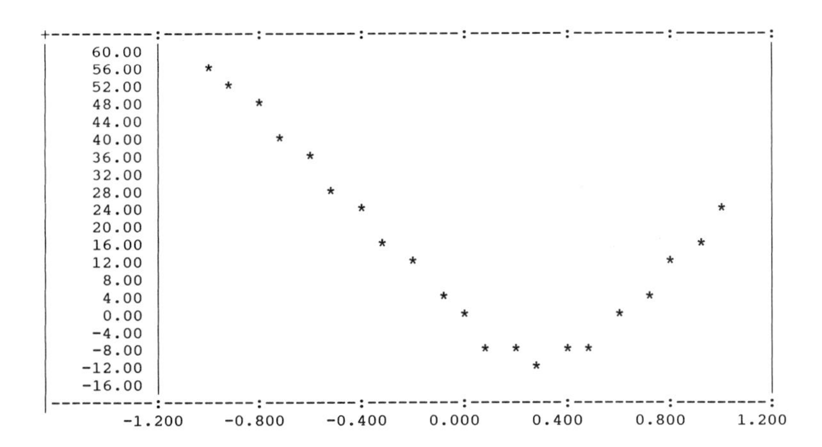

$DIN? .4

Increment?

$DIN? .01

Min. value of lambda?

$DIN? .2

which shows a minimum at $\lambda = 0.31$. The output also demonstrates that the deviance is roughly constant at -10.5 all the way from $\lambda = 0.28$ up to 0.33. Since the choice within this range will make no practical difference, we should accept either the simplest value, and work with $\lambda = 0.3$, or the most logically consistent (the cube root).

13.7 Overdispersion

Overdispersion is the polite statistician's version of Murphy's law: if something can go wrong, it will. Overdispersion can be a problem when working with Poisson or binomial errors, and tends to occur because you have not measured one or more of the factors that turn out to be important. It may also result from the underlying distribution being non-Poisson or non-binomial. This means that the probability you are attempting to model is not constant within each cell, but behaves like a random variable. This, in turn, means that the residual deviance is inflated. In the worst case, all the explanatory variables you have measured may turn out to be unimportant so that you have no information at all on any of the genuinely important predictors. In this case, the minimal adequate model is just the overall mean, and all your 'explanatory' variables provide no extra information.

One way to get round the problem of overdispersion is to *assume that the variance is proportional to the binomial variance rather than equal to it.* To do this, one calculates the ratio of scaled deviance to d.f. (or the ratio of Pearson's χ^2 (the system scalar %x2) to d.f.) and uses the **scale** directive to alter the scaled deviance. The parameter estimates are not affected by this procedure, but the standard errors are inflated, so that significance tests are more stringent (and, naturally, one gets fewer significant effects). The detailed procedures for dealing with overdispersion in models with Poisson errors are described in Section 14.9, and in models with binomial errors in Section 15.6.

The comparison of models is also complicated by overdispersion. Suppose, for example, that model 1 is simpler than model 2. Model 1 has p_1 parameters and model 2 has p_2 ($p_1 < p_2$), where the parameters of model 2 are the same p_1 parameters as in model 1 plus some extra parameters. The deviance of model 1 is D_1 and the deviance of model 2 is D_2. Because model 2 is more complex, then $D_2 \leq D_1$. Now there is overdispersion, so that D_2 is larger than the value of χ^2 in tables with $R - 1 - p_2$ d.f. To assess whether the extra parameters contribute significant explanatory power, we construct the analysis of deviance table (Manly, 1990) shown in Table 13.3.

Table 13.3 ANODEV with overdispersion

Source	Deviance	d.f.	Mean deviance
Extra parameters in model 2	$D_1 - D_2$	$p_2 - p_1$	$(D_1 - D_2)/(p_2 - p_1)$
Error (fit of model 2)	D_2	$R - 1 - p_2$	$D_2/(R - 1 - p_2)$
Total	D_1	$R - 1 - p_1$	

The more complex model 2 is worth retaining if the F-ratio

$$F = \frac{(D_1 - D_2)/(p_2 - p_1)}{D_2/(R - 1 - p_2)}$$

is larger than the value in F tables with $p_2 - p_1$ and $R - 1 - p_2$ d.f.

For example, suppose that we have already established that 18 different chemicals had a major effect on insect counts, but the effect of four different applicators was marginal based on a t-test using the adjusted standard errors. The final deviance of the full model with both factors (chemicals and applicators) was 165 with 17 d.f. There are four application methods, so the simpler model without applicators has 3 d.f. more than the complex model. When applicators were removed from the model the deviance increased by 35, so the analysis of deviance table looks like Table 13.4.

Table 13.4 ANODEV for overdispersed insect count data

Source	Deviance	d.f.	Mean deviance
Application methods	35	3	11.66
Full model	165	17	9.7
Chemicals only	200	20	

Thus the F-ratio is $11.66/9.7 = 1.2$, much less than the value in tables with 3 and 17 d.f. We conclude that application method had no significant impact on insect counts.

It is important to stress that these techniques introduce another level of uncertainty into the analysis. Overdispersion happens for real ecological reasons, and these reasons may throw doubt upon our ability to interpret the experiment in an unbiased way. It means that something we did not measure turned out to have an important impact on the results. If we did not measure this factor, then we have no confidence that our randomization process took care of it properly and we may have introduced an important bias into the results.

CHAPTER 14

Analysing count data: Poisson errors

Up to this point, the data have all been continuous measurements like weights, heights, lengths, temperatures, growth rates and so on. A great deal of the data collected by ecologists, however, is in the form of *counts* (whole numbers or integers) — the number of animals that died, the number of branches on a plant, the number of days of frost, or the number of insects on a leaf. With count data, the number 0 is often the value of a response variable (consider, for example, what a 0 would mean in the context of the four examples just listed).

For our present purposes, it is useful to think of count data as coming in three types.

1 Data on *frequencies*, where we count how many times something happened, but we have no way of knowing how often it did *not* happen; natural cases might be the number of lightning strikes per year, but this kind of data is perhaps most commonly met in population estimation (the number of fish in a trawl, the number of dormant seeds in a soil core or the number of wildebeest in an aerial photograph).

2 Data on *proportions*, where both the number doing a particular thing and the total group size are known, e.g. the response variable might be the number of insects dying in an insecticide bioassay, and the group size is the number of insects treated with the chemical.

3 *Category* data, in which the response variable is a score of some sort; for example, with binary response data the individuals are scored as dead or alive, pupated or not, hatched or not, set seed or not, etc.

Each of these types is dealt with in a chapter of its own. We begin, here, with a discussion of generalized linear models for analysing data on frequencies.

Straightforward linear regression methods are not appropriate for count data for two main reasons: (i) the linear model might lead to the prediction of negative counts; and (ii) the variance will not be constant. In GLIM, count data are handled very elegantly by specifying:

1 Poisson errors;

2 log link.

The log link ensures that all the fitted values are positive (see Section 10.4.1), while the Poisson errors take account of the fact that the data are integer and have variances equal to their means.

14.1 The Poisson distribution

The Poisson distribution is widely used for the description of count data that refer to cases where we know how many times something happened (e.g. kicks from cavalry horses, lightning strikes, bomb hits), but we have no way of knowing how many times it did not happen. This is in contrast to the binomial distribution (Chapter 15) where we know how many times something did not happen as well as how often it did happen (e.g. if we got six heads from 10 tosses of a coin, we must have got four tails).

The Poisson is a one-parameter distribution, specified entirely by the mean. The variance is identical to the mean, so the variance/mean ratio is equal to one. Suppose we are studying the ecology of a leaf miner on birch trees, and our data consist of the numbers of mines per leaf (x). Many leaves have no mines at all, but some leaves may have as many as five or six mines. If the mean number of mines per leaf is λ, then the probability of observing x mines per leaf is given by:

$$P_x = \frac{e^{-\lambda}\lambda^x}{x!} \tag{14.1}$$

This can be calculated very simply on a hand calculator because:

$$P_x = P_{x-1}\frac{\lambda}{x} \tag{14.2}$$

This means that if you start with the *zero term*:

$$P_0 = e^{-\lambda} \tag{14.3}$$

then each successive probability is obtained simply by multiplying by the mean and dividing by x.

Because the data (insects per leaf, plants per quadrat, droplets per plate, etc.) are integers, it means that the residuals (y−%fv) can take only a restricted range of values. If the estimated mean was 0.5, for example, then the residuals for counts of 0, 1, 2 and 3 could only be −0.5, 0.5, 1.5 and 2.5. The normal distribution assumes that the residuals can take any value (it is a continuous distribution).

Similarly, the normal distribution allows for negative fitted values. Since we cannot have negative counts, this is clearly not appropriate. GLIM deals with this by working with a logarithmic link function when Poisson errors are specified (see Section 10.4). Since the fitted values are antilogs, they can never go negative. Even if the linear predictor was $-\infty$, the fitted value would be $\exp(-\infty) = 0$.

A further difference from the examples in previous chapters is that GLIM uses maximum likelihood methods to estimate the parameters, rather than least squares methods. The notion of maximum likelihood will probably be unfamiliar, but it is a remarkably straightforward concept. This is how it works. Given a set of data and a particular model, then the maximum likelihood estimates of the parameter values are *those values that would make the observed data most likely* (hence the name maximum likelihood). You will see what this means by working through the maximum likelihood estimator for the Poisson distribution which is explained on p. 174.

Because the Poisson is a one-parameter distribution (the mean is equal to the variance), GLIM does not attempt to estimate the scale parameter (it is set to a default value of 1.0). If the error structure of the data really is Poisson, then the ratio of residual deviance to degrees of freedom after model-fitting should be 1.0. If the ratio is substantially greater than 1, then the data are said to show overdispersion, and remedial measures may need to be taken (e.g. the use of negative binomial errors; see Section 14.9).

With Poisson errors, the change in deviance attributable to a given factor is distributed asymptotically as χ^2. This makes hypothesis testing extremely straightforward. We simply remove a given factor from the maximal model and note the resulting change in deviance and in the degrees of freedom. If the change in deviance is larger than the value of χ^2 in tables, we retain the term in the model; if the change in deviance is less than the value of χ^2 in tables, the factor is insignificant and can be left out of the model.

Thus, the only important differences you will need to remember in modelling with Poisson rather than normal errors are the following:

1 the **error p** directive must be specified;

2 hypothesis testing involves χ^2 rather than variance ratios;

3 beware of overdispersion;

4 do not collapse contingency tables over explanatory variables.

These points are explained fully in the examples that follow.

14.2 An example of count data

Count data (c) have been obtained on the numbers of slugs found beneath 40 tiles placed in a stratified random grid over each of two different grasslands. This was part of a preliminary study in which the question was simply whether the mean slug density differs significantly between the two grasslands.

```
$units 80 $

$data c $

$dinput 6 $

File name? glex5.dat
```

```
$tab the c mean $

1.775
```

The overall mean density was 1.775 slugs per tile, and we can see how slugs were distributed over tiles using the **tab for** directive:

```
$tab for c $

        0.000   1.000   2.000   3.000  4.000  5.000  6.000
[ ]    34.000  14.000  10.000   7.000  4.000  5.000  2.000

        7.000   8.000   9.000  10.000
[ ]     1.000   1.000   1.000   1.000
```

These are typical Poisson data: whole numbers (integers) with lots of zeros (34 of them in this example), with the mode at the left (zero in this case) and a longish tail to the right. Most tiles had no slugs underneath them, while the maximum was 10 slugs, and this was observed just once.

We calculate a new vector A to represent the two fields and declare it to be a factor with two levels (the first 40 numbers refer to the first field).

```
$calc a=%gl(2,40) $

$factor a 2 $
```

Next we determine the mean slug densities in each field:

```
$tab the c mean for a $

      1      2
[ ]  1.275  2.275
```

so it appears that nearly twice as many slugs were caught in field 2 as in field 1. We need to determine how likely it is that a difference as large as this could occur by chance alone, if the fields really had the same slug densities. We shall say that the fields are significantly different only if such a difference could arise by chance with a probability of less than about 0.05. We assess the significance in the usual way, but the new ingredient is that we use the **error p** directive to tell GLIM that we have count data with Poisson errors:

```
$yvar c $

$error p $

$fit $

scaled deviance = 224.86 at cycle 4
           d.f. = 79

$disp e $
```

```
      estimate  s.e.      parameter
  1   0.5738    0.08375   1
  scale parameter taken as 1.000
```

The total deviance is 224.86 (note that this is not the total sum of squares as it was in the previous examples with normal errors and identity link), and the overall mean is 0.5738 with a standard error of 0.08375. Because we are using a log link, this mean value is in natural logs, so to obtain the observed sample mean of 1.775 slugs per tile, we would need to calculate the natural antilog (this is the function %exp):

```
$calc %exp(.5738) $

1.775
```

To determine the significance of the difference between the fields, we fit the factor A to the model:

```
$fit +a $

scaled deviance = 213.44 (change = −11.42) at cycle 4
           d.f. = 78     (change = −1)
```

This causes a change in deviance of 11.42 with one degree of freedom. With Poisson errors, the change in deviance can be compared directly with χ^2 tables to assess its significance. Since 11.42 is much larger than the value of χ^2 in tables with one degree of freedom (3.841 at 5%) we would conclude that the fields are indeed significantly different in their mean slug densities.

To see the parameter estimates we type **disp e** as usual:

```
$disp e $

      estimate  s.e.     parameter
  1   0.2430    0.1391   1
  2   0.5789    0.1742   A(2)
  scale parameter taken as 1.000
```

To see what these figures mean, we need to look up the antilogs of 0.243 and 0.8219 (= 0.234 + 0.5789):

```
$calc %exp(.243) $

1.275

$calc %exp(.243+.5789) $

2.275
```

This demonstrates that the parameter estimates are the natural logs of the arithmetic means. Note that they are *not* the average of the logs of the slug counts. If you think about it, you will realize that they could not be, since many of the counts were zeros, and you cannot take the log of zero. For more about the link function, see Section 10.4.

In the place where we are used to seeing the error variance, GLIM has printed 'scale parameter taken as 1.000'. This is because the Poisson is a one-parameter distribution. The variance is equal to the mean, and so the scale parameter is not estimated independently from the data.

A word of caution is necessary at this point. GLIM works on the assumption that if we say the errors are Poisson, then they really are Poisson. This means that once we have fitted the minimal adequate model to the data, the residual deviance divided by the residual degrees of freedom ought to be 1. Now, in the present example, the residual deviance is 213.44 on 78 d.f. (a ratio of 2.736). This is substantially larger than the assumed scale parameter of 1, and suggests that the data are overdispersed. We consider how to deal with this at the end of the present chapter (see Section 14.9). •

14.3 Contingency tables

An important use for Poisson errors is in the analysis of contingency tables. Ecologists analyse frequency data in four main ways:

1 as 2×2 tests of association;

2 in comparisons between observed and theoretical frequency distributions;

3 in pattern tests on counts, equivalent to ANOVA;

4 in multi-way tests of association (complex contingency tables).

14.3.1 G-test of independence

The preferred means of analysis for frequency data is the G-test. It is computationally more straightforward than the χ^2 test, and is amenable to more thorough analysis. The basis of the test is as follows.

Assume that we wish to test whether a sample of animals conforms to the expected mendelian ratio of phenotypes of 3:1. Out of 100 animals sampled at random, 69 had red eyes and 31 had white eyes. We refer to these observed frequencies as f_1 and f_2 respectively. If they had followed mendelian expectation exactly, we should have found 75 red-eyed individuals and 25 white-eyed. These expected frequencies are denoted by $\widehat{f_1}$ and $\widehat{f_2}$ respectively. Do the departures between f_1 and $\widehat{f_1}$ suggest that something more complicated is going on, or are these the kind of departures that could be expected due to chance alone?

Given that the animals had to have either red or white eyes (there were only two contingencies), the two probabilities can be computed from the binomial distribution, where p is the probability of red and $q = (1 - p)$ is the probability of white eyes.

$$C(n,f_1)p^{f_1}q^{f_2}$$

Similarly, the expected probabilities $\hat{p}$ and $\hat{q} = (1 - \hat{p})$ are obtained from the mendelian ratios (0.75 and 0.25):

$$C(n,f_1)\hat{p}^{f_1}\hat{q}^{f_2}$$

We can now calculate the ratio between these two likelihoods, which is greatly simplified by the fact that the combinatorial formula cancels out:

$$L = \frac{C(n,f_1)p^{f_1}q^{f_2}}{C(n,f_1)\hat{p}^{f_1}\hat{q}^{f_2}} = \left(\frac{p}{\hat{p}}\right)^{f_1} \left(\frac{q}{\hat{q}}\right)^{f_2}$$

Now, since $f_1 = np$ and $f_2 = nq$, the likelihood ratio can be written as:

$$L = \left(\frac{f_1}{\hat{f}_1}\right)^{f_1} \left(\frac{f_2}{\hat{f}_2}\right)^{f_2}$$

and the *log likelihood* is simply:

$$\ln L = f_1 \ln\left(\frac{f_1}{\hat{f}_1}\right) + f_2 \ln\left(\frac{f_2}{\hat{f}_2}\right) = \sum_{i=1}^{k} f_i \ln\left(\frac{f_i}{\hat{f}_i}\right) \qquad (14.4)$$

The test statistic, G, is given by $2 \ln L$ and is asymptotically χ^2 distributed with one degree of freedom.

Working through the example, we have $f_1 = 69$, $\hat{f}_1 = 75$, $f_2 = 31$ and $\hat{f}_2 = 25$. Thus:

$$G = 2\left[69 \ln\left(\frac{69}{75}\right) + 31 \ln\left(\frac{31}{25}\right)\right] = 1.83$$

This value is smaller than the value of χ^2 in tables with 1 d.f., so we conclude that there is no significant departure from mendelian expectation.

14.3.2 Two-by-two contingency tables

In the past, ecologists would have analysed 2×2 contingency tables using the χ^2 test of association. The row and column totals would have been computed, and the expected frequencies in each cell of the table obtained by multiplying the relevant row and column totals together, and dividing by the grand total. These *expected frequencies* (E) were then subtracted from the *observed frequencies* (O), and the test statistic calculated as follows:

$$\chi^2 = \sum\frac{(O - E)^2}{E} \qquad (14.5)$$

and the value compared with χ^2 tables with 1 d.f. (in general, rows − 1 times columns − 1). The trouble with this method is that the test statistic is not really a χ^2 value at all (it can't be, since χ^2 is supposed to be continuous, but the test statistic can take only discrete values). In fact, the test produces a *sample statistic* that is approximately distributed as χ^2 with 1 d.f. The mistake is so deeply ingrained, and in practice makes so little difference, that we shall pass on without further ado. The main reason for abandoning the χ^2 test in favour of GLIM is that it does not generalize readily to the analysis of more complicated tables of data.

Consider the following example from a study of the ecology of gall-forming cynipid wasps on oak trees. The presence or absence of two gall-forming species was noted on a sample of 111 different oak trees, and the data were presented as a 2×2 contingency table as shown in Table 14.1.

Table 14.1 2×2 contingency table: two species of gall wasp on oak

	Gall B		
Gall A	Present	Absent	Row totals
Present	13	44	57
Absent	25	29	54
Column totals	38	73	111

Thus 13 oaks had both galls on them and 29 trees had neither. Gall species A was somewhat more frequent than species B (found on 57 trees compared with 38).

We calculate the expected frequencies in each cell as follows:

$$\widehat{f}_{i,j} = \frac{R_i C_j}{T}$$

where $\widehat{f}_{i,j}$ is the expected frequency in row i and column j, R_i is the ith row total, C_j is the jth column total, and T is the grand total. We write the expected frequencies in a column next to the observed frequencies, calculate the differences, square the differences, divide by the expected frequencies, and add up the total of the scaled squared differences (see Table 14.2).

The sum of the scaled, squared differences is distributed approximately as χ^2 with 1 d.f. Tables give a χ^2 value of 3.841. Because the calculated value is larger than the value in tables we reject the null hypothesis that the distribution of the two gall species is independent. Note that the analysis does not tell us *how* the species are associated, and we must inspect the data to see whether the association is positive or negative. Because the two species are found on the same tree only 13 times –

Table 14.2 Calculations for the χ^2 test of association

O	E	$(O-E)$	$(O-E)^2$	$\frac{(O-E)^2}{E}$
13	19.51	6.51	42.38	2.172
44	37.49	6.51	42.38	1.130
25	18.49	6.51	42.38	2.292
29	35.51	6.51	42.38	1.193
			$\sum\frac{(O-E)^2}{E}$	6.787

compared with an expected 19.51 times if they were independently distributed — we conclude that there is a negative correlation to their distributions. This might be caused by different habitat preferences by the egg-laying females of the two species, by genetic or microhabitat differences between the oak trees, or by some other combination of factors. Significant values of χ^2 in such tests usually raise more questions than they answer.

Now we repeat the same example, but using the preferred G-test. We could use the formula $G = 2[\Sigma f \ln(f/\hat{f})]$, but there is an easier method that does not involve the prior calculation of all the expected frequencies, $\hat{f}$. First calculate $\Sigma f \ln f$ for the observed cell frequencies (see Table 14.3).

Table 14.3 The G-test for contingency tables: step 1

f	$\ln f$		$f \ln f$
13	2.5649		33.344
44	3.7842		166.50
25	3.2189		80.472
29	3.3673		97.652
		$\Sigma f \ln f$	377.968

Next, calculate $\Sigma f \ln f$ for the row and column totals (see Table 14.4).

Table 14.4 The G-test for contingency tables: step 2

Row and column totals	$\ln f$		$f \ln f$
57	4.0431		230.45
54	3.9890		215.41
38	3.6376		138.23
73	4.2905		313.20
		$\Sigma f \ln f$	897.29

Then calculate $n \ln n = 111 \times 4.7095 = 522.76$, where n is the grand total. Finally, compute:

$$G = 2(\Sigma f \ln f_{\text{cells}} - \Sigma f \ln f_{\text{totals}} + n \ln n) \quad (14.6)$$

$$377.968 - 897.29 + 522.76 = 3.438$$

$$G = 2 \times 3.438 = 6.876$$

The value of G is asymptotically distributed as χ^2 with 1 d.f. Since the calculated value is larger than the value in tables, we conclude that the galls are not independently distributed over trees (see above). Note that the value of G is close to the value of χ^2 obtained earlier (6.788), but not identical.

14.3.3 Two-by-two tables in GLIM

The calculation of G could not be more straightforward in GLIM, because with Poisson errors the scaled deviance gives the test statistic (twice the log likelihood) directly. For the example of gall wasps on oak, the **units** directive specifies the total number of frequencies in the table; rows × columns ($2 \times 2 = 4$ in this case). Two vectors are then calculated to represent the row subscript and the column subscript for each cell, and each is declared as a factor with two levels.

```
$units 4 $

$data n $

$read 13 44 25 29 $

$calc row=%gl(2,2) $

$calc col=%gl(2,1) $

$factor row 2 col 2 $
```

To check that the data are correctly configured, we can tabulate n using the **tprint** directive:

```
$tprint n row; col $

COL   1       2
ROW
1     13.00   44.00
2     25.00   29.00
```

We now declare the errors to be Poisson, and fit the row and column main effects in order to constrain the marginal totals (see below):

```
$yvar n $

$error p $

$fit row+col $
```

```
scaled deviance = 6.8783 at cycle 3
             d.f. = 1
```

The *scaled deviance gives the required value of G directly*. Since 6.8783 is larger than the value of χ^2 in tables with 1 d.f., we reject the null hypothesis of independence in the distribution of the two wasp species over different oak trees of the same species. The gall distributions are different, and more work appears to be justified to discover the mechanism underlying the association.

In order to see GLIM's estimates of the expected frequencies, based on the hypothesis of independence, we can tabulate the fitted values (%fv) along with our observed counts (%yv) by rows and columns:

```
$tprint %yv;%fv row;col $

         COL   1      2
ROW
1        N     13.00  44.00
         %FV   19.51  37.49

2        N     25.00  29.00
         %FV   18.49  35.51
```

Thus, if the row and column factors had been independent, 19.51 rather than 13 oak trees would have been expected to support both gall insects (cell 1,1) and 35.51 rather than 29 to support neither (cell 2,2). Note that these expected frequencies from GLIM are exactly the same as we calculated by hand during the χ^2 test, above.

This table of fitted values shows what we mean by *constraining the marginal totals*. By fitting ROW + COL to the data we ensure that the row totals in the data (57 and 54) and the column totals (38 and 73) are retained in the fitted values. In more complicated contingency tables (e.g. Section 14.8) we use the interaction between the explanatory factors to constrain the marginal totals. ●

Thus, if we were interested in parasite infection (P = 1 = uninfected, 2 = infected) as a function of category variables such as age (A = 1 = young, 2 = juvenile, 3 = adult), sex (S = 1 = male, 2 = female) and body condition (B = 1 = healthy, 2 = wasted), then we would begin the analysis by fitting the *response category* (parasitism in this case). As usual, the response variable is the vector containing the count data, say c. ○

```
$yvar c $

$error p $

$fit p $
```

Next, we fit *all the explanatory categories and their interactions*:

```
$fit +a*s*b $
```

This term constrains all the marginal totals correctly, which means that there is the correct number of individuals of each sex in each age class and each body condition. Then the relationship between parasite infection and the explanatory categories is investigated *by looking at their interaction effects*. For example, to see whether age affects parasite infection, we fit +P*A, and for body condition +P*B, and so on. This is explained in detail in Section 16.3.

14.3.4 Fisher's exact test

The contingency tables we have dealt with so far had expected frequencies greater than 5 in all of their cells. For smaller samples, where one or more of the expected frequencies is less than 5, the recommended test is Fisher's exact. Consider the example of an insect ecologist testing host-plant choice in insects. Ten trees of species *A* and 10 trees of species *B* were selected for use in the experiment. The trees were transplanted into a common garden and laid out in a grid, so that each point on the ground had equal numbers of neighbours of each tree species. The investigator then collected eight ants' nests and established them in a regular pattern amongst the transplanted trees, and allowed the ants to establish foraging trails into the different tree canopies. The hypothesis being tested was that the ants did not discriminate between the two tree species. The data are shown in Table 14.5.

Is the apparent preference for ants to forage in tree species *A* statistically significant?

If the frequencies in the four cells of a 2×2 table are denoted a, b, c and d as shown in Table 14.6, then the probability of any given configuration is given by:

Table 14.5 2×2 contingency table: ants' nests in two tree species

	Tree *A*	Tree *B*	Row totals
With ants	6	2	8
Without ants	4	8	12
Column totals	10	10	20

Table 14.6 Fisher's exact test for 2×2 contingency tables

	Column 1	Column 2	Row totals
Row 1	a	b	$a+b$
Row 2	c	d	$c+d$
Column totals	$a+c$	$b+d$	$n=a+b+c+d$

$$P = \frac{(a + b)!(c + d)!(a + c)!(b + d)!}{a!b!c!d!n!}$$

where the numerator is the product of the factorials of the two row totals and the two column totals, and the denominator is the product of the factorials of the four cell totals and the grand total, n.

To calculate the probability of a significant interaction between the row factor and the column factor, we need the likelihood of the case *plus the likelihood of all more extreme cases*. For the example in hand, the calculations are as follows:

$$P = \frac{8!12!10!10!}{6!2!4!8!20!} \qquad (14.7)$$

which, after lots of cancelling out, becomes:

$$P = \frac{3 \times 3 \times 7 \times 5}{19 \times 17 \times 13} = \frac{315}{4199} = 0.075$$

which is already greater than the acceptable 0.05. We could stop here and conclude that there is no significant preference for tree A, but we shall complete the calculation of the more extreme cases to demonstrate how the procedure works. If only one ant colony had foraged in tree B, the denominator would be (7!1!3!9!20!), and, again, after some cancelling out:

$$P = \frac{10 \times 4}{19 \times 17 \times 13} = \frac{40}{4199} = 0.00953$$

Likewise, for the most extreme case when no ants at all were found in tree B, we should have the same numerator, but a denominator of (8!0!2!10!20!). Note that 0! is *defined* as being 1, so that after cancelling we get:

$$P = \frac{3}{19 \times 17 \times 2 \times 13} = \frac{3}{8398} = 0.0003572$$

so the probability of the observed case or a more extreme result is $0.075 + 0.00953 + 0.0003572 = 0.0849$. However, we must also allow for the probability of a more extreme result in the opposite direction (i.e. the ants happen to exhibit an apparent preference for tree B), and a two-tail test therefore has a probability of $2 \times 0.0849 = 0.1698$. There is no evidence, therefore, that the ants exhibit a significant preference for tree A, and a result as extreme as this (one way or the other) would be expected by chance alone in about 17% of cases.

Let us see what GLIM would have made of this table with low frequencies in half of its cells.

```
$units 4 $

$data c $

$read 6 2 4 8 $

$calc row=%gl(2,2) $

$calc col=%gl(2,1) $

$factor row 2 col 2 $

$yvar c $

$fit row+col $

scaled deviance = 3.4522 at cycle 3
             d.f. = 1
```

To see the fitted values, we use **tprint**:

```
$tprint %yv;%fv row;col $

        COL   1       2
ROW
1       C     6.000   2.000
        %FV   4.000   4.000

2       C     4.000   8.000
        %FV   6.000   6.000
```

Notice, again, that the fitted values are constrained by fitting ROW + COL to have the same row totals (8 and 12) and column totals (10 and 10) as the data. The scaled deviance (3.4522) is less than the value of χ^2 in tables with 1 d.f. (3.841), so there is no evidence to suggest a preference. The GLIM analysis agrees with the interpretation of Fisher's exact test. For a full discussion of the issues involved, see Aitkin *et al.* (1989, pp. 195–200) where the notions of *conditional likelihoods* and *profile likelihoods* are explained. A GLIM macro to evaluate Fisher's exact test is explained in Exercise 19.1c.

Perhaps the most important lesson here concerns the design of the experiment. With only eight ants' nests, it is bound to be extremely difficult to demonstrate a significant preference, and the experiment really requires a much bigger sample size of ant colonies (and perhaps a larger number of trees, especially if the ants exhibit interference competition between colonies). •

14.4 Nested analysis of counts

Entomologists often obtain count data from plants or from plant parts. Their data might consist of the number of mines per leaf caused by a particular species of leaf-mining fly, and we are interested in understanding the factors that are responsible for causing variation in the number of mines per leaf.

Consider the following hierarchical sampling of holly leaf miner: 30 randomly selected leaves were taken from each of four branches on three trees at four different sites (a total of 1440 leaves). Many of the leaves scored 0 because they contained no mines.

```
$units 1440 $

$data x $

$dinput 6 $

File name? glex25.dat
```

We need to generate factor levels using %gl to label each leaf with the site, tree and branch number from which it was taken.

```
$calc branch=%gl(4,30) $

$calc tree=%gl(3,120) $

$calc site=%gl(4,360) $

$factor branch 4 tree 3 site 4 $
```

A total of 1029 mines were found, distributed over the 1440 sampled leaves as follows:

```
$tab for x $
```

	Mines per leaf					
	0	1	2	3	4	5
Frequency	735	488	145	40	29	3

The arithmetic mean numbers of mines per leaf on each sample branch on each tree at each site are found with:

```
$tab the x mean for site;tree;branch $
```

		1	2	3	4
1	1	1.0333	0.8667	0.9667	1.1000
	2	0.1000	0.2333	0.3667	0.1000
	3	0.5333	0.4667	0.6333	0.4667
2	1	1.1333	1.0333	1.0000	1.2000
	2	0.2333	0.3333	0.4667	0.2333
	3	0.6000	0.6000	0.7000	0.5000
3	1	1.3333	1.1333	1.0667	1.3000
	2	0.3000	0.5000	0.5000	0.2667
	3	0.6333	0.6000	0.8000	0.6667

```
4 1 1.3667 1.2333 1.1333 1.5333
  2 0.4667 0.5333 0.5333 0.3667
  3 0.7000 0.7000 0.9667 0.7667
```

There appears to be rather little variation in mine density from branch to branch within trees (the four columns), but substantial variation from tree to tree within sites. We now declare the response variable to be the counts of mines per leaf, and the error distribution to be Poisson:

```
$yvar x $

$error p $

$fit $

scaled deviance = 1727.2 at cycle 4
           d.f. = 1439
```

First, we fit site in order to determine the significance of the difference in the mean leaf miner density between sites. These fits take a long time to compute on older machines.

```
$fit +site $

scaled deviance = 1704.5 (change = -22.7) at cycle 4
           d.f. = 1436   (change = -3)
```

The sites are significantly different because the change in scaled deviance (22.7) is substantially greater than the value of χ^2 with 3 d.f. in tables (even allowing for the slight overdispersion; see below).

```
$tab the x mean for site $

        1       2       3       4
[ ] 0.5722  0.6694  0.7583  0.8583
```

The mean rate of mining varies from 0.5722 mines per leaf in site 1, up to 0.8583 mines per leaf in site 4.

Now we look at the differences between the tree means. Trees are nested within sites, and so it is not appropriate to fit a main effect for trees. Tree 1 in site 1 has nothing in common with tree 1 in site 2, and the numbering of the trees is entirely arbitrary (the tree labels could have been switched around without altering the sense of the experiment). The significance of differences in leaf mining between trees within sites is assessed by fitting the site-by-tree interaction term (this takes a long time to compute):

```
$fit +site.tree $

scaled deviance = 1474.7 (change = -229.72) at cycle 4
           d.f. = 1428   (change = -8)
```

Again, this is highly significant (the value of χ^2 in tables with 8 d.f. is only 15.507).

```
$disp e $

     estimate    s.e.      parameter
 1   -0.008368   0.09164   1
 2    0.09607    0.1266    SITE(2)
 3    0.1976     0.1237    SITE(3)
 4    0.2835     0.1214    SITE(4)
 5   -1.601      0.2215    SITE(1).TREE(2)
 6   -0.6360     0.1556    SITE(1).TREE(3)
 7   -1.238      0.1837    SITE(2).TREE(2)
 8   -0.5985     0.1466    SITE(2).TREE(3)
 9   -1.127      0.1676    SITE(3).TREE(2)
10   -0.5823     0.1386    SITE(3).TREE(3)
11   -1.020      0.1543    SITE(4).TREE(2)
12   -0.5193     0.1302    SITE(4).TREE(3)
scale parameter taken as 1.000
```

To assess the significance of the differences between individual trees within sites we need to obtain the standard errors of the differences:

```
$disp s $

S.E.s of differences of parameter estimates
 1  0.000
 2  0.2030  0.0000
 3  0.2012  0.1205  0.0000
 4  0.1998  0.1181  0.1150  0.000
 5  0.2725  0.2198  0.2181  0.2168  0.0000
 6  0.2223  0.1532  0.1507  0.1488  0.2377  0.000
 7  0.2053  0.2550  0.2214  0.2201  0.2878  0.2407
 8  0.1729  0.2297  0.1918  0.1903  0.2656  0.2138
 9  0.1910  0.2100  0.2391  0.2069  0.2778  0.2287
10  0.1662  0.1877  0.2198  0.1842  0.2613  0.2084
11  0.1795  0.1996  0.1978  0.2263  0.2700  0.2192
12  0.1592  0.1816  0.1796  0.2106  0.2570  0.2029
      1       2       3       4       5       6

 7  0.000
 8  0.1999  0.000
 9  0.2486  0.2226  0.000
10  0.2301  0.2017  0.1830   0.000
11  0.2399  0.2129  0.2278   0.2075   0.000
12  0.2251  0.1961  0.2122   0.1902   0.1677   0.000
      7       8       9        10       11       12
scale parameter taken as 1.000
```

So the standard error of the difference between trees 2 and 3 in site 3 (parameters 9 and 10) is 0.1830. The standard errors differ from comparison to comparison, but the biggest of them is about 0.25. This shows that any pair of trees differing in their parameter estimates by more than about $2 \times 0.25 = 0.5$ are likely to have significantly different leaf miner infestations. Reference to the **disp e** table shows that all three trees have significantly different infestation rates at each of the three sites.

It is important to understand the relationship between the arithmetic mean mine densities and the parameter estimates in the linear predictor. The arithmetic means are as follows:

```
$tab the x means for site;tree $

       1       2       3
1  0.9917  0.2000  0.5250
2  1.0917  0.3167  0.6000
3  1.2083  0.3917  0.6750
4  1.3167  0.4750  0.7833
```

so that tree 2 in site 3 has a mean of 0.3917 mines per leaf. To find the fitted value equivalent to this we must add together the intercept, the site 3 difference and the site 3/tree 2 interaction effect:

$$-0.008368 + 0.1976 - 1.127 = -0.9378$$

Now recall that because Poisson errors use the log link, this value is a natural logarithm, and so to find the fitted value for site 3/tree 2 we need to calculate the natural antilog, $\exp(-0.9378) = 0.3915$. This is quite close to the arithmetic mean, and the difference is due only to rounding errors in calculating the logarithms. If you found the output from the **disp e** statement hard to interpret, you should re-read Section 8.5.

What about branch-to-branch variation within trees?

```
$fit +site.tree.branch $
```

On some smaller or older machines it will be impossible to fit this interaction because of space limitations (you will get an error message in this case). GLIM is a small program, and not ideally suited to handling big nested designs (you would be better off using Genstat on a larger computer). The deviance reduction of 25.1 on 36 d.f. is not significant.

Testing the differences between branches within trees for each of the three sites separately is possible by using the **weight** directive to consider each of the four sites in turn (you should do this as an exercise). This shows that, not surprisingly, there is much less variation between branches within trees than between trees at any given site. TREE(2) in SITE(2) has branch means that differ by a factor of 2, but this is not typical of the data set as a whole.

In summary, when there is nested sampling, we do not estimate main effects (except for the factor at the top of the hierarchy — sites in this case). The significance of nested factors is assessed by their *interactions with all factors higher in the hierarchy*. Fitting complex nested designs can be slow in GLIM and a program like Genstat is likely to be more efficient. Be sure you understand the difference between nested sampling (this example) and split-plot experiments (see Section 8.8); this distinction is the cause of much confusion. ●

14.5 Bird ring recoveries

Certain kinds of survival analysis involve the periodic recovery of small numbers of dead animals. Bird ringing records are a good example of this kind of data. Out of a reasonably large, known number of ringed birds, a usually small number of dead birds carrying rings are recovered each year. The number of rings recovered each year declines because the pool of ringed birds declines each year as a result of natural mortality. There are two important variables in this case: (i) the probability of a bird dying in a given time period; and (ii) the probability that, having died, the ring will be discovered. It is possible that one or both of these parameters varies with the age of the bird (i.e. with the time elapsed since the ring was applied). We can use GLIM with Poisson errors to fit log-linear models to data like this, and to compare estimated survival and ring-recovery rates from different cohorts of animals.

The probability that a bird survives from ringing up to time t_{j-1} is the survivorship s_{j-1} (see Section 18.1). The number of ringed birds still alive at time t_{j-1} is therefore $N_0 s_{j-1}$, where N_0 is the number of birds initially ringed and released at time t_0. Thus, the number of birds that die in the present time interval D_j is:

$$D_j = N_0 s_{j-1} d_j$$

where d_j is the probability of an animal that survived to time t_{j-1} dying during the interval $(t_j - t_{j-1})$. Now, out of these D_j dead animals, we expect to recover a small proportion p_j, so the expected number of recoveries, R_j, is:

$$R_j = N_0 s_{j-1} d_j p_j$$

Maximum likelihood estimates of the death and recovery probabilities are possible only if assumptions are made about the way in which the two parameters change with age and with time after ringing (e.g. Seber, 1982). Common assumptions are that the probability of death is constant, once a given age has been reached, and that the probability of discovery of a dead bird is constant over time.

In order to analyse ring return data by log-linear modelling, we must combine the two probabilities of death and recovery into a single parameter z_j. This is the *probability of a death being recorded* in year j for an animal that was alive at the beginning of year j.

$$R_j = N_0 s_{j-1} z_j$$

Now, taking logs, rearranging and then taking antilogs we can write this expression as:

$$R_j = \exp[\ln(N_0) - \lambda_{j-1} t_{j-1} + \ln(z_j)]$$

replacing the survivorship term S_{j-1} by the proportion dying λt. Since we know the number of birds originally marked, we can use $\ln(N_0)$ as an offset. Then, using the log link, we are left with a graph of log of recoveries $\ln R_j$ against time t_j in which the intercept is given by the log of the recovery probability, $\ln(z_j)$, and the slope λ_{j-1} is the survival rate per unit time over the period 0 to t_{j-1}.

GLIM can now be used to analyse count data on the number of recoveries of dead ringed birds, and to compare models based on different assumptions about survivorship and recovery:

1 different cohorts have constant death rates and recovery rates, and these rates are the same in all cohorts (the null model);

2 different cohorts have the same death rates but different recovery probabilities;

3 different cohorts differ in both their death rates and their recovery rates;

4 the death rates may be time dependent;

5 a variety of more complex models.

14.5.1 An example with bird ringing data

Suppose that tawny owls were ringed in three different kinds of woodland — oak, birch and mixed woodland — in the same county. The numbers marked in the three habitats were 75, 49 and 128 respectively. The numbers of dead ringed birds recovered in subsequent years were as shown in Table 14.7.

Table 14.7 Recoveries of dead ringed tawny owls

Year	Oak	Birch	Mixed
1	5	3	8
2	3	3	5
3	1	2	5
4	2	0	1
5	1	0	3
6	0	2	0
7	1	0	0

```
$units 21

$data t r $

$dinput 6 $

File name: glex29.dat

$calc wood=%gl(3,7) $

$factor wood 3 $

$tprint r t;wood $

WOOD 1       2       3
T
1.000  5.000  3.000  8.000
2.000  3.000  3.000  5.000
3.000  1.000  2.000  5.000
4.000  2.000  0.000  1.000
5.000  1.000  0.000  3.000
6.000  0.000  2.000  0.000
7.000  1.000  0.000  0.000
```

Next, we create a vector N_0 which contains the initial numbers of owls ringed in each wood, then calculate the natural log of this vector. Note the use of the logical functions (e.g. wood==1) in setting up the vector of initial numbers:

```
$calc n0=(wood==1)*75+(wood==2)*49+(wood==3)*128 $

$calc logn=%log(n0) $
```

Now set the response variable to the number of ring recoveries, the errors to Poisson, and the offset to the log of the number of birds ringed initially:

```
$yvar r $

$error p $

$offset logn $
```

Recall that the log link is the default for Poisson errors. We fit the maximal model first, which assumes that both survival and recovery rates vary from wood to wood:

```
$fit $

scaled deviance = 42.839 at cycle 4
           d.f. = 20
```

```
$fit +wood+wood.t $

scaled deviance = 15.670 (change = -27.17) at cycle 4
           d.f. = 15     (change = -5)
```

The residual deviance is well behaved with a scaled deviance of only 15.67 on 15 d.f. The model has explained a highly significant amount of the deviance (the value of χ^2 in tables with 5 d.f. is 11.07), but perhaps the model is unnecessarily complicated. We begin by inspecting the parameter estimates:

```
$disp e $

     estimate    s.e.     parameter
 1   -2.432      0.5116   1
 2    0.08609    0.7787   WOOD(2)
 3    0.2295     0.6421   WOOD(3)
 4   -0.3874     0.1644   WOOD(1).T
 5   -0.3584     0.1830   WOOD(2).T
 6   -0.4832     0.1371   WOOD(3).T
scale parameter taken as 1.000
```

Neither of the recovery rate differences for WOOD(2) or WOOD(3) comes anywhere near their standard errors, so it looks as if differences between woods might be removed. We try this:

```
$fit -wood $

scaled deviance = 15.807 (change = +0.136) at cycle 4
           d.f. = 17     (change = +2)
```

which is a trivial change. Perhaps the survival rates could be equated between the different woods as well? We inspect the three different rates in the current model:

```
$disp e $

     estimate  s.e.     parameter
 1   -2.301    0.2733   1
 2   -0.4234   0.1184   T.WOOD(1)
 3   -0.3702   0.1228   T.WOOD(2)
 4   -0.4544   0.1081   T.WOOD(3)
scale parameter taken as 1.000
```

The highest rate in WOOD(3) differs from the lowest in WOOD(2) by only 0.0842 which is less than one standard error. It looks as if a single estimate for survival would be just as good, so we remove the interaction term, and fit the same slope for all three woods:

```
$fit -wood.t+t $

scaled deviance = 16.250 (change = +0.444) at cycle 4
           d.f. = 19     (change = +2)
```

Again, the deviance has changed by only 0.444 on 2 d.f., so this model looks to be minimal adequate.

```
$disp e $

   estimate  s.e.     parameter
1  -2.305    0.2731   1
2  -0.4252   0.09109  T
scale parameter taken as 1.000
```

The intercept gives the log of the recovery rate and the slope gives the log of the survival rate. We calculate the antilogs:

```
$calc %exp(-2.305) $

0.09976

$calc %exp(-0.4252) $

0.6536
```

and conclude that the recovery probability of an owl that died in a given year is about 10% and that the annual survival rate is about 65%. There is no evidence that either of these parameters differs between the three kinds of woodland.

To obtain the fitted values, we need to calculate:

$$R_i = \exp[\ln(N_0) - 2.035 - 0.4252t]$$

so, for time 1 in wood 1 (where the offset is log 75) we get:

$$R_{1,1} = \exp[\ln(75) - 2.035 - 0.4252] = 4.890$$

which, using the calculator in GLIM, is:

```
$calc %exp(%log(75)-2.305-0.4252) $

4.890
```

To compare all the observed and fitted values, write:

```
$look wood r %fv $

    WOOD R      %FV
 1  1    5.000  4.8881
 2  1    3.000  3.1950
 3  1    1.000  2.0883
 4  1    2.000  1.3650
 5  1    1.000  0.8922
 6  1    0.000  0.5832
```

```
 7  1     1.000  0.3812
 8  2     3.000  3.1935
 9  2     3.000  2.0874
10  2     2.000  1.3644
11  2     0.000  0.8918
12  2     0.000  0.5829
13  2     2.000  0.3810
14  2     0.000  0.2490
15  3     8.000  8.3423
16  3     5.000  5.4528
17  3     5.000  3.5641
18  3     1.000  2.3296
19  3     3.000  1.5227
20  3     0.000  0.9953
21  3     0.000  0.6505
```

which shows the fitted value that we calculated, long-hand, in row 1 (as 4.8881) and demonstrates what a good description of the data is achieved by this remarkably simple, two-parameter model.

14.6 The danger of contingency tables

In observational studies we quantify only a limited number of explanatory variables. It is inevitable that we shall fail to note (or to measure) a number of factors that have an important influence on the ecological behaviour of the system in question. That's life, and given that we make every effort to note the important factors, there is little we can do about it. The problem comes when we ignore factors that have an important influence on ecological behaviour. This difficulty can be particularly acute if we *aggregate data over important explanatory variables*. An example should make this clear.

Suppose we are carrying out a study of induced defences in trees. A preliminary trial has suggested that feeding on a leaf by aphids may cause chemical changes in the leaf which reduce the probability of that leaf being attacked later in the season by hole-making insects. To this end we mark a large cohort of leaves, then score whether they were infested by aphids early in the season and whether they were holed by insects later in the year. The work was carried out on two different trees and the results were as shown in Table 14.8.

The data are unequivocal. Aphid infestation had absolutely no effect on the probability of a leaf being holed by caterpillars later in the season. What is equally clear is that the two trees differed greatly in their rates of leaf-holing; on tree 1 about 2% of the leaves were holed, whereas on tree 2 the rate was four times higher.

The problems arise if we make the mistake of *collapsing the data over one of our categories*. Let's say we ignored the trees, and presented all the data in a single 2×2 contingency table. It would look like Table 14.9.

Table 14.8 Study of induced defences: frequencies of leaves with and without holes which had and had not been attacked previously by aphids; aphid attack had no effect on the proportion of leaves subsequently holed (2% holed on tree 1; 8% holed on tree 2)

Tree	Aphids	Holed	Not	Total leaves	Proportion holed
Tree 1	Without	35	1750	1785	0.0196
	With	23	1146	1169	0.0197
Tree 2	Without	146	1642	1788	0.0817
	With	30	333	363	0.0826

Table 14.9 The contingency table collapsed over trees

Aphids	Holed	Not	Total leaves	Proportion holed
Without	181	3392	3573	0.0507
With	53	1479	1532	0.0346

The results are quite different. It now appears that early infestation by aphids does indeed reduce the probability of a leaf being holed from about 5% to under 3.5%. This difference is highly significant (binomial test; $p < 0.05$). But the result is entirely an artefact of our having *aggregated the data over an important explanatory variable.*

In observational studies there is no randomization of individuals to treatments, and we shall always run the risk of this kind of misinterpretation. The moral of the example is clear: write down as many explanatory variables as possible at the data gathering stage, and do not aggregate the data at the analysis stage until (and if) GLIM has demonstrated that this is permissible (i.e. do not throw away explanatory variables). A GLIM analysis of the full data set would prevent us from misinterpreting the between-tree differences as induced defences caused by aphid-feeding (see Exercise 14.1).

Of course, an observational study can never tell us *why* the trees had different rates of leaf-holing. It could be historical, they might support different densities of caterpillar-feeding birds, the tree genotypes may be different in their constitutive chemical defences, the water-status of the trees may differ, rendering their leaves more or less palatable, and so on.

The GLIM analysis of these data demonstrates that there is no interaction between leaf-holing and aphid infestation, and highlights the significant difference between trees in both their aphid infestation rates and their mean levels of leaf-holing. The 'evidence' for induced defences in the collapsed data set is entirely spurious.

14.7 Constructing a contingency table from continuous data

In many cases it is useful to turn a set of continuous data into a contingency table by grouping certain values into a small number of factor levels. To demonstrate this we use data from an experiment on competition between two parasitoid species. Preliminary study had suggested that the two parasite species differed in their host-finding behaviour (one was a low-density specialist and the other a high-density generalist). The data consist of percentage parasitism, host density, parasite species 1 present (true/false) and parasite species 2 present (true/false).

```
$units 150 $

$data perc den p1 p2 $

$dinput 6 $

File name? glex33.dat
```

The data were gathered primarily to look at patterns of density dependence. If we restrict plotting to parasite species 2, for example, we can look for evidence of density dependence by plotting percentage parasitism against host density:

```
$calc %re=p2 $

$plot perc den $
```

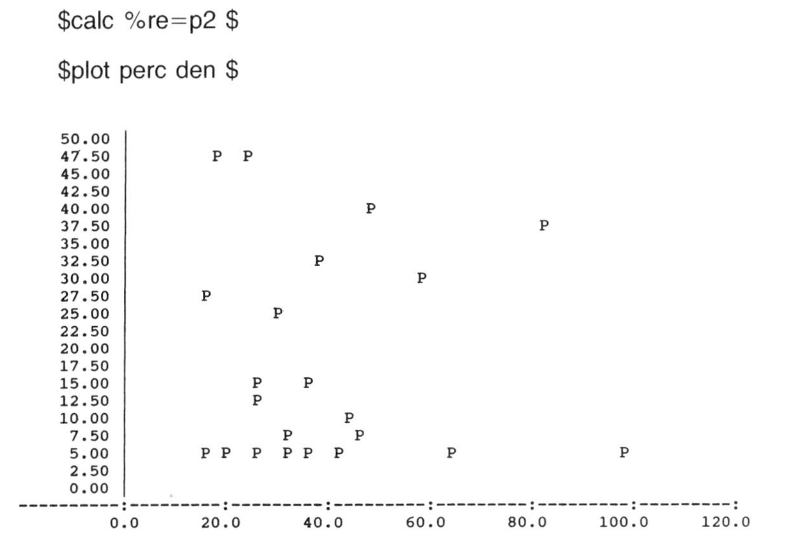

There is a slight hint of inverse density dependence (i.e. percentage parasitism declining with increasing host density), and this could be assessed by regression analysis (see Exercise 14.4).

The present aim is to demonstrate how data like this can be condensed into a contingency table, and then subjected to analysis using Poisson errors. The objective in this case is to test whether there is an association between host density and the relative importance of the two parasite species. Preliminary studies had suggested that species 1 was more important at high host densities and species 2 at low.

The first thing to do is to turn the continuous explanatory variable den (host density) into a category variable that we can use in contingency tables. We define low host densities as those with fewer than 25 hosts, medium host densities as 25–49.99 hosts, and high host densities as 50 or over.

```
$assign int=0,25,50 $

$assign val=3,2,1 $
```

The **group** directive can now be used to replace host density by a three-level code in a new factor called N:

```
$group n=den interval int * values val $
```

The asterisk means 'create a third category from 50 up to plus infinity'; this gets the value 1 (the third value in the list). Notice that in using the group directive, there needs to be one more number in the **intervals** list than in the **values** list. In the present case, this extra number (plus infinity) was specified in the group directive itself, rather than in the assign directive.

We can see what the **group** directive has done by tabulating the values of n:

```
$tab for n $

       1      2      3
[ ] 25.00  71.00  54.00
```

There were 25 high-, 71 medium- and 54 low-density samples.

Because we intend to use the identities of the parasite species P1 and P2 as factors, we need to change them from vectors containing 0's and 1's (absent and present) to vectors containing 1's and 2's. This is simple; we just add 1 to all the values:

```
$calc p1=1+p1 $

$calc p2=1+p2 $

$factor p1 2 p2 2 $
```

Now we can inspect the contingency table that we intend, eventually, to analyse:

```
$tab for n;p1;p2 $
```

```
   1                 2
   1        2        1        2
1  15.000    2.000    6.000  2.000
2  39.000   10.000   19.000  3.000
3  33.000    3.000   16.000  2.000
```

Now repeat the **tabulate** procedure, but save the contingency table in a new, shorter vector called table, and get GLIM to produce new, shorter vectors to classify this table by host density (CN), parasite 1 (CP1) and parasite 2 (CP2), where the prefix 'C' stands for contingency. You can use whatever variable names you like, so long as the names have not been used before (this would cause an error because they would have the wrong number of units).

```
$tab for n;p1;p2 using table by cn;cp1;cp2 $
```

Use **tprint** to check that the table and the new classifying vectors have been properly set up:

```
$tprint table cn;cp1;cp2 $
```

```
CP1  1                  2
CP2  1        2         1        2
CN
1     15.000    2.000    6.000  2.000
2     39.000   10.000   19.000  3.000
3     33.000    3.000   16.000  2.000
```

The next step is to count the number of cells in the table (12 in this case) and reset the units directive to this value (for larger or more complex examples you would use **env d** to inspect the length of CN; this would tell you the new value for **units**):

```
$units 12 $
```

```
[w] -- model re-initialized
```

Now, declare the new table to be the response variable, and the errors to be Poisson:

```
$yvar table $
```

```
$error p $
```

Fit the full model to reproduce the frequencies exactly:

```
$fit cn*cp1*cp2 $
```

```
scaled deviance = 0.0000000 at cycle 6
            d.f. = 0
```

If the relative abundance of the two parasites changes with host density, then removal of the three-way interaction term (host density × parasite 1 × parasite 2) will cause a significant increase in deviance:

```
$fit -cn.cp1.cp2 $

scaled deviance = 1.2643 (change = +1.264) at cycle 3
           d.f. = 2      (change = +2)
```

Evidently, the relative attack rate of the two parasites does not change with host density (the deviance change would need to be greater than 6 for significance; this is the value of χ^2 in tables with 2 d.f.).

We can get a visual impression of the fit of the model by comparing the observed and expected frequencies:

```
$tprint table;%fv cn;cp1;cp2 $

      CP1    1               2
      CP2    1        2        1        2
CN
1     TABL   15.000   2.000    6.000    2.000
      %FV    14.285   2.715    6.715    1.285

2     TABL   39.000   10.000   19.000   3.000
      %FV    40.042   8.958    17.958   4.042

3     TABL   33.000   3.000    16.000   2.000
      %FV    32.673   3.327    16.327   1.673
```

The overall fit is excellent and, as demonstrated by the deletion test, there is no hint of a three-way interaction.

```
$disp e $

      estimate    s.e.     parameter
 1     2.659      0.2590   1
 2     1.031      0.3000   CN(2)
 3     0.8274     0.3101   CN(3)
 4    -0.7548     0.4360   CP1(2)
 5    -1.660      0.5678   CP2(2)
 6    -0.04715    0.4996   CN(2).CP1(2)
 7     0.06105    0.5178   CN(3).CP1(2)
 8     0.1628     0.6252   CN(2).CP2(2)
 9    -0.6242     0.7190   CN(3).CP2(2)
10     0.006274   0.4986   CP1(2).CP2(2)
scale parameter taken as 1.000
```

None of the two-way interactions looks significant either, but we shall test these more thoroughly in Exercise 14.4.

To summarize, a contingency table is created from the full data set, using the **tab for using** directive to save a new, shorter vector containing the frequencies, along with a set of new classifying vectors. The response variable is then redefined as the frequency vector and analysed using Poisson errors. The technique is extremely powerful and very useful with ecological data, but it does take lots of practice to understand how it works. ●

14.8 A complex contingency table

In this example, animal carcases were collected by wildlife managers and classified according to their sex (S; male = 1, female = 2), the habitat (H) in which they were collected (woodland = 1, scrub = 2, open = 3), the aspect (A) of the slope (north and east = 1, south and west = 2), the season (W) the carcase was found (1 = spring, 2 = summer, 3 = autumn, 4 = winter) and the age (J) of the corpse (1 = juvenile, 2 = adult), so the full contingency table has $2 \times 3 \times 2 \times 4 \times 2 = 96$ cells. The data were counts with a low mean, and so Poisson errors were considered appropriate. All habitats were searched at all seasons and many of the data were 0's. The aim of the analysis is to look for significant interactions between the explanatory variables (see Agresti, 1990).

```
$units 96 $

$data c s h a w j $

$dinput 6 $

File name? glex28.dat
```

Before making any attempt at modelling, it is most important when working with complicated contingency tables to carry out a thorough data inspection. The **tab with** directive is valuable for this. First we look at the marginal totals (equivalent to main effects in ANOVA); they show the numbers of dead animals in each level of a factor.

```
$tab with c for s $

        1.000    2.000
[ ]   32.00    82.00
```

so there were 32 male and 82 female carcases in all:

```
$tab with c for h $

        1.000    2.000    3.000
[ ]   27.00    23.00    64.00

$tab with c for a $

        1.000    2.000
[ ]   41.00    73.00
```

```
$ tab with c for w $

        1.000    2.000    3.000    4.000
[ ]    30.00    30.00    28.00    26.00

$tab with c for j $

        1.000    2.000
[ ]    54.00    60.00
```

Thus, out of the 114 carcases recovered, 54 were juveniles and 60 were adults; and so on. With observational data like these, the main effects do not mean very much because we do not know whether they reflect differences in collecting effort or in any number of other potentially confounding variables. Our interest is with the *interactions between classifying factors*.

Next, all the two-way tables should be inspected. Here are three as examples:

```
$tab with c for s;h $

          1.000    2.000    3.000
1.000   17.000    7.000    8.000
2.000   10.000   16.000   56.000
```

which suggests that male carcases were found disproportionately in habitat 1 (woodland) whereas females were found most frequently in habitat 3 (open);

```
$tab with c for s;a $

          1.000    2.000
1.000    16.00    16.00
2.000    25.00    57.00
```

which suggests that females were overrepresented on south-westerly aspects; and

```
$tab with c for s;w $

          1.000    2.000    3.000    4.000
1.000   12.000    3.000    4.000   13.000
2.000   18.000   27.000   24.000   13.000
```

which suggests that males carcases tend to be found in the cold seasons and females in the warm seasons. The modelling is time consuming but not difficult. The steps involved are these:

1. declare the counts of dead animals to be the response variable;
2. declare the errors to be Poisson (the default is the log link function);
3. fit the maximal model;

4 remove the high-order interaction terms, one at a time;
5 continue with model simplification;
6 determine the minimal adequate model;
7 ensure that all the marginal totals are properly constrained;
8 test for overdispersion.

```
$yvar c $

$factor s 2 h 3 a 2 w 4 j 2 $

$error p $

$fit $

scaled deviance = 172.09 at cycle 4
           d.f. = 95
```

The object of the exercise is to find a simple model for the data that explains as much of the scaled deviance of 172.09 as possible. We also require that the residual deviance does not show evidence of significant overdispersion. You may find that the first stages of the analysis are rather slow. Fitting the full model and removing the four-way interactions may take hours rather than minutes on older computers (like XT or AT PC's). If you are willing to take it from me that none of the four-way interactions is significant, you can fit the main effects:

```
$fit +s+h+a+w+j $

scaled deviance = 114.42 (change = −57.68) at cycle 4
           d.f. = 87     (change = −8)
```

two-way interactions:

```
$fit +s.h+s.a+s.w+s.j $

scaled deviance = 70.755 (change = −43.66) at cycle 4
           d.f. = 80     (change = −7)

$fit +h.a+h.w+h.j+a.w+a.j+w.j $

scaled deviance = 59.813 (change = −10.942) at cycle 4
           d.f. = 63     (change = −17)
```

and three-way interactions:

```
$fit +s.h.a+s.h.w+h.a.w $

scaled deviance = 43.589 (change = −16.225) at cycle 9
           d.f. = 49     (change = −14)
```

from the bottom up. The ratio of scaled deviance to degrees of freedom is very close to 1, so there is no evidence of overdispersion (see Section 14.9).

Inspection of the raw data suggested that there was little difference in the numbers of juvenile and adult carcases discovered (54 and 60), so we begin the process of model simplification by removing the two-way interactions involving this factor.

```
$fit −s.j $

scaled deviance = 43.742 (change = +0.153) at cycle 9
            d.f. = 50     (change = +1)

$fit −h.j $

scaled deviance = 44.151 (change = +0.409) at cycle 9
            d.f. = 52     (change = +2)

$fit −a.j $

scaled deviance = 44.194 (change = +0.044) at cycle 9
            d.f. = 53     (change = +1)

$fit −w.j $

scaled deviance = 44.288 (change = +0.094) at cycle 9
            d.f. = 56     (change = +3)
```

Even the largest (the interaction with habitat) gives a χ^2 value of only 0.409 on 2 d.f., and so we can leave out all the two-way interactions involving age structure. We shall remove the main effect (J) in due course.

Next we test whether any of the three-way interactions is required. Some of them had looked as if they might be significant (e.g. males appeared to be overrepresented on northern slopes in woodland and females on southern slopes in open habitat).

```
$fit −h.a.w $

scaled deviance = 49.863 (change = +5.57) at cycle 8
            d.f. = 62     (change = +6)

$fit +h.a.w−s.h.w $

scaled deviance = 50.622 (change = +0.760) at cycle 8
            d.f. = 62     (change = 0)

$fit +s.h.w−s.h.a $

scaled deviance = 50.136 (change = −0.486) at cycle 8
            d.f. = 58     (change = −4)

$fit−s.h.w−h.a.w $
```

```
scaled deviance = 60.513 (change = +10.38) at cycle 4
           d.f. = 70     (change = +12)
```

This shows that the habitat/aspect/season, sex/habitat/season and sex/habitat/aspect interactions account for approximately equal but insignificant fractions of the deviance. They can all be left out of the model.

Now we can test each of the two-way interactions by a series of deletion tests, removing them from the current model (i.e. the model with all the main effects and the two-way interactions involving sex, aspect, habitat and season). Note that the factors are added back into the model in case there are correlations with other factors.

```
$fit −a.w: +a.w $

scaled deviance = 65.139 (change = +4.63) at cycle 4
           d.f. = 73     (change = +3)

scaled deviance = 60.513 (change = −4.627) at cycle 4
           d.f. = 70     (change = −3)
```

and so on, for the other terms:

```
$fit −h.w:+h.w $

$fit −h.a:+h.a $

$fit −s.w:+s.w $

$fit −s.a:+s.a $

$fit −s.h:+s.h $
```

It is reasonably clear that we need retain only two two-way interactions: sex/season (scaled deviance of 11.66 on 3 d.f.) and sex/habitat (18.65 on 2 d.f.). This gives the following model:

```
$disp e $

     estimate   s.e.     parameter
 1    0.1366    0.3555   1
 2   −1.066     0.5041   S(2)
 3    0.5769    0.1950   A(2)
 4   −0.8873    0.4475   H(2)
 5   −0.7538    0.4283   H(3)
 6   −1.386     0.6412   W(2)
 7   −1.099     0.5757   W(3)
 8    0.08004   0.4001   W(4)
 9    1.357     0.6022   S(2).H(2)
10    2.477     0.5488   S(2).H(3)
11    1.792     0.7097   S(2).W(2)
12    1.386     0.6547   S(2).W(3)
13   −0.4055    0.5408   S(2).W(4)
scale parameter taken as 1.000
```

The main effect of age structure was removed without a significant increase in deviance, but it is clear that all the other main effects need to be retained to constrain the marginal totals.

We finish the analysis by tidying up the factor levels wherever it is sensible to do so. For example, spring (W = 4) is not significantly different from winter (W = 1) in main effect (0.08004, s.e. = 0.4001) or in its interaction terms (with sex −0.4055, s.e. = 0.5408). Nor do summer (W = 2) and autumn (W = 3) appear to differ significantly, or to have different interaction terms. We can try a new model that has just two season categories, one for bad weather (winter and spring) and one for good weather (summer and autumn). We compute a new factor W1 to do this. We can use either the **group** directive or **calc** with logical functions to do this. Both give the same result; 1's for W = 1 and W = 4 and 2's for W = 2 and W = 3.
either

```
$calc w1=(w==1)+(w==4)+2*(w==3)+2*(w==2) $

$factor w1 2 $
```

or

```
$assign codes=1,2,2,1 $

$group w1=w values codes $
```

The **group** directive does not require a **factor** statement, because all **group** directives produce factors by default. We fit the new model, omitting W but adding W1 and the interaction between W1 and sex:

```
$fit s+h+a+w1+s.h+s.w1 $

scaled deviance = 76.197 at cycle 4
           d.f. = 87
```

The increase in deviance is negligible (1.17 on 4 d.f.), and the simplification of reducing the season levels from 4 to 2 appears to be justified. The model now looks like this:

```
$disp e $

    estimate   s.e.     parameter
1    0.1774   0.2883    1
2   -1.256    0.4330    S(2)
3   -0.8873   0.4475    H(2)
4    0.7538   0.4284    H(3)
5    0.5769   0.1950    A(2)
6   -1.273    0.4258    W1(2)
```

```
7   1.357   0.6022  S(2).H(2)
8   2.477   0.5489  S(2).H(3)
9   1.771   0.4828  S(2).W1(2)
scale parameter taken as 1.000
```

You will notice that all the parameters are now significant, so this represents a minimal adequate model for describing the pattern of recoveries of carcases in relation to sex, habitat, aspect and season. The significance of the interactions between sex and habitat and sex and season would justify further field work on the behavioural ecology of the two sexes in relation to their habitat selection at different seasons.

Finally, it is good practice to inspect the residuals, and to check for systematic departures or outliers. This is left to the reader as an exercise (see Exercise 14.5).

14.9 Overdispersion in Poisson data

When the scale factor is significantly greater than 1 after the minimal adequate model has been fitted to the data, then there is overdispersion (see Section 13.7). Perhaps the real error distribution was negative binomial rather than Poisson, and that

$$s^2 = \bar{x} + \frac{\bar{x}^2}{k}$$

For a reasonable range of values of $\bar{x}$ we may be able to approximate the relationship so that $s^2 = \lambda\bar{x}$ where $(\lambda > 1)$. Thus, Poisson errors combined with an empirical scale parameter can act as a model with negative binomial errors.

If sampling errors really do follow a Poisson distribution, but all the frequencies are multiplied by λ, then we can correct for the extraneous variance by scaling the sample sizes. If the sample sizes are scaled by dividing by λ, then the standard errors should be increased by multiplying by $\sqrt{\lambda}$ since standard errors are generally proportional to $1/\sqrt{n}$. Variances and covariances should be multiplied by λ.

Suppose, for example, that we have fit a model to insect count data that involved a factorial experiment with four novel pesticide-application technologies and three different pesticide preparations. After the maximal model has been fit, the deviance is 200 and the degrees of freedom are 20. Ideally, the deviance ought to be 20, so the data are overdispersed by a factor of $\lambda = 10$. This means that the standard errors should be multiplied by $\sqrt{10} = 3.16$. If both application and pesticide are still significant, then all well and good. But if this increase in the standard errors has made one or both of the factors insignificant, then model simplification should be attempted, as described in Chapter 12.

14.9.1 Overdispersion in the slug data

We now return to the analysis of slug catches, considered earlier (see Section 14.2). The residual deviance was 2.76 times the residual degrees of freedom, suggesting significant overdispersion, probably caused by an aggregated spatial distribution of slugs within each field (perhaps because of clumping of progeny around parents, or because of within-field spatial heterogeneity in habitat quality). One way to take account of this is to adjust the scale parameter, then refit the model, and see whether our conclusions are materially affected. Rather than estimate a scale parameter from the residual deviance, it is recommended to use Pearson's χ^2 instead (Aitkin *et al.*, 1989). GLIM holds the value of Pearson's χ^2 in a system scalar called %x2, which we can see by typing:

```
$look %x2 $

245.3
```

Notice that this is rather larger than the scaled deviance of 213.44. The scale parameter (%s) is then calculated by dividing Pearson's χ^2 by the residual degrees of freedom:

```
$calc %s=245.3/78 $

$look %s $

3.145
```

We can then change the scale parameter with the **scale** directive:

```
$scale %s $

[w] -- model changed
```

If we now refit the model we can see what difference has been made by adjusting the scale parameter.

```
$fit $

scaled deviance = 71.500 at cycle 4
           d.f. = 79

$fit +a $

scaled deviance = 67.869 (change = −3.632) at cycle 4
           d.f. = 78     (change = −1)

$disp e $

   estimate  s.e.    parameter
1  0.2430    0.2467  1
2  0.5789    0.3088  A(2)
scale parameter taken as 3.145
```

The parameter estimates are unaffected, but the difference between the means now falls short of significance both by the deletion test ($\chi^2 = 3.632$) and by the t-test ($t = 0.5789/0.3088 = 1.87$).

A third option is to transform the data and carry out the analysis with the identity link and normal errors. The appropriate transformation for Poisson data is the square root, so we proceed as follows:

```
$calc rc=%sqrt(c) $

$yvar rc $

[w] -- model changed

$error n $

$fit $

deviance = 69.834
      d.f. = 79

$fit +a $

deviance = 62.386 (change = −7.448)
      d.f. = 78      (change = −1)

$disp e $

    estimate  s.e.     parameter
1   0.6447    0.1414   1
2   0.6103    0.2000   A(2)
scale parameter taken as 0.7998
```

On the basis of this test, the mean slug densities are significantly different in the two fields. You will recall that with normal errors and the identity link that the t-test and deletion test are exact, and give identical results ($t = 0.6103/0.2 = 3.05$; $F = 7.448/0.7998 = 9.31$; $F = t^2$).

A fourth option involves specifying negative binomial errors and using our **own** model. This is dealt with in Chapter 19 and Exercise 19.2.

Thus two of the tests suggest that the differences between the mean slug densities are highly significant, while a third test suggests that the difference is insignificant. What is clear is that there is more to this than mere differences between the fields. Differences between the means explain only 5% of the variation in slug counts from tile to tile (deviance change of 11.42 out of 224.86; see p. 230). Within fields it is clear that the data are overdispersed. Variance/mean ratios for each field can be calculated as follows:

```
$tab the c mean for a into cm by aa $

$tab the c variance for a into cv by aa $
```

```
$calc vm=cv/cm $

$look vm $

     VM
 1  4.183
 2  2.164
```

Both fields have variance/mean ratios greater than 2, which suggests that there is substantial within-field spatial heterogeneity in the numbers of slugs beneath sampling tiles. Because of this, it may be better to consider a different kind of error structure (say a negative binomial; see Section 19.10). It is very common for field data to have this kind of overdispersed structure, and it simply reinforces the point that ecological data reflect complex and interacting processes. It would be naive of us to assume that simple error structures will always work perfectly.

In summary, it does look as if the fields support different slug population densities, but we need to do more work on the within-field variation in slug catches, perhaps using different sampling methods (e.g. separation of soil cores), to be confident about the significance of these differences.

14.10 Negative binomial distribution

Count data from ecological studies are often well described by the negative binomial, a two-parameter distribution with a variance mean ratio > 1. The probability of getting x individuals in a sample with mean μ and aggregation parameter k (low values of k describing high degrees of aggregation) is:

$$P_{(x)} = \left(1 + \frac{\mu}{k}\right)^{-k} \frac{(k + x - 1)!}{x!(k - 1)!} \left(\frac{\mu}{\mu + k}\right)^{x}$$

In fact, the terms of the negative binomial are much easier to calculate than might appear from the full formula. Setting $x = 0$ allows us to calculate the zero-term:

$$P_{(x=0)} = \left(1 + \frac{\mu}{k}\right)^{-k}$$

Subsequent terms can then be obtained from the recursion relationship:

$$P_{(x)} = P_{(x-1)} \left(\frac{k + x - 1}{x}\right) \left(\frac{\mu}{\mu + k}\right)$$

The recursion equation is simple to evaluate on a calculator, and avoids the need to calculate gamma functions for the fractional factorials. Examples of the use of these formulas are provided in Exercise 14.3.

CHAPTER 15

Analysing proportion data: binomial errors

An important class of ecological problems involves data on proportions:
1 data on percentage mortality;
2 infection rates of diseases;
3 toxicological assay;
4 data on proportional response to an experimental treatment.
What all these have in common is that we know how many of the experimental objects are in one category (say dead or infected) and we know how many are in another (say alive or uninfected). This contrasts with Poisson data in the previous chapter where we knew how many times an event occurred, but not how many times it did not occur.

We model processes involving proportional response variables in GLIM by specifying *binomial errors*. The only complication is that whereas with Poisson errors we could simply say **error p**, with binomial errors we must specify the total sample from which each observed proportion has been drawn. To do this we need to specify another vector, say n, that contains the sample sizes. This vector is called the *binomial denominator*, and the error term is expressed like this:

```
$error b n $
```

The old-fashioned way of modelling this sort of data was to use the percentage mortality as the response variable. There are three problems with this: (i) the variance is not normally distributed; (ii) the variance is not constant; and (iii) by calculating the percentage, we lose information of the size of the sample from which the proportion was estimated. In GLIM, we use the *number dying* as the response variable, with the total number of animals in the original experiment as the binomial denominator. GLIM then carries out weighted regression, using the individual sample sizes as weights, and the *logit link function* to ensure linearity (as described below).

If the response variable takes the form of a percentage change in some continuous measurement (such as the percentage change in weight on receiving a particular artificial diet), then the percentage data should be arcsine transformed prior to analysis, especially if many of the percentage changes are smaller than 20% or larger than 80%. Note, however, that data of this sort are probably better treated by analysis of covariance (see Chapter 9), using final weight as the response variable and initial weight as a covariate, or by specifying the response variable to be the logarithm of the ratio of final weight to initial weight.

15.1 Data on proportions

The traditional transformations of proportion data were arcsine and probit. The arcsine transformation took care of the error distribution, while the probit transformation was used to linearize the relationship between percentage mortality and log dose in bioassay. There is nothing wrong with these transformations, and they are available within GLIM, but a simpler approach is often preferable, and is likely to produce a model that is easier to interpret.

The major difficulty with modelling proportion data is that the responses are *strictly bounded*. There is no way that the percentage dying can be greater than 100% or less than 0%. But if we use simple techniques like regression or analysis of covariance, then the fitted model could quite easily predict negative values, especially if the variance was high and many of the data were close to zero.

The *logistic* curve is commonly used to describe data on proportions, because, unlike the straight-line model, it asymptotes at 0 and 1 so that negative proportions, and responses of more than 100%, cannot be predicted (see Fig. 15.1). Throughout this discussion we shall use p to

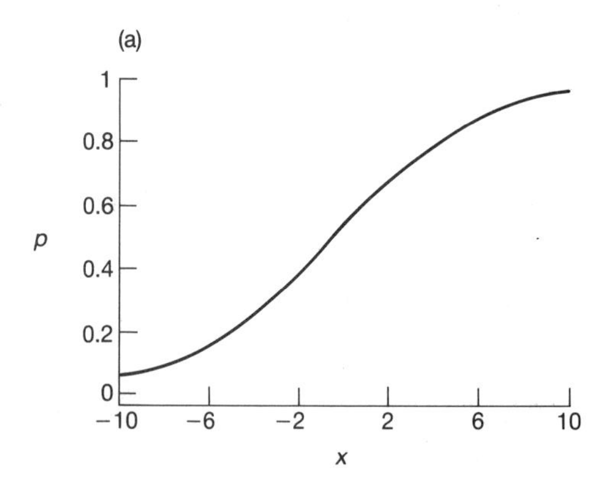

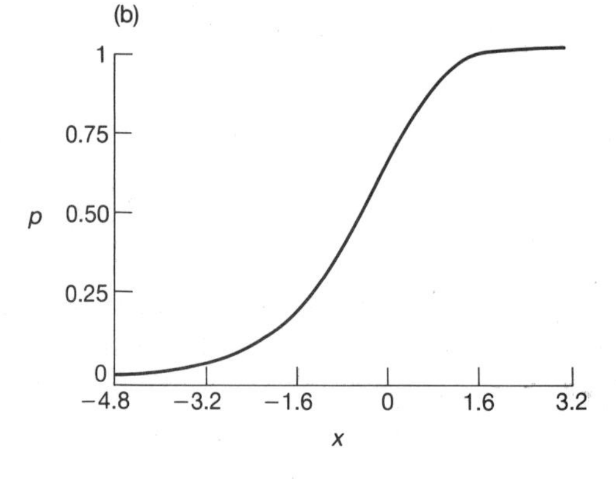

Fig. 15.1 (a) The logistic curve is the basis of the logit, or log-odds, link function. It is symmetrical about $p = 0.5$. (b) The complementary log-log link is an asymmetrical link function (see text for details).

describe the proportion of individuals observed to respond in a given way. Because much of their jargon was derived from the theory of gambling, statisticians call these *successes*, though this may seem somewhat macabre to an ecologist measuring death rates. The individuals that respond in other ways (the statistician's *failures*) are therefore $(1-p)$ and we shall call the proportion of failures q. The third variable is the size of the sample, n, from which p was estimated (it is the binomial denominator, and the statistician's *number of attempts*).

An important point about the binomial distribution is that the variance is not constant. In fact, the variance of the binomial distribution is:

$$s^2 = npq \qquad (15.1)$$

so that the variance changes with the mean as shown in Fig. 15.2. The variance is low when p is very high or very low, and the variance is greatest when $p = q = 0.5$. As p gets smaller, so the binomial distribution gets closer and closer to the Poisson distribution. You can see why this is so by considering the formula for the variance of the binomial (equation 15.1). Remember that for the Poisson, the variance is equal to the mean: $s^2 = np$. Now, as p gets smaller, so q gets closer and closer to 1, so the variance of the binomial converges to the mean:

$$s^2 = npq \approx np \qquad (q \approx 1)$$

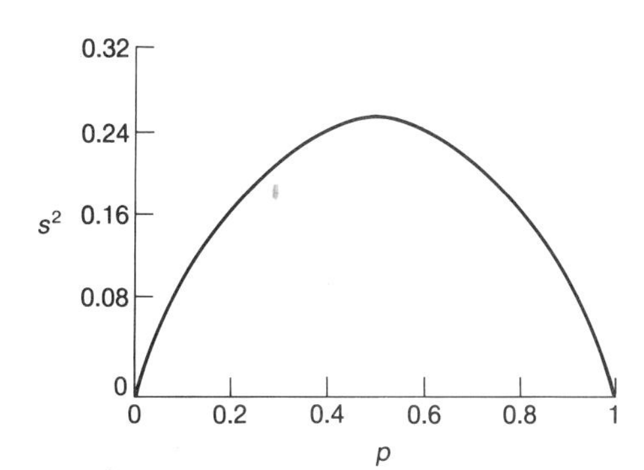

Fig. 15.2 The variance of the binomial distribution is a function of the mean since $s^2 = npq$ and $\bar{x} = np$. The variance peaks at $p = q = 0.5$.

15.2 Odds

The logistic model for p as a function of x looks like this:

$$p = \frac{e^{(a+bx)}}{1 + e^{(a+bx)}} \qquad (15.2)$$

and there are no prizes for realizing that the model is not linear. But if $x = -\infty$, then $p = 0$; if $x = +\infty$, then $p = 1$, so the model is strictly bounded. When $x = 0$ then $p = \exp(a)/[1 + \exp(a)]$. The trick of linearizing the logistic actually involves a very simple transformation. You may have come across the way in which bookmakers specify probabilities by quoting the *odds* against a particular horse winning a race (they might give odds of 2:1 on a reasonably good horse or 25:1 on an outsider). This is a rather different way of presenting information on probabilities than scientists are used to dealing with. Thus, where the scientist might state a proportion as 0.666 (2 out of 3), the bookmaker would give odds of two-to-one (2 successes and 1 failure). In symbols, this is the difference between the scientist stating the probability p, and the bookmaker stating the odds, p/q.

Now if we take the *odds* p/q and substitute this into the formula for the logistic, we get:

$$\frac{p}{q} = \frac{e^{(a+bx)}}{1 + e^{(a+bx)}}\left(1 - \frac{e^{(a+bx)}}{1 + e^{(a+bx)}}\right)^{-1}$$

which looks awful. But a little algebra shows that:

$$\frac{p}{q} = \frac{e^{(a+bx)}}{1 + e^{(a+bx)}}\left(\frac{1}{1 + e^{(a+bx)}}\right)^{-1} = e^{(a+bx)}$$

Now, taking natural logs, and recalling that $\ln(e^x) = x$, will simplify matters even further, so that:

$$\ln\left(\frac{p}{q}\right) = a + bx \qquad (15.3)$$

This gives a *linear predictor*, $a + bx$, not for p but for the *logit* transformation of p, namely $\ln(p/q)$. In the jargon of GLIM, the logit is the *link function* relating the linear predictor to the value of p.

You might ask at this stage 'why not simply do a linear regression of $\ln(p/q)$ against the explanatory x-variable'? GLIM has three great advantages here:

1 it allows for the non-constant binomial variance;

2 it deals with the fact that logits for p's near 0 or 1 are infinite;

3 it allows for differences between the sample sizes by weighted regression.

15.3 GLIM commands for proportional data

In addition to specifying that the y-variable contains the counted data on the number of successes, GLIM requires the user to provide a second vector containing the *binomial denominator* (the sample size from which each different sample was drawn). There are two constraints on the data in the binomial denominator. GLIM will stop and give an error message when you type:

```
$error b n $
```

if any of the rows in the data matrix contains a binomial denominator of 0 (if no animals are measured, we cannot possibly know how many died). Also, the response variable is not allowed to be larger than its binomial denominator (you cannot have 26 animals dying out of a sample of 24).

Suppose that we have gathered data on the weevil infestation of acorns from five different oak trees. The question is simply whether the proportion of acorns infected by weevils varies significantly from tree to tree. The data are to be stored in three vectors: one to identify the TREE, one to contain the number of weevily acorns and a third to contain the total number of acorns from the tree in question. Type in the data, using the **read** directive:

```
$units 5 $

$data tree weevil acorns $

$read

$REA? 1 10 17

$REA? 2 52 101

$REA? 3 21 40

$REA? 4 9 17

$REA? 5 36 60
```

We begin by declaring TREE to be a factor with five levels:

```
$factor tree 5 $
```

As ever, the first object is to get to know the data. With binomial data it is useful to turn the counts into proportions, and to work out the mean proportions:

```
$calc p = weevil/acorns $

$tab the p mean for tree $

          1        2        3        4        5
[ ]  0.5882   0.5149   0.5250   0.5294   0.6000
```

Thus, tree 5 has the highest mean with 60% of its acorns attacked and tree 2 has the lowest with 51.5%. Are these differences statistically significant?

To test this, the number of weevily acorns is defined to be the *response variable*:

```
$yvar weevil $
```

and the *error structure* is defined as binomial, with the total number of acorns as the *binomial denominator*:

```
$error binomial acorns $
```

Next, model-fitting proceeds exactly as usual. In the present example we simply fit the factor TREE and inspect the parameter estimates:

```
$fit $

scaled deviance = 1.3174 at cycle 3
            d.f. = 4

$fit + tree $

scaled deviance = 0.00000 (change = −1.317) at cycle 4
            d.f. = 0      (change = −4)
```

Adding TREE to the model caused a reduction in deviance of only 1.317 on 4 d.f., much less than the value of χ^2 in tables. Thus we conclude that there is no significant difference in the rate of weevil attack between the different trees. Inspection of the parameter values and their standard errors confirms this view; the largest *t*-ratio is for TREE(2), but this is much less than 2.

```
$disp e $

   estimate   s.e.     parameter
1   0.3567    0.4928   1
2  −0.2973    0.5315   TREE(2)
3  −0.2566    0.5858   TREE(3)
4  −0.2389    0.6921   TREE(4)
5   0.04879  0.5588   TREE(5)
scale parameter taken as 1.000
```

It is important to understand how to use the parameter estimates to obtain the mean proportions for the different trees. Remember that the parameters are *logits*; they are the natural logs of the odds ratio, %log(p/q). Thus, to get p we need first to take the antilogs. Take tree 1 as an example. Its logit is 0.3567. We work out its antilog using %exp as follows:

```
$calc %exp(.3567) $

1.429
```

Now, we need to use the formula for obtaining p from the odds 1.429:

$$p = \frac{1}{1 + \dfrac{1}{\text{odds}}}$$

```
$calc 1/(1+1/1.429) $

0.5883
```

which is the mean for tree 1 we obtained earlier. To find the mean proportion for TREE(4) we need to add its logit (−0.2389) to the logit of TREE(1):

```
$calc %exp(0.3567−0.2389) $

1.125

$calc 1/(1+1/1.125) $

0.5294
```

Note that the model including TREE is a *full model*; it fits the data perfectly (the residuals are zero). ●

15.4 A simple bioassay

Insects in Petri dishes were treated with different concentrations of insecticide and the numbers dead after 24 h were recorded. Most Petri dishes had 40 insects at the beginning, but because of accidents and other unforeseen calamities, a few of the Petri dishes started out with smaller numbers than this. We define a vector called dead, which contains the number of insects that died, and a vector called initial, which contains the number of insects that were treated with insecticide initially (usually 40, but sometimes fewer). The vector called dose contains the concentration of insecticide applied to each Petri dish. The **units** directive specifies the number of rows of data, which, in this case, is the total number of Petri dishes (one dish for each of seven different doses of insecticide). Note that **units** is *not* the total number of insects at the beginning of the experiment.

```
$units 7 $

$data dose dead initial $

$dinput 6 $

File name? glex13.dat
```

To see the shape of the data, we calculate the percentage dead, then plot percentage dead against dose:

```
$calc perc=100*dead/initial $

$plot perc dose $
```

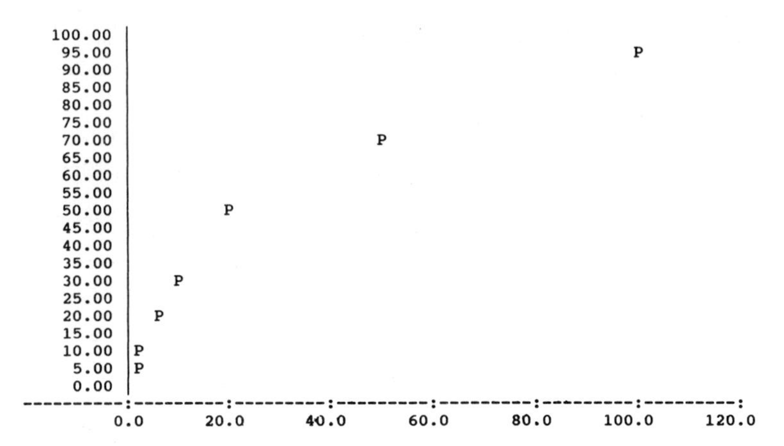

The relationship is obviously non-linear, but reasonably smooth; there are no conspicuous outliers. We begin by declaring the response variable and the error structure:

```
$yvar dead $

$error b initial $
```

So for each of the seven Petri dishes GLIM knows that dead individuals came from an initial sample. The process of model-fitting and interpretation is exactly the same with binomial errors as it was with normal errors and Poisson errors. The only difference from modelling with normal errors is that (like Poisson errors) the change in deviance gives us a value of χ^2, rather than a sum of squares. First, we do a linear regression of log-odds against dose:

```
$fit $

scaled deviance = 125.34 at cycle 3
            d.f. = 6

$fit +dose $

scaled deviance = 10.764 (change = −114.6) at cycle 4
            d.f. = 5      (change = −1)
```

Fitting dose reduced the scaled deviance from 125.34 to 10.764, so it is absolutely plain that increasing the dose increased the percentage killed. In this case, we are more interested in the parameter estimates:

```
$disp e $

    estimate    s.e.       parameter
1  −1.737      0.2074      1
2   0.05338    0.007143    DOSE
scale parameter taken as 1.000
```

Notice that the scale parameter is taken as 1.0. Just as with Poisson errors in the previous chapter, this is because GLIM is only estimating p from the data and is not attempting to find an independent estimate of the sample variance. Also, note that there is some evidence of overdispersion because the residual deviance is substantially larger than the residual degrees of freedom ($10.764/5 = 2.155$).

In order to see what the parameter estimates mean, let us use the equation to work out the predicted percentage kill at a dose of 20. We simply substitute $x = 20$ into the equation of the straight line, with its intercept of -1.737 and slope of 0.05338 (see the **disp e** table, above):

$$y = -1.737 + 20 \times 0.05338 = -0.6694$$

This number is a *logit*, the natural log of p/q. We want to work out p, so the first step is to calculate the antilog of -0.6694 which is 0.5120:

```
$calc %exp(-0.6694) $

0.5120
```

This means that p/q is 0.512 and we need to find p. Remember that $q = 1 - p$, so:

$$\frac{p}{1 - p} = 0.512$$

and we solve for p by taking reciprocals:

$$\frac{1 - p}{p} = \frac{1}{p} - 1 = \frac{1}{0.512}$$

and finally:

$$p = \frac{1}{1 + \frac{1}{0.512}} = 0.3386$$

So we predict about 34% kill at a dose of 20. Note that the fitted values (%fv) are *counts* (numbers of dead animals in this case; they are not logits). Let us continue by expressing the fitted values as percentages (pfit) so that we can plot the data and the model on the same axes, in order to assess the goodness of fit:

```
$calc pfit=100*%fv/%bd $
```

where %bd is the binomial denominator (it gets its values from the **error b** directive; it is known as a *system pointer* and currently refers to the values in the vector called initial). We can now plot the observed (*) and fitted (+) percentages on the same axes:

```
$plot perc pfit dose '*+' $
```

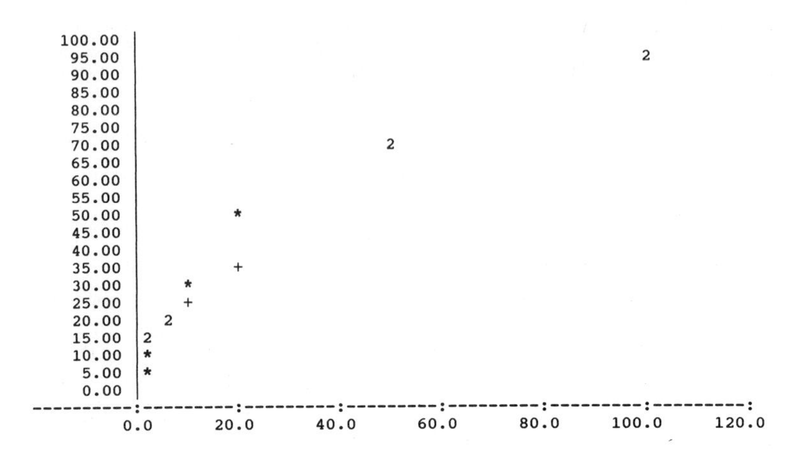

The model fits very well for the high values of dose, but we can see our predicted kill of 34% at dose = 20 marked by +. It is pretty clear that the model is a poor fit at these intermediate doses, because we actually measured a kill of almost 50% at dose = 20. This prompts us to inspect the residuals:

```
$calc resid=dead−%fv $

$plot resid %fv $
```

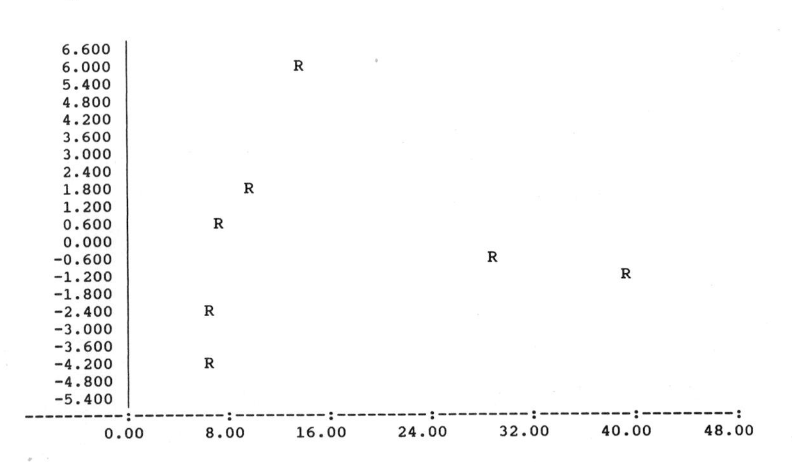

There is a clear n-shaped pattern in the residuals, so something needs to be done to the model (or perhaps to the error structure or link function) to rectify this. The simplest modification might be to transform the explanatory variable, i.e. to regress the logit mortality against the logarithm of dose rather than against dose. To do this, we calculate a new variable logdose, and repeat the fit:

```
$calc logdose=%log(dose) $

$fit $

scaled deviance = 125.34 at cycle 3
            d.f. = 6

$fit +logdose $

scaled deviance = 4.9425 (change = -120.4) at cycle 4
            d.f. = 5       (change = -1)
```

Notice that the residual deviance is much lower now (4.9425) than it was with a regression against dose (10.764), indicating that the model with log dose is a much better fit. Also notice that the ratio of residual scaled deviance to degrees of freedom is roughly 1. This is most satisfactory and shows no sign of overdispersion (see Section 15.6 and Exercise 15.3).

You should check that there is no pattern in the residuals:

```
$calc resid=dead-%fv $

$plot resid %fv $
```

This shows substantial scatter, but with a hint of increasing variance (see p. 287 for scaled residuals). The model with log dose appears to be acceptable. You should recalculate pfit and check the correspondence between observed and predicted mortalities:

```
$calc pfit=100*%fv/%bd $

$plot perc pfit logdose '*+' $
```

The fit is reasonably good, and we can now proceed to use the maximum likelihood equation for predictive purposes:

```
$disp e $

    estimate  s.e.    parameter
1   -3.269    0.3907  1
2    1.118    0.1309  LOGD
scale parameter taken as 1.000
```

which means that:

$$\log\left(\frac{p}{q}\right) = -3.269 + 1.118 \times \log \text{dose}$$

15.5 Estimating LD_{50}

We begin by working out the LD_{50}; this is the dose that is lethal to 50% of the individuals. When 50% die, this means that $p = q = 0.5$, so the logit is:

$$\ln\left(\frac{p}{q}\right) = \ln(1) = 0$$

In order to find the log dose that gives $y = 0$, we simply rearrange the equation to find:

$$\text{log dose} = \frac{3.269}{1.118} = 2.924$$

and we need the antilog of this to find the dose that kills 50% of the individuals. We use GLIM's calculator for this:

```
$calc 3.269/1.118 $

2.924

$calc %exp(2.924) $

18.62
```

So our estimate of the LD_{50} is a dose of about 18.62 (check this against our earlier plot of raw percentages).

We can obtain an estimate of the standard error in the death rate inflicted by a dose of 18.62 as follows. The **disp e** directive gives an estimate of the standard error of the intercept (the logit when log dose is 0), but the standard error of the predicted logit will be lower than 0.3907 at log doses in the middle of the range (see Section 7.4). A simple programming trick allows us to get the standard error we want from the **disp e** directive: we do this *by shifting the x-axis along until the origin is at 18.62*. The y-axis will cut the x-axis at a dose of 18.62 if we subtract 2.924 (the log of 18.62) from each of the log dose values:

```
$calc ld50=logdose−2.924 $
```

Now we fit the new, shifted x-axis (called ld50) to the model:

```
$fit ld50 $

scaled deviance = 4.9425 at cycle 4
            d.f. = 5

$disp e $

    estimate    s.e.     parameter
1   0.0008025   0.1587   1
2   1.118       0.1309   LD50
scale parameter taken as 1.000
```

The intercept is now equal to zero (almost) and the standard error is 0.1587. In order to establish a 95% confidence interval for the logit mortality at a dose of 18.62 we find t from tables with 5 d.f. (2.571) and multiply this by the standard error:

$$CI = t.SE = 2.571 \times 0.1587 = 0.408$$

If we add this logit on to the logit at LD_{50} (zero) we obtain a 95% upper bound, and we can subtract 0.408 to obtain a lower bound. Next we convert the logits to proportional mortality in the usual way:

```
$calc %exp(0.408) $

1.504

$calc 1/(1+1/1.504) $

0.601

$calc %exp(−0.408) $

0.665

$calc 1/(1+1/0.665) $

0.399
```

which means that, if we were to repeat the experiment, we could be 95% certain that the mortality rate inflicted by a dose of 18.62 would lie between 40 and 60%.

More usually, we shall be interested in establishing a confidence interval on the estimate of the LD_{50}. This involves establishing an interval on the x-axis; for this we need to find *the variance of a function of the parameters*, because LD_{50} is calculated as:

$$LD_{50} = \frac{-P(1)}{P(2)} = \frac{3.269}{1.118} = 2.923$$

where $P(1)$ is the first parameter in the list (the intercept) and $P(2)$ is the second (the slope). This is achieved by employing Fieller's theorem (see Collett, 1991, p. 97); a macro to carry this out is explained in Section 19.12, and here we simply use the macro called FIELLER to obtain the confidence interval directly:

```
$input 7 $

File name? fieller.mac

$fit logdose $

$use fieller $
```

and GLIM responds by printing:

```
Estimated LD50 = 2.923

95% Limits 2.648 and 3.220
```

This gives the upper and lower bounds on log dose; we then calculate the antilogs:

```
$calc %exp(3.220) $

25.03

$calc %exp(2.648) $

14.13
```

Thus, we can be reasonably sure that a repeat of the experiment would give an LD_{50} lying between doses of about 14 and 25. •

15.6 Overdispersion and hypothesis testing

All the different statistical procedures that we have met in earlier chapters can also be used with data on proportions. Factorial analysis of variance, multiple regression and a variety of mixed models in which different regression lines are fit in each of several levels of one or more factors can be carried out. The only difference is that we assess the significance of terms on the basis of χ^2 values, the increase in scaled deviance that results from removal of the term from the current model.

The important point to bear in mind is that hypothesis testing with binomial errors is less clear-cut than with normal errors. While the χ^2 approximation for changes in scaled deviance is reasonable for large samples (i.e. bigger than about 30), it is poorer with small samples. Most worrisome is the fact that the degree to which the approximation is satisfactory is itself unknown. This means that considerable care must be exercised in the interpretation of tests of hypotheses on parameters, especially when the parameters are marginally significant or when they explain a very small fraction of the total deviance. With binomial or Poisson errors we cannot hope to provide exact *P*-values for our tests of hypotheses (see Section 11.4).

As with Poisson errors, we need to address the question of overdispersion (see Section 13.7). When we have obtained the minimal adequate model, the residual scaled deviance should be roughly equal to the residual degrees of freedom. The present example was OK (look back to the last **fit** directive), but when the residual deviance is larger than the residual degrees of freedom there are two possibilities: either the model is misspecified, or the probability of success, p, is not constant within a given treatment level. The effect of randomly varying p is to increase the binomial variance from npq to:

$$s^2 = npq + n(n - 1)\sigma^2$$

leading to large residual deviance. This occurs even for models that would fit well if the random variation were correctly specified.

One simple solution is to assume that the variance is not npq but $npqs$, where s is an unknown *scale parameter* ($s > 1$). We obtain an estimate of the scale parameter by dividing the Pearson χ^2 value (the system scalar %x2) by the degrees of freedom (%df), and use this estimate of s to set the **scale** directive:

```
$calc %s=%x2/%df $

$scale %s $
```

then compare the resulting scaled deviances for terms in the model using an F-test instead of χ^2 (just as in conventional ANOVA; see Section 13.7).

While this procedure may work reasonably well for small amounts of overdispersion, it is no substitute for proper model specification. For example, it is not possible to test the goodness of fit of the model. A macro for carrying out Williams' adjustment for overdispersion is explained in Section 19.13.

15.7 Density-dependent parasitism: ANCOVA with binomial errors

To demonstrate the procedures involved in analysing a more complicated model involving data on proportions, we examine a field study on insect parasitism. The proportion of lepidopteran caterpillars that were parasitized was estimated by dissection of larvae from six independent random samples from each of three different habitats. At the time the insects were collected, a separate estimate of insect population density was made by measuring the amount of frass (insect faeces) in special traps beneath the host plants. Previous study had shown that this technique gave a reliable index of insect density (numbers per unit ground area).

Because we have a mixture of factor (habitat) and continuous variables (insect density) the data should be analysed by analysis of covariance, and we should normally fit the maximal model first, then proceed by model simplification (see Section 12.5.4). But for the purposes of demonstration, we begin by supposing that only population density had been measured as an explanatory variable, and treat the analysis as if it were a simple regression. Then we shall assume that the only explanatory information we have is the habitat from which the samples were gathered, and treat the analysis as a simple one-way classification. Finally, we shall use information on habitat and population density together to carry out a full analysis of covariance.

```
$units 18 $

$data x d n h $

$dinput 6 $

File name? glex24.dat

$factor h 3 $
```

```
$yvar d $

$error b n $
```

First, we test for density dependence in the rate of parasitism:

```
$fit $

scaled deviance = 34.868 at cycle 3
            d.f. = 17

$fit +x $

scaled deviance = 25.154 (change = -9.715) at cycle 4
            d.f. = 16     (change = -1)
```

The decrease in deviance of 9.715 is much larger than the value of χ^2 in tables with 1 d.f., so there is little doubt that population density affects the proportion of insects that were parasitized. To see *how* density affects the rate of parasitism we inspect the parameter estimates:

```
$disp e $

    estimate    s.e.      parameter
1  -0.2974     0.1930     1
2   0.03269    0.01112    X
scale parameter taken as 1.000
```

The intercept is not significantly different from zero, but it does appear that there is a significant positive relationship between population density and parasitism (perhaps because the parasites aggregate in patches of high host density).

It is important to know how well the model describes the data. The best way to see this is to calculate the proportion parasitized from the raw data (perc), and the proportion from the fitted values of the model (fitted), then to plot both these against x, using different symbols to represent the data (+) and the model (F):

```
$calc perc=d/n $

$calc fitted=%fv/n $

$plot perc fitted x '+F' $
```

While there may be a trend in percentage parasitism with population density, the fit of the model (F) to the data (+) is extremely poor. At low densities, the model overestimates the proportion parasitized, and at high densities underestimates it (except for the single outlying point in the lower right-hand corner of the graph).

Next, suppose that only the habitat factor had been recorded, and ask

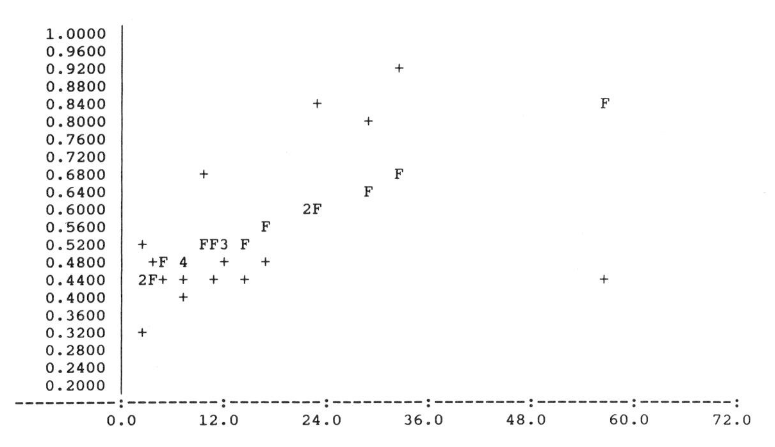

whether there are significant differences between habitats in the proportion of insects parasitized.

$fit $

scaled deviance = 34.868 at cycle 3
d.f. = 17

$fit +h $

scaled deviance = 25.730 (change = −9.138) at cycle 4
d.f. = 15 (change = −2)

The addition of habitats to the model brings about a reduction in deviance of 9.138 on two degrees of freedom (this represents 26.2% of the total deviance). The habitat deviance of 9.138 is highly significant (compare with χ^2 with 2 d.f. from tables = 5.99). Next, we investigate the parameter estimates:

$disp e $

	estimate	s.e.	parameter
1	−0.1054	0.2055	1
2	0.7985	0.2961	H(2)
3	0.08827	0.2764	H(3)

scale parameter taken as 1.000

The mean proportions parasitized in habitats H(1) and H(3) are similar, but the mean in H(2) appears to be significantly higher than the other two ($t = 0.7985/0.2961 = 2.70$). Note that the standard errors are different in H(2) and H(3) because of the unequal numbers of caterpillars collected from the two habitats.

With category data like this, it is often easier to understand what is going on if we calculate the proportions, and tabulate the mean proportion for each treatment:

```
$tab the perc means for h $

        1        2        3
[ ]  0.4847  0.6269  0.5041
```

Compare this with the **disp e** table where the parasitism estimates are in logits and differences between logits.

Finally, let us combine the factors and the continuous variables in an analysis of covariance. To do this, we add x to the model:

```
$fit +x $

scaled deviance = 20.659 (change = -5.071) at cycle 4
           d.f. = 14     (change = -1)
```

The change in deviance that results from adding local population density to the model as a covariate is statistically significant, because 5.071 is large compared with the χ^2 value in tables with 1 d.f. (3.841). To see the nature of the effect of population density on parasitism we examine the parameter estimates.

```
$disp e $

   estimate   s.e.      parameter
1  -0.4797    0.2666    1
2   0.6509    0.3081    H(2)
3   0.2285    0.2853    H(3)
4   0.02519   0.01157   X
scale parameter taken as 1.0002
```

which suggests that the rate of parasitism is higher at higher local densities, but the effect is of marginal significance ($t = 0.02519/0.01157 = 2.18$). It appears from the main effects that habitats H(1) and H(3) have similar means, and that the average parasitism in habitat H(2) is significantly higher.

In order to see whether a model with different slopes in each habitat would be an improvement, we fit the interaction term H.X:

```
$fit +h.x $

scaled deviance = 5.7507 (change = -14.91) at cycle 4
           d.f. = 12     (change = -2)
```

Inclusion of the interaction term causes a substantial reduction in scaled deviance, and the χ^2 value of 14.91 is much larger than the value in tables with 2 d.f. This means that one or more of the slopes is significantly different from the others. We inspect the parameter estimates to see what is going on:

```
$disp e $

     estimate      s.e.       parameter
1   -0.1146       0.2817     1
2   -1.361        0.6447     H(2)
3   -0.08775      0.6278     H(3)
4    0.0006091    0.01276    X
5    0.1087       0.03050    H(2).X
6    0.01932      0.05840    H(3).X
scale parameter taken as 1.000
```

Notice, first of all, that the two interaction terms H(2).X and H(3).X are *differences between slopes* because the slope in habitat 1 (X) appears in row 4 of the table. The slope in habitat H(1) is clearly density independent (i.e. it is not significantly different from zero). The first interaction term is the difference in slope between habitat H(2) and H(1); this suggests that the proportion parasitized is significantly density dependent in habitat H(2) ($t = 0.1087/0.0305 = 3.56$). There is no significant difference between the slopes in habitats H(3) and H(1), so we conclude that parasitism is density independent in H(3) ($t = 0.01932/0.05840 = 0.33$). As for the mean rate of parasitism in the three habitats, it is clear that H(1) and H(3) have very similar means ($t = 0.08775/0.6278 = 0.14$), but interpretation of the main effect of H(2) needs to be made more carefully. The estimate is somewhat lower than zero ($t = 2.111$, tables = 2.179), but the increasing trend with density is so pronounced that, unless the density within the habitat was known, it would be quite wrong to use the lower intercept to conclude that parasitism was generally lower in this habitat than the others (cf. the mean proportions calculated earlier).

Let us see how this maximal model fits the data. First we recalculate the fitted proportions, then plot both data and fitted values against population density:

```
$calc fitted=%fv/n $

$plot perc fitted x 'abcABC' h $
```

Here, the lower-case letters a, b and c show the proportion parasitized from habitat H(1), H(2) and H(3), and the upper-case letters shows the matching fitted values. The fit is generally good, and the pattern of the responses is clear. Neither habitat H(1) nor H(3) shows any relationship between parasitism and density, and their mean levels of parasitism are virtually identical. The response in habitat H(2), however, is dramatically different. The mean value is higher than in the other two habitats, but the important feature of the data is the strong and apparently linear dependence of the rate of parasitism upon population density. More work

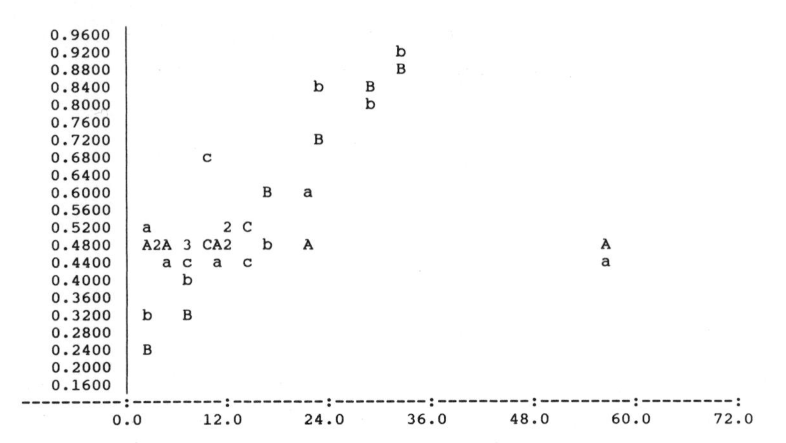

would now need to be done to understand the mechanism underlying the pattern. Was there some other factor that caused parasitism to be responsive to changes in density in habitat H(2), but not in habitats H(1) and H(3)? Conversely, why is it that percentage parasitism fails to respond to increases in population density in habitats H(1) and H(3)? Is parasitism, perhaps, regulated by a different factor than in habitat H(2)? As ever, clear patterns lead to a proliferation of new questions. The lesson to be learned from this exercise is that a thorough analysis of covariance has produced a model that explains 83.5% of the scaled deviance in *y*, whereas simple regression analysis of the density effects explained only 27.9% and ANOVA of habitat differences only 26.2%.

15.7.1 Model simplification

There is a certain amount of model simplification that can be done with this model. The intercepts for habitats 1 and 3 do not look significantly different from zero, and the slope of the regression line looks significantly different from zero only for habitat 2. The first task is to combine habitats 1 and 3 into a new factor H2 that has a value 1 for habitats 1 and 3, and 2 for habitat 2.

```
$calc h2=1+(h==2) $

$factor h2 2 $
```

Next, we remove the old habitat factor from the model and replace it with the new one:

```
$fit h2+h2.x $

scaled deviance = 5.9760 at cycle 4
           d.f. = 14
```

This simplification has caused an insignificant increase in deviance

(5.976 − 5.7507 = 0.225), and therefore appears to be justified. We can see how the simplification has affected the parameter estimates:

```
$disp e $

     estimate      s.e.       parameter
  1  -0.06209      0.1985     1
  2  -1.414        0.6129     H2(2)
  3   0.0004604    0.01204    H2(1).X
  4   0.1093       0.02770    H2(2).X
  scale parameter taken as 1.000
```

Next, we should give a slope of zero to the regression of parasitism against density in habitats 1 and 3 (i.e. the H2(1).X interaction). We do this by intentionally aliasing the parameter, calculating a new x2 vector that has 0's in all the habitat 1 and 3 locations and the original value of x in habitat 2. This is done as follows:

```
$calc x2=x*(h2==2) $
```

which says x2 has the value of x where H2 is equal to 2 but zero otherwise (because (H2==2) is zero (false) for habitats 1 and 3, where H2 is 1).

We fit the new model as follows:

```
$fit h2+h2.x2 $

scaled deviance = 5.9775 at cycle 4
           d.f. = 15
```

This, again, causes an insignificant increase in deviance (5.9775 − 5.976 = 0.0015). The parameters of the new model are:

```
$disp e $

     estimate    s.e.       parameter
  1  -0.05662    0.1374     1
  2  -1.419      0.5960     H2(2)
  3   0.000      aliased    H2(1).X2
  4   0.1093     0.02770    H2(2).X2
  scale parameter taken as 1.00
```

What does the intercept of −0.05662 predict about the proportion parasitized in habitats 1 and 3? To go from the logit to the proportion we first antilog the parameter estimate to get the odds ratio:

```
$calc %exp(-.05662) $

0.9450
```

then work out 1/(1 + 1/odds) as in Section 15.3:

```
$calc 1/(1+1/0.945) $
0.4859
```

The predicted parasitism is 48.6%, but since the intercept is not significantly different from 0 we might as well simplify this to 50% parasitism (logit(0.5) = ln(1) = 0). This is consistent with the raw mean proportions for habitats H(1) and H(3) we calculated earlier (see Exercise 15.4).

The slope of 0.1093 in habitat 2 means that for every increase in host density of 10 caterpillars per unit area, the odds of being parasitized increase by a factor of roughly 3 (= exp(1.093) = 2.983). Thus, increasing density from 20 to 30 caterpillars leads to an increase in the proportion parasitized from 0.683 to 0.865. This is the minimal adequate model; it contains the minimal number of explanatory variables and all the parameters are statistically significant.

15.7.2 Model criticism

Next, we need to assess whether the standardized residuals are normally distributed and whether there are any trends in the residuals, either with the fitted values or with the explanatory variables. It is necessary to deal with *standardized residuals* because with error distributions like the binomial, Poisson or gamma distributions, the variance changes with the mean (see Section 5.11). The simplest test is to use:

```
$disp r $
```

unit	observed	out of	fitted	residual
1	7	14	6.802	0.106
2	10	22	10.689	−0.294
3	9	21	10.203	−0.525
4	8	17	8.259	−0.126
5	6	10	4.858	0.722
6	5	11	5.344	−0.208
7	3	10	2.214	0.599
8	2	5	1.647	0.336
9	15	31	18.421	−1.251
10	17	20	14.766	1.137
11	9	11	9.290	−0.242
12	20	22	19.663	0.233
13	8	17	8.259	−0.126
14	10	22	10.689	−0.294
15	7	15	7.288	−0.149
16	6	9	4.373	1.085
17	22	43	20.892	0.338
18	5	11	5.344	−0.208

and to inspect the table of standardized residuals. To obtain plots of the standardized residuals, however, we need to calculate:

$$r_s = \frac{y - \mu}{\sqrt{V}} = \frac{(\%yv - \%fv)}{\sqrt{\%fv\left(1 - \frac{\%fv}{\%bd}\right)}} \qquad (14.4)$$

where the formula for the *variance function* of the binomial is obtained from Table 19.2.

```
$calc rs=(%yv−%fv)/%sqr(%fv*(1−%fv/%bd)) $
```

```
$look %yv %bd %fv rs $
```

	D	N	%FV	RS
1	7.000	14.000	6.802	0.1059
2	10.000	22.000	10.689	−0.2938
3	9.000	21.000	10.203	−0.5252
4	8.000	17.000	8.259	−0.1259
5	6.000	10.000	4.858	0.7222
6	5.000	11.000	5.344	−0.2077
7	3.000	10.000	2.214	0.5986
8	2.000	5.000	1.647	0.3362
9	15.000	31.000	18.421	−1.2512
10	17.000	20.000	14.766	1.1367
11	9.000	11.000	9.290	−0.2415
12	20.000	22.000	19.663	0.2335
13	8.000	17.000	8.259	−0.1259
14	10.000	22.000	10.689	−0.2938
15	7.000	15.000	7.288	−0.1486
16	6.000	9.000	4.373	1.0853
17	22.000	43.000	20.892	0.3382
18	5.000	11.000	5.344	−0.2077

which, you will note, is exactly the output obtained by **disp r**. Other kinds of standardized residuals (like Pearson residuals or deviance residuals) can be calculated from the system vector containing the variances of the linear predictor, %vl. This is achieved by using the **extract** directive (see Exercise 15.1):

```
$ext %vl $
```

or obtained from model-checking macros like mcheck (see below). We can use our calculated vector of standardized residuals rs to form various model-checking plots. To plot the residuals against the fitted values, we might arcsine transform the x-axis (because the fitted data are proportions):

```
$calc fitted=%ang(%fv/%bd) $
```

Finally, we plot the residuals against the fitted values and inspect the graph for patterns:

```
$plot rs fitted '+' $
```

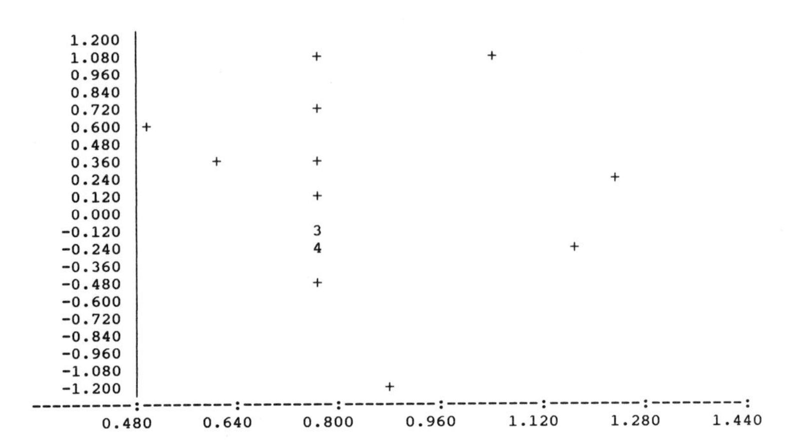

There is no clear pattern to the residuals and the scatter suggests that the model and the error distribution are reasonably well behaved.

There was only one large residual (−1.2512) which occurred in sample 9 where only 15 out of 31 caterpillars were parasitized. A check of the field notebook shows that these figures were indeed correct, and so the analysis stands as it is. If, for example, we had found that 15 had been transcribed in error instead of 25, then the data would have been edited, the model refitted and the residual analysis carried out afresh. It is extremely bad practice, however, to rely on the analysis of the residuals to show up errors in data transcription. Not only is it unscientific, but it will not work if the erroneous data points are influential because, as we have seen, influential values have small residuals (see Section 5.14).

To demonstrate the advantages of using model-checking macros, we can employ mcheck. There is no extra calculation involved and the macro called mcheck will automatically produce two plots of standardized residuals when it is read into GLIM:

```
$input 7 $

File name? mcheck.mac
```

First, the standardized residuals are plotted against the fitted values (see top graph, p. 289) and then the ranked residuals are plotted against the standard normal deviates (see bottom graph, p. 289).

As we saw earlier, the scatter of the residuals against the fitted values shows no pattern suggestive of mis-specification of the model, and the plot of standardized residuals against normal order statistics makes a reasonably straight line. We conclude that the error specification and the model structure are satisfactory.

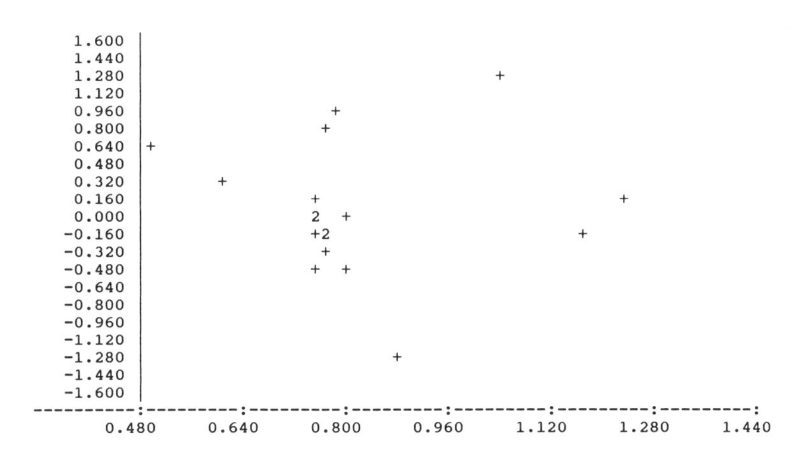

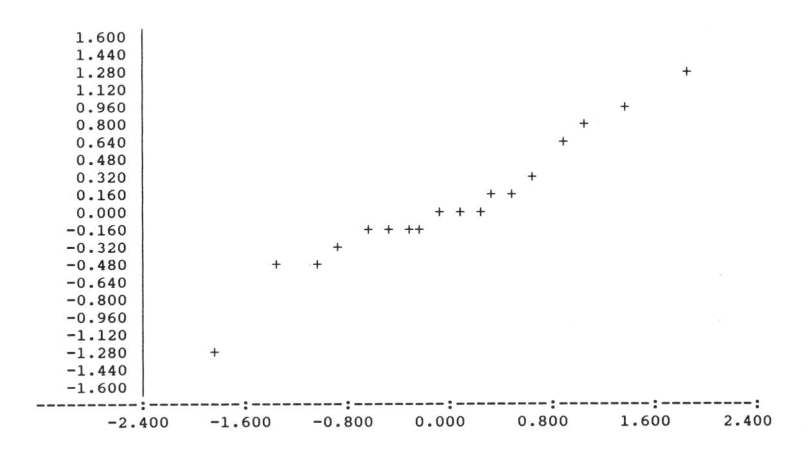

Just as in Chapter 14 on Poisson errors, there is a great deal of new material in this chapter, and it is certainly worth repeating the exercises until the steps involved have become completely familiar. The most important points to emphasize in modelling with binomial errors are:

1 make sure that the binomial denominator contains no zeros, and that none of the counts is greater than its binomial denominator;

2 check for overdispersion, and correct for it by using the **scale** directive or Williams' procedure if necessary;

3 remember that you do not obtain exact *P*-values with binomial errors; the χ^2 approximations are sound for large samples, but small samples present a problem;

4 the fitted values are counts, like the response variable;

5 the linear predictor is in logits (the log of the ratio p/q);
6 the proportion is estimated from the odds by $p = 1/(1 + 1/\text{odds})$;
7 use F-tests rather than χ^2 tests after adjusting the scale parameter for overdispersion.

CHAPTER 16

Binary response variables

Many ecological problems involve binary response variables. For example, we often classify individual organisms as:

1 dead or alive;
2 parasitized or not;
3 infected or not;
4 wilted or turgid;
5 male or female;
6 or patches as occupied or vacant;

and it is interesting to understand the factors that are associated with an individual being in one class or the other (Cox & Snell, 1989). In studying mortality attributable to a given factor, for instance, the data tend to consist of a list of measurements made on the dead individuals (their age, sex, weight, body condition, location, habitat, and so on) and a similar list for the live individuals. The question then becomes which, if any, of the factors increase the probability of an individual being dead?

16.1 Binary data in GLIM

The response variable contains only 0's or 1's; 0 to represent dead individuals and 1 to represent live ones. The way GLIM treats this kind of category data is to assume that the 0's and 1's come from *a binomial trial with sample size 1*. If the probability that an animal is dead is p, then the probability of obtaining x (where x is either 0 or 1) is given by an abbreviated form of the binomial distribution with $n = 1$, known as the Bernoulli distribution (see Section 15.1):

$$P(x) = p^x(1 - p)^{(1-x)} \tag{16.1}$$

The random variable x has a mean of p and a variance of pq, and the object is to determine how the explanatory variables influence the value of p.

In order to carry out linear modelling on a binary response variable we take the following steps:

1 create a vector of 0's and 1's as the response variable;
2 use **calc** to produce a vector, n, full of 1's for the binomial denominator;
3 declare the **error** to be binomial and the binomial denominator to be n;
4 chose an appropriate **link** function (usually either logit or complementary log-log);

5 **fit** the model in the usual way;
6 test significance by deletion of factors from the maximal model, and compare the change in deviance with χ^2 tables;
7 there is no test for overdispersion with a binary response variable. Choice of link function is generally made by trying both links and selecting the link that gives the lowest deviance. The logit link that we used in the previous chapter is symmetric in p and q, but the complementary log-log link is asymmetric.

In order to see how this kind of analysis fits in with the analysis of other kinds of categorical data using contingency tables, it is worth knowing that Poisson with the log link is exactly equivalent to binomial with the logit link when the response is two-level (in the absence of structural zeros; for details see Aitkin *et al.*, 1989, pp. 225–55). When we deal with multi-category response variables in Section 16.3 later in this chapter, you will see an example of the use of log-odds with Poisson rather than binomial errors.

16.2 Death and reproductive effort

We shall use a binary response variable to analyse questions concerning the costs of reproduction in a perennial plant. Suppose that we have a set of measurements on plants that survived through the winter to the next growing season, and another set of measurements on otherwise similar plants that died between one growing season and the next. We are particularly interested to know whether, for a plant of a given size, the probability of death is influenced by the number of seeds it produced during the previous year (i.e. is there a trade-off between reproductive effort and mortality?).

There were 59 plants, and for each plant it was recorded whether the plant was dead or alive (d = 0 or 1), the number of capitula (flower-heads) it had produced the previous summer (cap) and the volume of its rootstock (root).

```
$units 59 $

$data d cap root $

$dinput 6 $

File name? glex26.dat
```

Some calculations are necessary; the product of the number of flower-heads and the rootstock volume (cxr) will be used to test for interaction effects, and the binomial denominator, n, needs to be set to 1 for each plant.

```
$calc cxr=cap*root $
```

```
$calc n=1 $
```

The response variable is set to the binary variate d (0 for dead, 1 for alive), and the errors are declared to be binomial, with binomial denominator equal to 1:

```
$yvar d $

$error b n $
```

From this point, it is a simple exercise in multiple regression analysis. We fit the null model first:

```
$fit $

scaled deviance = 78.903 at cycle 4
           d.f. = 58
```

Note that the scaled deviance is substantially larger than the degrees of freedom, so there is plenty of variation in the data for the model to explain. Next we fit both capitula and rootstock to the model:

```
$fit+cap+root $

scaled deviance = 54.068 (change = -24.8352) at cycle 5
           d.f. = 56     (change = -2)
```

The change in deviance is much larger than the value of χ^2 in tables with two degrees of freedom, so this has caused a significant reduction in deviance.

```
$disp e $

    estimate   s.e.      parameter
1    0.9615    0.6151    1
2   -0.1064    0.03320   CAP
3    6.600     2.087     ROOT
scale parameter taken as 1.000
```

Inspection of the parameter estimates demonstrates that both flower-head production and rootstock volume are required in the model. You should check this by removing each variable in turn from the current model.

The graphical output from binary response models takes some getting used to. Remember that the data consist entirely of 0's and 1's, so these will make two lines, parallel to the x-axis. The fitted values are based on the estimated probability of death at each value of x. If there is a sudden threshold, then the fitted values will show an S-shaped step function, whereas if the effect of the explanatory variable is subtle, then the fitted values will show a more gradual rise. Since we only have two-dimensional graphics, we need to fit one variable at a time to the model in order to get an intelligible graph (if we plot the fitted values from the full model against capitulum numbers, they will be scattered all over the place because of the uncontrolled variation in root size, and vice versa). To look at the effects of reproduction on the risk of death we fit capitulum numbers on their own:

```
$fit cap $

scaled deviance = 75.008 at cycle 4
             d.f. = 57

$plot d %fv cap '*+' $
```

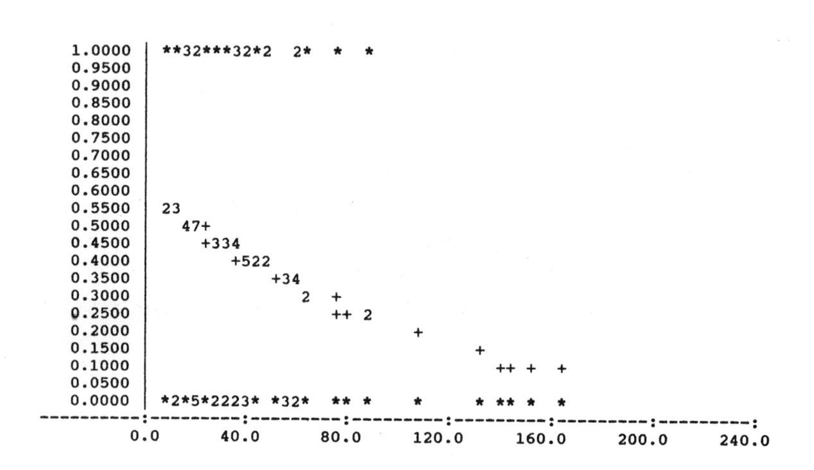

Note the concentration of live plants (1 on the *y*-axis) at low capitulum numbers, and the much wider scatter of dead plants (0 on the *y*-axis) indicating that nearly all the plants with very large flower production died. The fitted values (+) show a smooth decline in the probability of surviving as the number of flower-heads increases, from about a 55% chance of survival with a small flower crop, down to a 10% chance if more than 120 flower-heads were produced.

Likewise, to see the impact of rootstock volume, we fit this on its own:

```
$fit root $

scaled deviance = 78.136 at cycle 4
             d.f. = 57

$plot d %fv root '*+' $
```

Here you see the positive effect of a large rootstock on the probability of

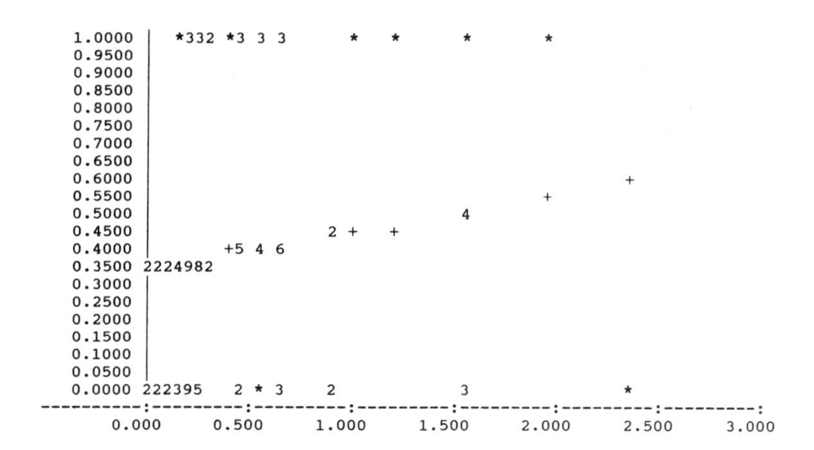

survival: for a given amount of flower production, a plant with a larger rootstock is more likely to survive the winter.

It is possible, of course, that there is an interaction between fecundity and rootstock size. To investigate this we have created a new variable cxr which is the product of capitulum numbers and rootstock volume (the notation cap.root is not allowed for continuous variables in GLIM 3.77). The maximal model is now:

```
$fit cap+root:+cxr $
```

the last term of which produces warning messages before printing:

```
scaled deviance = 37.128 (change = −16.94) at cycle 7
           d.f. = 55     (change = −1)
```

The warning messages just mean that two of the parameters have been held at their limits (0 or 1) while other maximum likelihood estimates are obtained. It is clear from the large reduction in deviance that the interaction term is indeed important. The new parameter values are:

```
$disp e $
```

	estimate	s.e.	parameter
1	−2.960	1.525	1
2	−0.07888	0.04164	CAP
3	25.15	7.809	ROOT
4	−0.2091	0.08840	CXR

scale parameter taken as 1.000

which suggests that large plants suffer disproportionately high mortality for a given level of flower production. To get a graph on this effect, we fit cxr on its own:

```
$fit cxr $

scaled deviance = 78.092 at cycle 4
           d.f. = 57

$plot d %fv cxr '*+' $
```

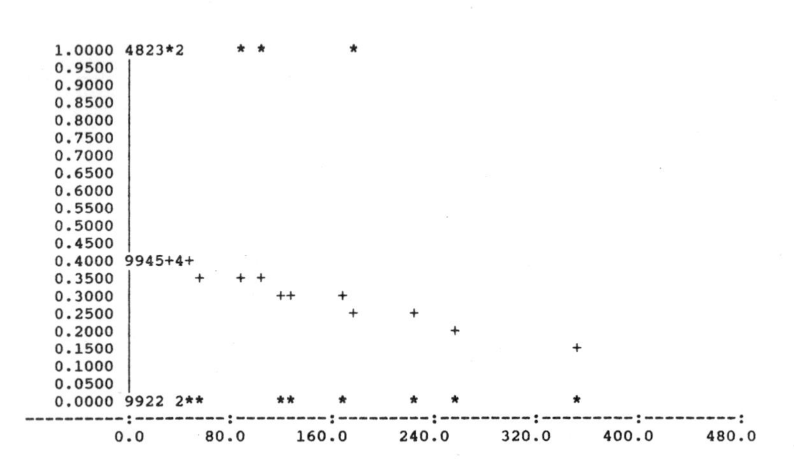

In order to make predictions we need to specify the root volume and the capitulum production. Thus, with a root of 0.5 and a capitulum number of 70 (so the interaction term cxr is 35) we work out:

$$-2.96 - (70 \times 0.07888) + (0.5 \times 25.15) - (35 \times 0.2091) = -3.2251$$

then convert this logit to a value of p (see Section 15.2) to get $p = 0.0382$. This should be compared with the prediction from the simpler, two-parameter model without the interaction term, from which p is predicted to be 0.0397.

Model criticism for this example leads us to question the link function and the need for transformation of the explanatory variables. By trying a range of options, you should convince yourself that log transformation of the explanatory variables does not improve the fit, but the complementary log-log transformation produces a marginally lower residual deviance than the logit link with which we have been working here (35.46 against 37.13). In no case, however, are the conclusions altered. The risk of death is greater for plants of a given rootstock volume that produce more flowers. The data support the hypothesis that there is a cost of reproduction in terms of reduced survival of adult plants. ●

16.3 Multi-category data

It is quite common to have three or more categories against which an individual might be classified; for example,

1 parasite-free, light or heavy infestation;
2 turgid, partially wilted or wilted;
3 symptom-free, ambiguous or clear symptoms;
4 very light, light, moderate, heavy or very heavy response;

and we should like a single model for the probabilities of all the categories. GLIM achieves this with the *multinomial logit model* as follows.

Assume that the response variable has r distinct categories and that the probability of each of these categories is p_j. There are only $r-1$ distinct probabilities because $\Sigma p_j = 1$. The multinomial logit Θ_j is simply:

$$\Theta_j = \log\left(\frac{p_j}{p_1}\right), \qquad j = 1,\ldots,r$$

where the first category is the reference category and the probabilities of all other categories are compared to this. Because GLIM does not handle the multinomial distribution, we cannot fit simultaneous multinomial logistic models directly. Instead, we estimate Poisson means λ_j for each category, and use:

$$\Theta_j = \log\left(\frac{\lambda_j}{\lambda_1}\right), \qquad j = 1,\ldots,r$$

The contingency table has three important components:

1 a category variable representing the response;
2 classifying factors representing the explanatory variables;
3 a vector of counts containing the frequencies in each cell of the table.

Thus, in an investigation of viral disease incidence, we might have a three-level response variable based on symptoms – none, mild and severe – with the classifying factors of age (four levels) and sex (two levels). The question is to what degree do age and sex influence the probability of expressing a particular category of symptom? Since there are three symptom categories, two sexes and four age classes, the data matrix has $3 \times 2 \times 4 = 24$ rows. There is a column for the frequency (count) data containing the number of animals showing a particular symptom (this is declared as the response variable with Poisson errors), plus three other columns containing factors: one for the response category (SYMPT: none = 1, mild = 2, severe = 3), one for SEX (male = 1, female = 2) and one for AGE (1 = young, 2 = sub-adult, 3 = adult, 4 = senescent).

As with the complex contingency tables in Section 14.8, the important point is to specify the *null model* correctly; this is necessary to reproduce the marginal totals in the contingency table. In general, if SYMPT is our response category (viral infection symptoms in this case), and AGE and

SEX are the explanatory variables, then we begin by fitting the null model:

```
$fit sympt + age*sex $
```

where all three have been declared as factors. We then carry out the modelling by asking which of the *interactions involving* SYMPT are statistically significant. As usual, hypothesis testing is based on the change in deviance resulting from a given interaction (see Section 11.4).

```
$units 24 $

$data sympt count sex age $

$dinput 6 $

File name? glex42.dat

$factor sympt 3 sex 2 age 4 $
```

The response variable is the count, containing the number of animals expressing each of the symptom categories:

```
$yvar count $

$error p $

$link l $
```

We begin by looking at the raw data table; the three symptoms (none, mild and severe) are the rows, males are the left-hand table and females the right, and the ages 1–4 are the columns. There is a strong indication that symptom severity depends upon age (both mild and severe symptoms peak in the sub-adult and adult classes), but no obvious effect of sex, nor of any interaction between age and sex.

```
$tprint count sympt;sex;age $
```

SEX	1				2			
AGE	1	2	3	4	1	2	3	4
SYMP								
1	17.000	25.000	32.000	3.000	22.000	36.000	28.000	15.000
2	3.000	10.000	21.000	1.000	10.000	12.000	31.000	10.000
3	1.000	4.000	8.000	0.000	5.000	2.000	8.000	10.000

To constrain the marginal totals, we begin by fitting the response category (SYMPT) and the interaction of all the classifying factors (AGE * SEX in this case):

```
$fit sympt+age*sex $

scaled deviance = 29.084 at cycle 4
           d.f. = 14
```

The interactions between SYMPT and the explanatory factors are now added one at a time:

```
$fit +sympt.age $

scaled deviance = 9.3434 (change = −19.74) at cycle 3
            d.f. = 8      (change = −6)

$fit +sympt.sex $

scaled deviance = 7.1751 (change = −2.168) at cycle 3
            d.f. = 6      (change = −2)
```

The symptom/age interaction is significant (χ^2 table with 6 d.f. = 12.592) but the sex effect is insignificant. The residual deviance is not overdispersed (ratio = 1.196). Can you see how the six residual degrees of freedom are made up?

	estimate	s.e.	parameter
1	2.648	0.2366	1
2	0.6382	0.2946	AGE(2)
3	0.7055	0.2850	AGE(3)
4	−2.034	0.5773	AGE(4)
5	0.5664	0.2731	SEX(2)
6	−1.099	0.3202	SYMP(2)
7	−1.872	0.4381	SYMP(3)
8	−0.3179	0.3467	AGE(2).SEX(2)
9	−0.4726	0.3255	AGE(3).SEX(2)
10	1.603	0.5921	AGE(4).SEX(2)
11	0.07878	0.4054	AGE(2).SYMP(2)
12	−0.4473	0.6118	AGE(2).SYMP(3)
13	0.9555	0.3720	AGE(3).SYMP(2)
14	0.5500	0.5207	AGE(3).SYMP(3)
15	0.6061	0.4988	AGE(4).SYMP(2)
16	1.284	0.5893	AGE(4).SYMP(3)

scale parameter taken as 1.000

The only parameters that appear significant are the increased prevalence of mild symptoms in adults (2.6 times the odds; row 13) and perhaps of severe symptoms in senescent females (3.6 times the background odds; row 16).

We use the macro **mcheck** to assess the error structure and residual distribution (see top graph, p. 300):

```
$input 10 $

File name? mcheck.mac
```

The scatterplot of scaled residuals against fitted values shows no important patterns (see bottom graph, p. 300) and the ranked residuals appear to be normally distributed (the plot is reasonably straight). We conclude that the model with only three parameters for viral symptoms is

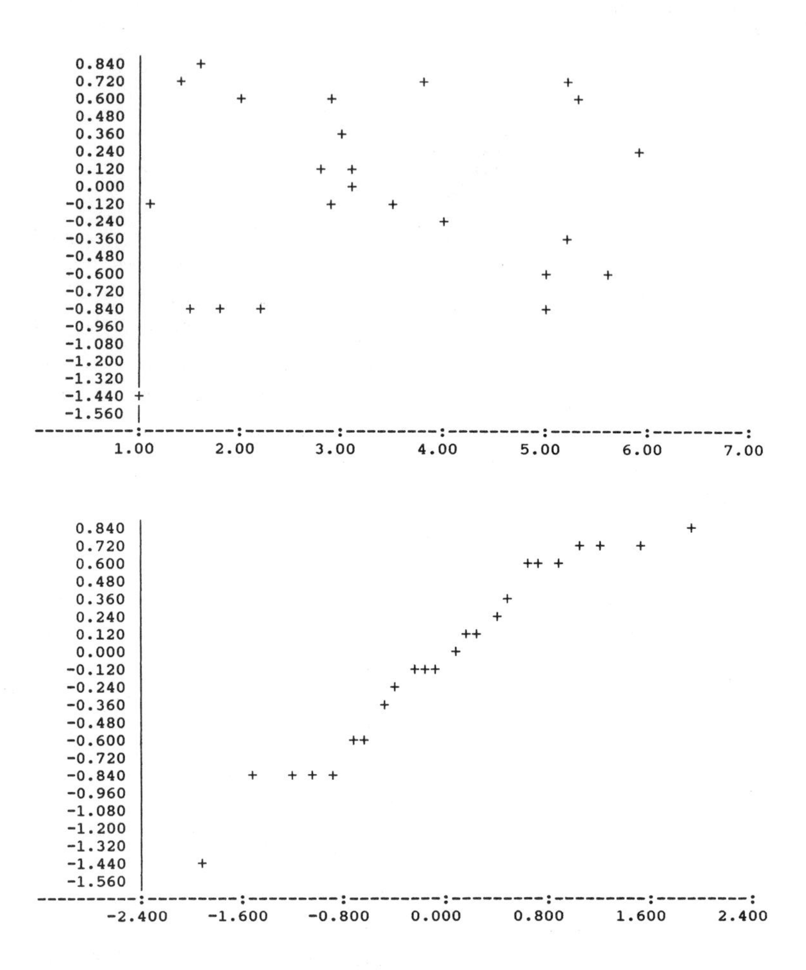

minimal adequate (the background rate, plus additional parameters for AGE(3).SYMPT(2) and AGE(4).SYMPT(3)).

In summary, we can analyse a two-category response variable using binomial errors, and a multi-category response variable using Poisson errors. The first is ideal when we have one or more continuous explanatory variables for each individual that we classify as dead or alive, parasitized or not, etc. The second is ideal when our explanatory variables are (or can easily be made into) factors. More examples of the analysis of contingency tables are to be found in Sections 14.3, 14.6, 14.7 and 14.8. •

CHAPTER 17

Gamma errors

With continuous response variables, the traditional models of regression and ANOVA assume constant variance and normal errors. For many ecological purposes, however, this structure is inappropriate because: (i) variance often increases with the mean; and (ii) the distribution of errors is often skewed (sometimes highly skewed). One way out of this dilemma is to assume log-normal errors in y and to carry out the analysis on the log-transformed response variable using least squares with normal errors and the identity link (see Section 3.4). An alternative is to use GLIM's capacity to deal with gamma errors.

17.1 The gamma distribution

The gamma distribution is appropriate when the response variable has a constant coefficient of variation rather than a constant variance. If s is the standard deviation, and m is the mean, then

$$CV = \frac{s}{m} = \text{constant} \tag{17.1}$$

whereas with normal errors we assume that it is the standard deviation, s, which is constant. Another measure of trend in variability used by ecologists is the *variance mean ratio*. For the Poisson distribution, the variance is defined as being equal to the mean, so the variance mean ratio is 1. For the binomial distribution the mean is np and the variance is npq. Since q must be less than 1 ($q = (1 - p)$), the variance must be less than the mean, and the variance mean ratio will be less than 1. For the kind of aggregated distributions that are so common in ecological studies (e.g. negative binomial or log normal distributions), the variance changes with the mean (see Taylor's power law; Section 4.19), and the variance mean ratio is often much greater than 1. If the variance is proportional to the square of the mean, as is commonly the case in ecological applications, then the coefficient of variation is constant:

$$CV = \frac{\sqrt{s^2}}{m} = \frac{\sqrt{a.m^2}}{m} = \sqrt{a} = c \tag{17.2}$$

and gamma errors are appropriate. The density function of the gamma distribution is given by:

$$f(x) = \frac{1}{\beta^{\alpha}\Gamma(\alpha)}\, x^{\alpha-1}e^{-x/\beta} \tag{17.3}$$

where α is known as the shape parameter. When $\alpha \leq 1$ the distribution declines monotonically, while for larger values of α (>10) the distribution becomes normal (Fig. 17.1). Thus, all gamma distributions are more or less skewed with a tail to the right, and distributions with $\alpha > 1$ have a mode to the right of $x = 0$ and pass through the origin.

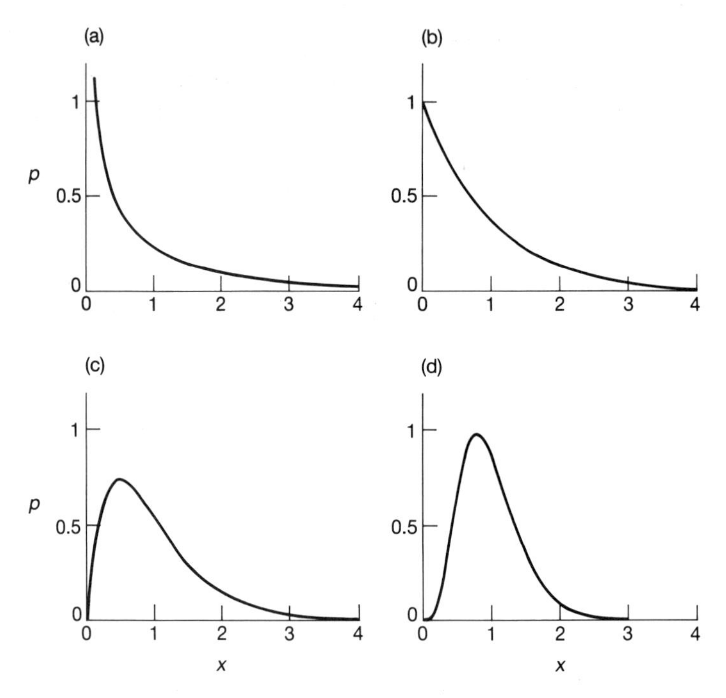

Fig. 17.1 The gamma distribution can take a variety of forms depending upon the value of the shape parameter: (a) $\alpha = 0.5$, (b) $\alpha = 1$, (c) $\alpha = 2$, (d) $\alpha = 5$. All these distributions have the same mean, $\alpha\beta = 1.0$.

There are two special cases of the gamma distribution that are of interest to ecologists: the exponential distribution and the χ^2 distribution. For the exponential the density function is:

$$f(x) = \frac{1}{\beta} e^{-x/\beta}$$

and for χ^2 the density function is:

$$f(x) = \frac{1}{2^{\nu/2}\Gamma(\nu/2)} x^{\nu/2-1} e^{-x/2}$$

where ν (pronounced 'new') is an integer (the degrees of freedom). The means and variances of the three distributions are shown in Table 17.1.

Table 17.1 First and second moments of three related distributions

Distribution	Mean	Variance
Gamma	$\alpha\beta$	$\alpha\beta^2$
Exponential	β	β^2
χ^2	ν	2ν

An important class of models for which gamma errors can be used are the *inverse polynomials* (see Fig. 17.2). The simplest of these is the inverse linear:

$$\frac{1}{y} = a + \frac{b}{x} \tag{17.4}$$

which is an asymptotic curve familiar to biologists under a variety of names: Holling's functional response, Briggs–Haldane and Michaelis–Menten.

$$y = \frac{x}{b + ax} \tag{17.5}$$

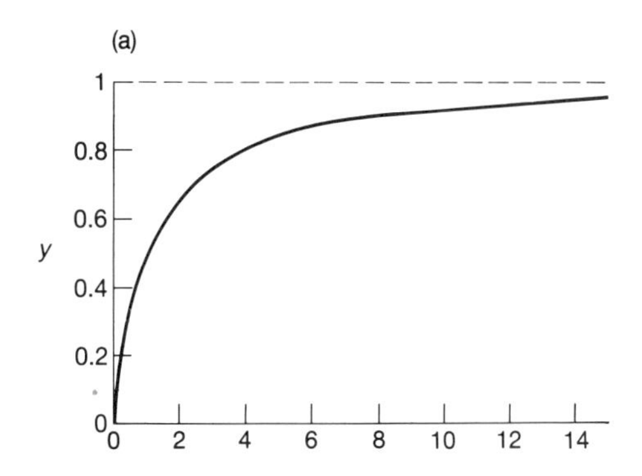

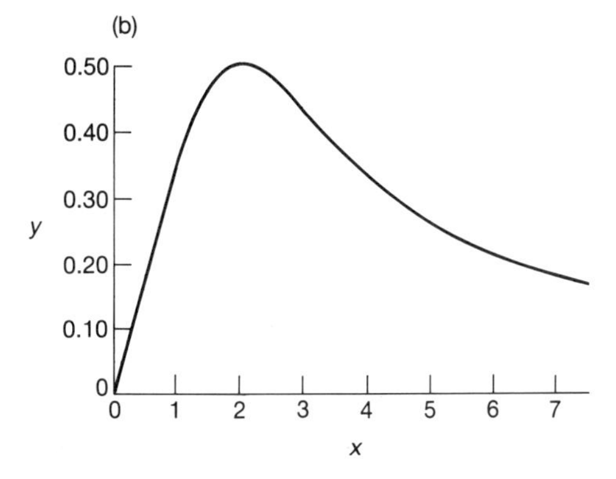

Fig. 17.2 Inverse polynomials: (a) the Michaelis–Menten or Holling functional response equation; (b) the n-shaped curve $1/y = a + bx + c/x$.

The curve rises at an initial slope of $1/b$, and asymptotes at $1/a$ (Fig. 17.2). The general formulation of the inverse polynomial looks like this:

$$\frac{x}{y} = a + bx + cx^2 + dx^3 + \ldots$$

and can have a wide range of shapes, including asymmetrical humped curves (e.g. try plotting y against x for the function $1/y = x - 2 + 4/x$).

It is clear that a reciprocal link function is appropriate for these models and, indeed, the reciprocal is the canonical link function for the gamma error distribution.

17.2 An example of gamma errors and the reciprocal link

Suppose that we wanted to estimate the parameters of a functional response curve. This shows the number of prey eaten per predator per time period (y) as a function of the number of prey available per unit area (x). Initial inspection of the data suggests that the variance in prey capture increases with the mean. This, coupled with the clearly asymptotic shape of the relationship, indicates that gamma errors with a reciprocal link might be appropriate.

```
$units 30 $

$data x y $

$dinput 6 $

File name: glex14.dat

$plot y x $
```

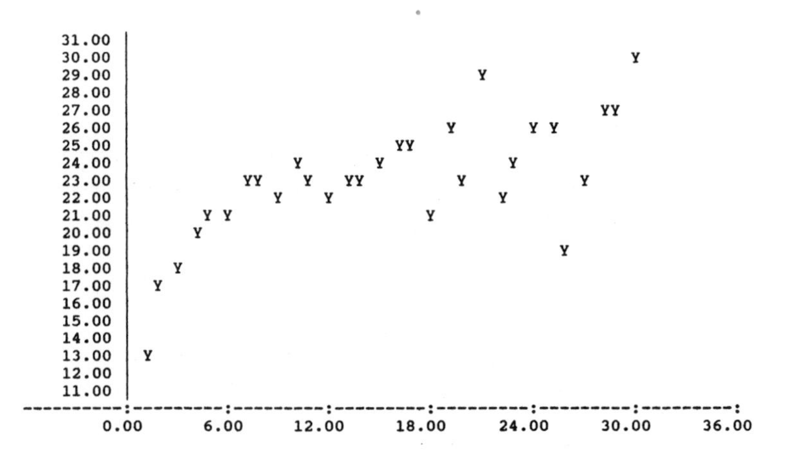

We begin by calculating the reciprocal of prey density ($1/x$), which will be the explanatory variable in the model:

```
$calc x1=1/x $

$yvar y $

$error g $

$link r $

$fit $

deviance = 0.79263 at cycle 3
      d.f. = 29

$fit+x1 $

deviance = 0.20955 (change = -0.5831) at cycle 3
      d.f. = 28       (change = -1)
```

This change in deviance is highly significant ($F = 0.5831/0.007484 = 77.9$), but this is hardly surprising given the trend in prey capture rate which we saw in the plot. Note that with gamma errors we use variance ratio tests and F tables in testing hypotheses (not χ^2 as with Poisson or binomial errors). The parameter estimates and their standard errors are:

```
$disp e $

   estimate  s.e.       parameter
1  0.03862   0.0008760  1
2  0.04405   0.005731   X1
scale parameter taken as 0.007484
```

which gives as the maximum likelihood estimates of the functional response curve:

$$y = \frac{x}{0.04405 + 0.03862x}$$

We should see how well these parameters describe the data, and then inspect the standardized residuals to ensure that the systematic trend in the residuals in original data has been dealt with by the gamma error distribution.

```
$plot y %fv x'*+' $
```

The fit is quite satisfactory, considering the wide scatter in prey capture rates at high prey densities. The standardized residuals show no obvious pattern:

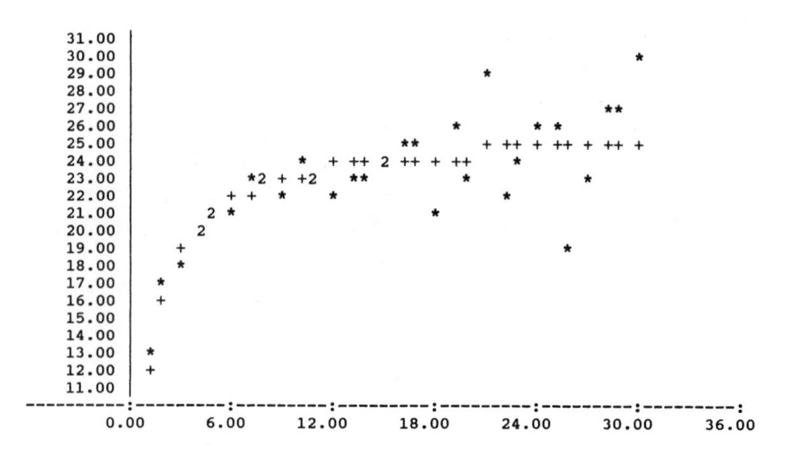


```
$disp r $
```

unit	observed	fitted	residual
1	12.52	12.10	0.035
2	16.76	16.49	0.016
3	18.43	18.76	−0.018
4	19.65	20.15	−0.025
5	20.54	21.09	−0.026
6	21.04	21.76	−0.033
7	23.00	22.27	0.033
8	22.50	22.67	−0.007
9	21.58	22.98	−0.061
10	23.66	23.25	0.018
11	23.19	23.46	−0.012
12	22.42	23.65	−0.052
13	23.23	23.81	−0.024
14	23.23	23.95	−0.030
15	23.83	24.07	−0.010
16	25.42	24.17	0.052
17	25.31	24.27	0.043
18	20.85	24.35	−0.144
19	26.22	24.43	0.073
20	23.48	24.50	−0.042
21	28.60	24.56	0.164
22	21.62	24.62	−0.122
23	23.96	24.67	−0.029
24	26.49	24.72	0.072
25	26.31	24.77	0.062
26	19.15	24.81	−0.228
27	23.27	24.85	−0.063
28	26.91	24.88	0.082
29	26.99	24.92	0.083
30	29.73	24.95	0.192

As an exercise, you might use the mcheck macro on the model in order to see whether we are justified in accepting this as a reasonable description of the data. •

17.3 Yield–density relationships

An important use of gamma errors is in the analysis of yield–density experiments (Connolly, 1987). Here, the dry weight at maturity of individual plants declines as density is experimentally increased, because of intraspecific competition for light and/or soil resources. The variance in yield is greater when mean plant size is greater, so the assumption of constant normal errors is not appropriate. Both these features of the data can be dealt with by specifying gamma errors with a reciprocal link. In the simplest case, with one plant species, the model is:

$$\frac{1}{Y} = a + b.D$$

where Y is the mean yield per plant (dry weight at maturity) and D is the planting density (plants per square metre).

An important class of experiments involves the analysis of patterns of yields in species mixtures. For example, it may be possible, by a judicious mix of plant growth forms and growth phenologies, to obtain a higher yield from a mixture of two species than from a monoculture of the higher yielding of the two species. The expectation would be that over-yielding is not possible, because in the mixture you are replacing an individual of the higher yielding species with an individual that will give a lower yield. It is possible, however, that biological interactions between the species might mean that the yield per individual was greater in the mixture than in the monoculture for one or both of the species. This might result from beneficial effects on pests or pollinators, enhanced nutrient uptake, nitrogen fixation by one of the species, or weed suppression. In these multispecies experiments, the model becomes:

$$\frac{1}{Y_i} = a_i + b_i D_1 + c_i D_2$$

Suppose we have an experiment with a cereal and a legume, each grown at a range of densities. If D_1 is the density of wheat and D_2 is the density of beans, then the ratio c_i/b_i describes the relative effect of individuals of beans and wheat in depressing the yield of plant species i. It may sometimes be necessary to include terms for D_1^2 and/or D_2^2 in the model, but we shall work with the simpler model without quadratic terms.

The same density trial was repeated in seven blocks. We measured the mean weight of individual plants of wheat and of beans that were grown in monoculture and in various mixtures, and the aim was to determine the relative impact of bean density on individual plant yields. There are seven density combinations for each crop (49 yields in all) and the data for wheat and beans are stored in separate files. We begin with the wheat yield data: the yield per plant of wheat (yw) and the densities of wheat (dw) and beans (db) are stored in the three columns of glex21.dat:

```
$units 49 $

$data yw dw db $

$dinput 6 $

File name: glex21.dat
```

To see the effect of wheat density on wheat yield we plot one against the other:

```
$plot yw dw $
```

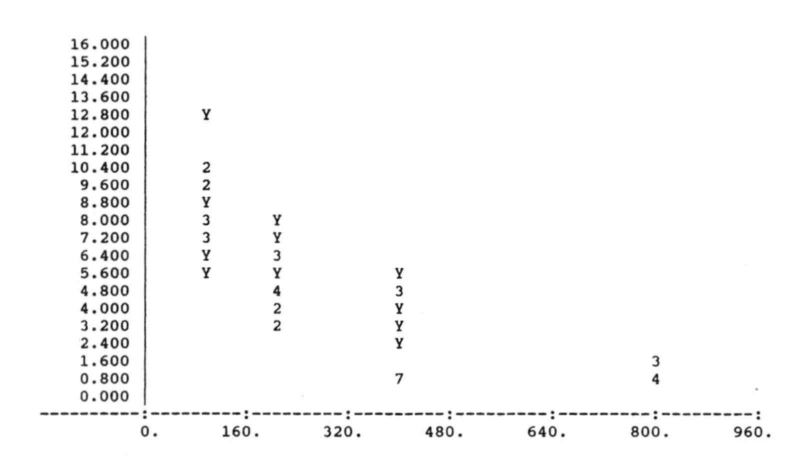

There is a good deal of scatter, but a hyperbolic relationship between wheat yield and wheat density looks reasonable. You can do a similar plot for the effect of bean density on wheat yield (see p. 309):

```
$plot yw db $
```

Note the degree of scatter along the *y*-axis at zero bean density; this shows the variation in wheat plant size with density in the wheat monocultures (i.e. at zero bean densities). As with wheat density, the mean size of wheat plants declines hyperbolically as the density of bean plants is increased. The big difference in the scales of the *x*-axes reflects the fact that individual bean plants are substantially bigger than individual wheat plants.

The analysis is straightforward. The response variable is wheat yield and we have two explanatory variables, wheat density and bean density. We declare the error structure to be gamma and the link to be reciprocal:

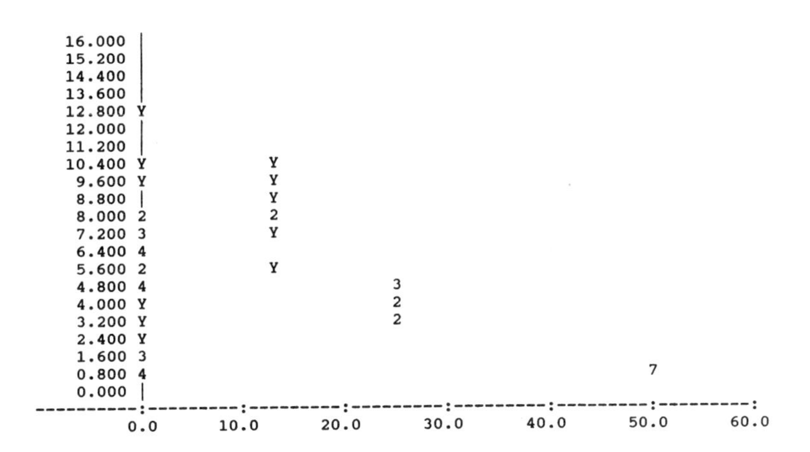


```
$yvar yw $

$error g $

$link r $

$fit $

deviance = 34.575 at cycle 5
      d.f. = 48
```

The total deviance is 34.575, and we can add the two densities to the model:

```
$fit+dw+db $

deviance = 9.487 (change = −25.09) at cycle 4
      d.f. = 46    (change = −2)
```

The deviance declines by 25.09, and

```
$calc 25.09/34.58 $

0.7256
```

This shows that the model explains 73% of the deviance. The parameter estimates are obtained in the usual way:

```
$disp e $

     estimate      s.e.         parameter
1   −0.005048    0.01997       1
2    0.0008863   0.0001213    DW
3    0.005731    0.001390     DB
scale parameter taken as 0.2062
```

It is evident that both wheat and bean densities need to be retained in the model (removing either one causes a massive increase in deviance).

In order to see the relative impact of the two species on mean wheat yield we simply calculate the ratio of the parameter estimates:

```
$calc 0.005731/0.0008863 $

6.466
```

This suggests that a bean plant is equivalent to about 6.5 wheat plants in terms of its effect on reducing mean wheat plant size. Given their relative sizes when freely grown, this seems perfectly reasonable. You can repeat the analysis on the bean plant data in file glex22.dat. Note that the data are in the form:

```
$data yb db dw $
```

The fit is less good, but the effect of a bean plant on mean bean weight is equivalent to 12.17 wheat plants. This is nearly double the relative impact of a bean plant on wheat plant size.

As an exercise, you might like to read both files into GLIM at the same time, calculate biomass (mean size × density) for each species, and then investigate whether total biomass of wheat plus beans is greater than the highest biomass of the monocultures (i.e. whether there is any evidence for overyielding).

17.4 Gamma errors and the log link

When the variance increases with the square of the mean, but the model structure is thought to be multiplicative rather than additive, then gamma errors with the log link are appropriate. Suppose that we wish to understand how the number of female bark beetles attracted to an experimentally induced wound on a pine tree is influenced by the rate of resin production ($ml\,h^{-1}$) and the concentration of volatile oils ($\mu g\,ml^{-1}$) in a standard sample of resin. It is plausible to assume, since beetles are attracted to the volatiles, that a multiplicative model might be suitable. Once some females have arrived at a wound, pheromones produced by the insects themselves might increase the subsequent rate of arrival of other females (we shall discuss this later).

```
$units 30 $

$data y vol conc $

$dinput 6 $

File name? glex40.dat
```

Plot beetle numbers (y) against resin volume then against the concentration of volatiles in the resin; you will see a good relationship for volume and a positive association, but much greater scatter, for concentration. Next, plot resin volume against the concentration of volatiles to see how (and whether) the explanatory variables are correlated. Now, declare beetle numbers as the response variable, set the errors to gamma and the link to log:

```
$yvar y $

$error g $

$link l $
```

Next, fit both volume and resin concentration to the model, and inspect their parameter estimates and standard errors:

```
$fit $

deviance = 19.958 at cycle 4
      d.f. = 29

$fit +vol+conc $

deviance = 3.3353 (change = −16.62) at cycle 4
      d.f. = 27      (change = −2)

$disp e $

   estimate  s.e.      parameter
1  0.1160    0.3379    1
2  0.8409    0.07659   VOL
3  0.1078    0.03282   CONC
scale parameter taken as 0.1235
```

Now remove each term from the model and assess its significance using an *F*-test.

```
$fit −conc $

deviance = 4.5860 (change = +1.251) at cycle 3
      d.f. = 28      (change = +1)
```

This gives an *F*-ratio of more than 10 (1.251/0.1235) and is clearly significant. Add it back into the model and then assess whether there are patterns in the residuals using mcheck.

```
$fit + conc $

$input 7 $

File name? mcheck.mac
```

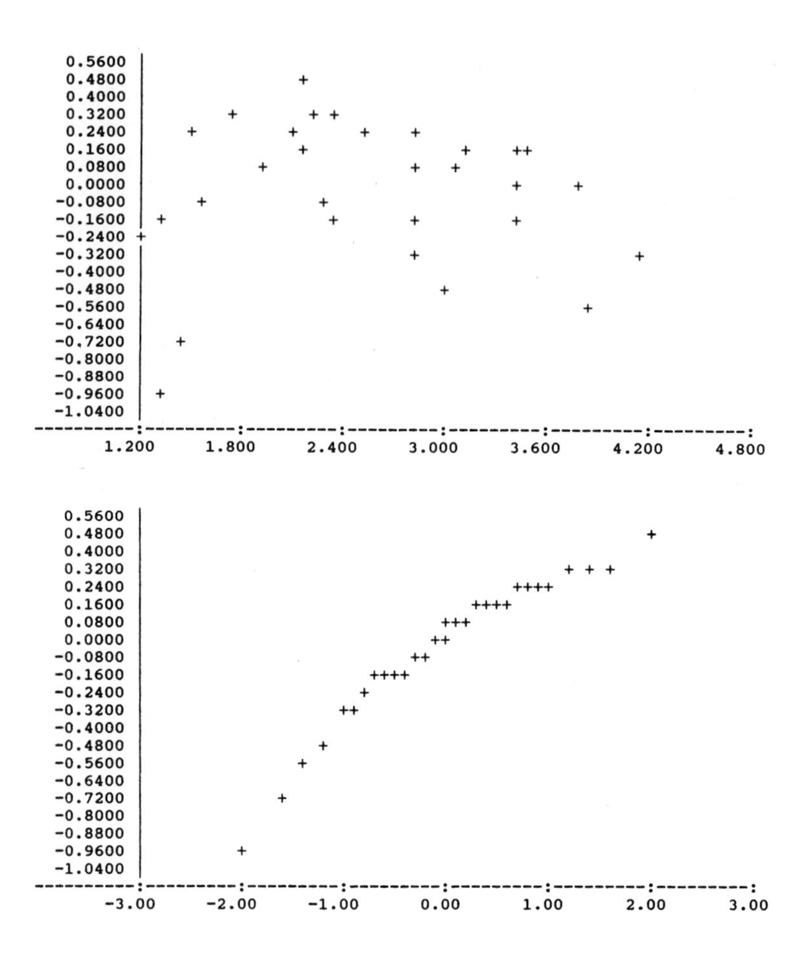

You will see that there is a distinct n-shaped pattern in the plot of standardized residuals against log fitted values, and the probability plot is obviously curved. Perhaps these problems would be alleviated by transformation of the explanatory variables? Since we are currently using the log link, it may be that the identity link is more appropriate.

```
$link i $

[w] -- model changed

$fit:+vol+conc$disp e $

deviance = 19.96 at cycle 3
      d.f. = 29

deviance = 2.2092 (change = −17.75) at cycle 4
      d.f. = 27       (change = −2)
```

```
    estimate  s.e.     parameter
1   -4.743    1.965    1
2   10.42     0.6819   VOL
3    0.4674   0.1950   CONC
scale parameter taken as 0.08182

$input 8 $

File name? mcheck.mac
```

The residual plots are much improved; the probability plot is linear and the n-shaped residuals are improved (although they are now somewhat fan-shaped). The residual deviance is somewhat lower and the residuals are better behaved with the identity link than the log link. We therefore accept the model with gamma errors and the identity link as minimal adequate, and we reject our initial hypothesis that the properties of resin volume and concentration act in a multiplicative way in attracting beetles.

What about the pheromone attraction hypothesis? It should be obvious that we cannot use the number of beetles present on the tree in order to explain the number of beetles arriving (this is equivalent to using the response variable as an explanatory variable in the same model). The question we want to answer is whether the number of females already on a tree influences the subsequent rate of arrival of other beetles. We can say nothing about this from the present data, because all we have is a single number of beetles determined at the end of the experimental period. Any response to pheromones of the kind hypothesized is completely confounded with the differential attractiveness of tree resins.

It would take an extra manipulative experiment to test the pheromone hypotheses. We could cage different numbers of females (B) on a range of trees (T) known in advance to differ in their resin production. Then we could investigate the model:

$$\log(\text{beetles attracted}) = \beta_0 + \beta_1 T + \beta_2 B + \beta_3 T.B$$

to look for tree effects (differences in resin production), beetle density effects (the pheromone hypothesis) and interactions between trees and beetle density (whether pheromones were more attractive on some trees than on others). ●

CHAPTER 18

Survival analysis

A great many ecological studies deal with deaths: the numbers of deaths, the timing of death, and the risks of death to which different classes of organisms are exposed (categorized by age, size, local microhabitat, predation risk, disease susceptibility and so on). The analysis of survival data is a major focus of the statistics business (see Cox & Oakes, 1984), and here we shall look at some of the tests that GLIM can do. The main theme of this chapter is the analysis of data that take the form of measurements of the *time to death*. Up to now, we have dealt with mortality data by considering the *proportion* of animals that were dead *at a given time*. In this chapter each individual is followed until it dies, then the time of death is recorded (this will be the response variable). Individuals that survive to the end of the experiment will die at an unknown time in the future; they are said to be *censored* (see below).

With data on time to death, the most important decision to be made concerns the error distribution. We are unlikely to know much about the error distribution in advance of the study, except that it will certainly not be normal. In GLIM we are offered several choices of error distribution for the analysis of survival data:

1 gamma;
2 exponential;
3 piece-wise exponential;
4 extreme value;
5 log-logistic;
6 log normal;
7 Weibull;

and, in practice, it is often difficult to choose between them. In general, the best solution is to try several distributions and to pick the error structure that produces the minimum error deviance.

18.1 Ecological background

Since everything dies eventually, it is often not possible to analyse the results of survival experiments in terms of the proportion that were killed (as we did in Chapter 15); in due course, they *all* die. Look at Fig. 18.1. It is clear that the two treatments caused different patterns of mortality, but both start out with 100% survival and both end up with zero. We could pick some arbitrary point in the middle of the distribution at which to compare the percentage survival, but this may be difficult in practice,

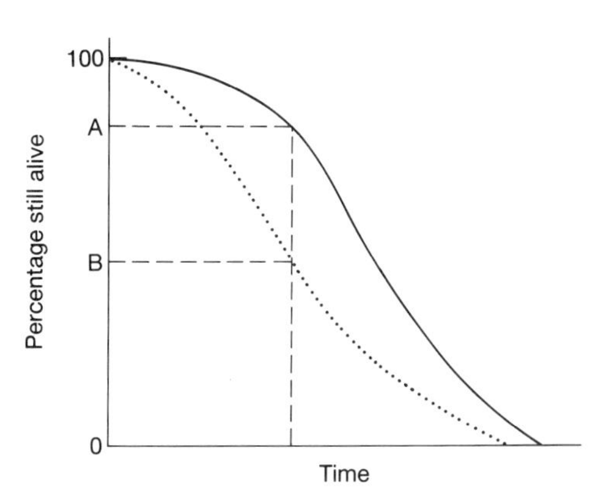

Fig. 18.1 Survival curves for two cohorts of individuals. Rather than selecting a subjective time at which to compare percentage mortality between the two treatments and using binomial errors, it is better to analyse the age at death of each individual, using an appropriate error distribution like the exponential or Weibull.

because one or both of the treatments might have few observations at the same location. Also, the choice of when to measure the difference is entirely subjective and hence open to bias. It is much better to use GLIM's powerful facilities for the analysis of survival data.

Ecologists use three interchangeable concepts when dealing with data on the timing of death:

1 survivorship;
2 age at death;
3 instantaneous risk of death.

Figure 18.2 shows graphs of these three mathematical ideas. The survivorship curve plots the natural log of the proportion of a cohort of individuals that started out at time 0 and is still alive at time t. For the so-called Deevey Type II survivorship curve, we have a linear decline in log numbers with time. This means that a constant proportion of the individuals alive at the beginning of a time interval will die during that time interval (i.e. the proportion dying is density independent and constant for all ages). When the death rate is highest for the younger age classes we get a steeply descending Type III survivorship curve, and when it is the oldest animals that have the highest risk of death, we obtain the Type I curve (characteristic of human populations in affluent societies where there is low infant mortality).

The age at death shows the proportion of a cohort dying at a given age. For the Type I curve this is a negative exponential. Because the fraction of individuals dying is constant with age, the number dying declines exponentially as the number of survivors (the number of individuals at risk of death) declines exponentially with the passage of time.

The instantaneous death rate is the derivative of the survivorship curve; it is the instantaneous rate of change in the log of the number of survivors per unit time (it is the slope of the line in Fig. 18.2a). Thus, for the Type II survivorship we have a horizontal line, because the risk of

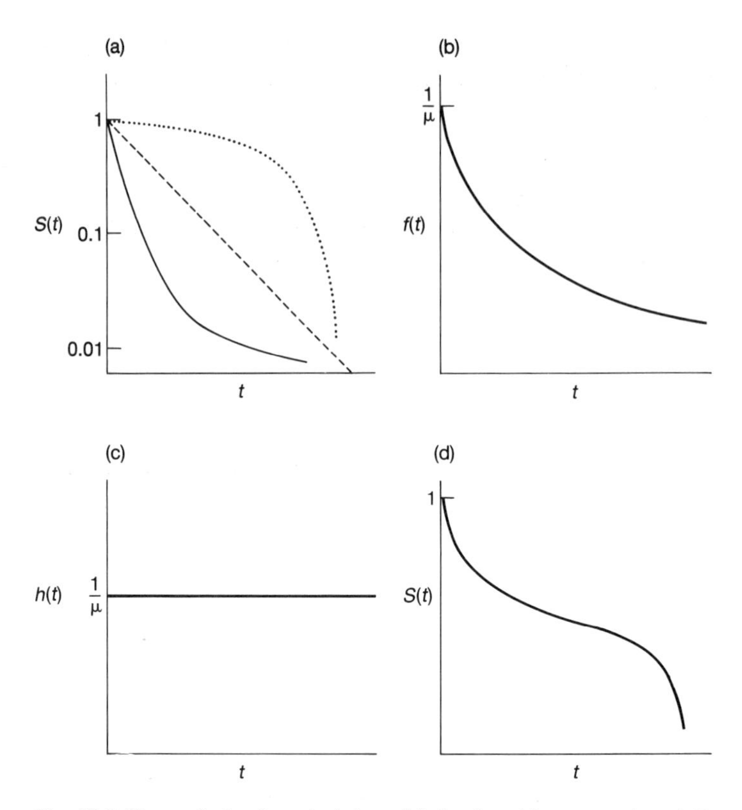

Fig. 18.2 The analysis of survival data. (a) Survivorship curves: (····) Type I = mortality concentrated in the oldest age classes; (– – –) Type II = constant risk of death; (——) Type III = mortality concentrated in the youngest age classes. Note the log scale. (b) Age at death for a Type II survivorship curve; the numbers dying per unit time decline exponentially as the size of the cohort declines. (c) The hazard function or age-specific instantaneous death rate is constant for the Type II curve. (d) When age-specific survival is low for both the youngest and the oldest individuals, we get a survivorship curve like an S on its side.

death is constant with age. Although this sounds highly unrealistic, it is a remarkably robust assumption in certain ecological applications. It also has the substantial advantage of parsimony. In many cases, however, it is clear that the risk of death changes substantially with the age of the individuals, and we need to be able to take this into account in carrying out our statistical analysis. In the case of Type I survivorship, the risk of death declines with age, while for Type III survivorship (as in humans) the risk of death increases with age.

18.2 The exponential distribution

This is a one-parameter distribution in which the hazard function is independent of age (i.e. it describes a Type II survivorship curve). The exponential is a special case of the gamma distribution in which the shape parameter α is equal to 1 (see Section 17.1). We are interested in estimating the value of μ, the mean age at death.

18.2.1 Density function

The density function is the probability of a death occurring in the small interval of time between t and $t + \delta t$; a plot of the number dying in the interval around time t as a function of t (i.e. the proportion of the original cohort dying at a given age) declines exponentially:

$$f(t) = \frac{e^{-t/\mu}}{\mu}$$

where both μ and $t > 0$. Note that the density function has an intercept of $1/\mu$ (remember that e^0 is 1). The numbers dying decline exponentially with time and a fraction $1/\mu$ of the survivors dies during each subsequent time interval.

18.2.2 Survivor function

This shows the proportion of individuals from the initial cohort still alive at time t:

$$S(t) = e^{-t/\mu}$$

The survivor function has an intercept of 1 (i.e. all the cohort is alive at time 0), and shows the probability of surviving longer than t.

18.2.3 Hazard function

This is the statistician's equivalent of the ecologist's *instantaneous death rate*. It is defined as the ratio between the density function and the survivor function, and is the conditional density function at time t, given survival up to time t. In the case of Type II curves this has an extremely simple form:

$$h(t) = \frac{f(t)}{S(t)} = \frac{e^{-t/\mu}}{\mu e^{-t/\mu}} = \frac{1}{\mu}$$

because the exponential terms cancel out. Thus, with the exponential distribution the hazard is constant with age; it is the reciprocal of the mean time to death, and vice versa. For example, if the mean time to death is 3.8 weeks, then the hazard is 0.2632; if the hazard were to increase to 0.32, then the mean time of death would decline to 3.125 weeks.

Of course, the death rate may not be a linear function of age. For example, the death rate may be high for very young as well as for very old animals, in which case the survivorship curve is like an S-shape on its side (Fig. 18.2d).

18.3 Survival analysis in GLIM

There are three cases that concern us here:

1 constant hazard and no censoring;
2 constant hazard with censoring;
3 age-specific hazard, with or without censoring.

The first case is dealt with in GLIM by specifying exponential errors. This involves using gamma errors with the scale factor fixed at 1:

```
$error g $

$scale 1 $
```

The second case involves the use of Poisson errors and a log link, with the censoring indicator as the response variable. The third case uses the Weibull distribution, in which the shape parameter α allows for the risk of death to increase ($\alpha > 1$) or decrease ($\alpha < 1$) with time.

18.4 An example of survival analysis

To see how the exponential distribution is used in modelling we take an example from plant ecology, in which individual seedlings were followed from germination until death. We have the times to death measured in weeks for two cohorts, each of 30 seedlings. The plants were germinated at two times, in early September (treatment 1) and mid October (treatment 2). We also have data on the size of the gap into which each seed was sown (a covariate x). The questions are these.

1 Is an exponential distribution suitable to describe these data?
2 Was survivorship different between the two planting dates?
3 Did gap size affect the time to death of a given seedling?

We begin by reading in the 60 pairs of data (survival time and relative gap size):

```
$units 60 $

$data y x $

$dinput 6 $

File name? glex7.dat
```

Next we create the factor levels for the two germination cohorts; the first 30 numbers relate to cohort 1 and the second 30 to cohort 2. We define a factor A with two levels to describe each group of seedlings:

```
$calc a=%gl(2,30) $

$factor a 2 $
```

With data on survival times and the exponential distribution there are three things to remember: (i) we use **gamma** errors; (ii) the **scale** parameter is set to 1; and (iii) we need to think about an appropriate link function. The **scale** parameter is set to 1 because the exponential is a one-parameter distribution, and we do not want GLIM to estimate a scale parameter from the data (as it would otherwise do with gamma errors; see Section 17.1). The choice of link function depends upon the data. In this example it is natural to choose the reciprocal link (the default for gamma errors) because this estimates the expected hazard directly (the reciprocal of the mean time to death), but in other cases the log or identity links may be more appropriate. If there is a risk of negative fitted values, then the log link is to be preferred, since it is obvious that we cannot have negative times to death. We begin as follows:

```
$yvar y $

$error g $

$scale 1 $

$fit $

scaled deviance = 45.239 at cycle 5
             d.f. = 59

$disp e $

    estimate   s.e.      parameter
1   0.1863     0.02404   1
scale parameter taken as 1.000
```

This means that GLIM estimates the hazard function $h(t)$ to be constant at 0.1863. This implies a mean time to death of 5.368 weeks (recall that $\mu = 1/h(t)$ for the exponential distribution):

```
$calc 1/0.1863 $

5.368
```

and a mean weekly survival rate of 83% for all seedlings taken together (the survivor function $S(1) = \exp(-h)$):

```
$calc %exp(-0.1863) $

0.8300
```

Notice that the ratio of scaled deviance to degrees of freedom is not too bad ($45.239/59 = 0.767$) and that the scale parameter is constrained to be 1. Let us see whether planting time had any significant effect on time to death:

```
$fit +a $
```

```
scaled deviance = 44.783 (change = −0.46) at cycle 5
            d.f. = 58     (change = −1)

$disp e $

     estimate    s.e.      parameter
1    0.2041      0.03724   1
2   −0.03265     0.04863   A(2)
scale parameter taken as 1.000
```

Evidently not (the change in scaled deviance is much less than the χ^2 value in tables). The two cohorts seem to have the same average time to death. We remove planting date from the model and add gap size:

```
$fit −a+x $

scaled deviance = 44.803 (change = +0.02) at cycle 5
            d.f. = 58     (change = 0)

$disp e $

   estimate  s.e.      parameter
1  0.1599    0.04476   1
2  0.05245   0.07921   X
scale parameter taken as 1.000
```

Again, there is no appreciable effect of gap size on survival for these seedlings. It is possible, however, that there was an interaction between planting date and gap size, and that the two effects cancelled out to give an insignificant main effect. For instance, bigger gaps may have been a disadvantage early in the season because they were drier, but an advantage late in the season because there was more light. To test this, we remove the overall regression on gap size and fit separate regression slopes for each cohort:

```
$fit −x + a.x $

scaled deviance = 44.036 (change = −0.77) at cycle 5
            d.f. = 57     (change = −1)

$disp e $

   estimate  s.e.      parameter
1  0.1606    0.04484   1
2  0.09147   0.09486   X.A(1)
3  0.01681   0.08585   X.A(2)
scale parameter taken as 1.000
```

There is no evidence of an interaction, and so we conclude that *the null model is minimal adequate for this experiment*. We refit the overall mean and inspect the standardized residuals:

```
$fit $

scaled deviance = 45.239 at cycle 5
           d.f. = 59

$disp r $
```

Only two of the 60 points have large residuals: unit 49 had 21 weeks to death ($r = 2.91$) and unit 58 had 16 weeks ($r = 1.981$). Neither of these is incompatible with a mean time to death of 5.367 weeks from such a large sample. •

18.5 Censoring

Censoring occurs when we do not know the time to death for all of the individuals. This comes about principally because some individuals outlive the experiment. We can say they survived for the duration of the study but we have no way of knowing at what age they will die. These individuals contribute something to our knowledge of the survivor function, but nothing to our knowledge of the age at death. Another reason for censoring occurs when individuals are lost from the study; they may be killed in accidents, they may emigrate, or they may lose their identity tags.

In general, then, our survival data may be a mixture of times at death and times after which we have no more information on the individual. We deal with this by setting up an extra vector called the *censoring indicator* to distinguish between the two kinds of numbers. If a time really is a time to death, then the censoring indicator takes the value 1. If a time is just the last time we saw an individual alive, then the censoring indicator is set to 0. Thus, if we had the time data T and censoring indicator W:

T	4	7	8	8	12	15	22
W	1	1	0	1	1	0	1

this would mean that all the data were genuine times at death except for two cases, one at time 8 and another at time 15, when animals were seen alive but never seen again.

With repeated sampling in survivorship studies, it is usual for the degree of censoring to decline as the study progresses. Early on, many of the individuals are alive at the end of each sampling interval, whereas few if any survive to the end of the last study period.

18.6 The likelihood function for censored data

A given individual contributes to the likelihood function depending upon whether it is alive or dead at time t. If it has died, then the censoring indicator is 1 and we learn more about $f(t)$; if it is still alive, then w_i is 0 and we learn only about $S(t)$:

$$L(\beta) = \prod_{i=1}^{n} [f(t_i)]^{w_i} [S(t_i)]^{1-w_i}$$

Now recall that the hazard function $h(t)$ is given by $f(t)/S(t)$. Thus we have $S(t)^w$ in the denominator and $S(t)^{1-w}$ in the numerator, so the likelihood function becomes:

$$L(\beta) = \prod_{i=1}^{n} [h(t_i)]^{w_i} S(t_i)$$

which involves data only from the hazard function of the uncensored individuals. If we replace $1/\mu_i$ by λ_i, then we can write the likelihood function for the exponential distribution as follows:

$$L(\beta) = \prod_{i=1}^{n} \lambda_i^{w_i} e^{-\lambda_i t_i}$$

For reasons that will become clear in a moment, it is convenient to multiply both the numerator and the denominator by $\Pi t_i^{w_i}$. This gives:

$$L(\beta) = \frac{\prod_{i=1}^{n} (\lambda_i t_i)^{w_i} e^{-\lambda_i t_i}}{\prod_{i=1}^{n} t_i^{w_i}}$$

Because the denominator is not a function of the estimated parameters β, it can be omitted from the likelihood formula, leaving only a term for the likelihood of a set of n observations w_i, having independent Poisson distributions, with means $\lambda_i t_i$, where w_i is either 0 or 1.

The model can be fitted to the hazard rate λ in one of two ways. Let θ_i represent the Poisson mean $\lambda_i t_i$. Using the *linear hazard model*:

$$\lambda_i = \beta' x_i$$

and

$$\theta_i = \beta'(t_i x_i)$$

The modelling proceeds as follows:

1 use the censoring indicator w_i as the response variable;
2 declare the error as Poisson;
3 declare the link function as the identity link;
4 multiply each of the explanatory variables, including the unit vector, by t_i;
5 fit the model as usual.

This is somewhat long-winded, and the fitting is easier if the *log linear hazard model is employed*, because:

$$-\log \mu_i = \log \lambda_i = \beta' x_i$$

and so

$$\log \theta_i = \log \lambda_i + \log t_i = \beta' x_i + \log t_i$$

This is easier to fit because we do not need to multiply through all the explanatory vectors by t_i. Instead, we use $\log t_i$ as an offset, and proceed as follows:

1 use the censoring indicator w_i as the response variable;
2 declare the error as Poisson;
3 declare the link function as the log link;
4 declare $\log t_i$ as an offset;
5 fit the model as usual.

This rather curious procedure of using the censoring indicator full of 0's and 1's as the response variable should become clearer with an example.

18.7 An example of censored survival data

When there is censoring, a simple technique is to use the censoring indicator w as the response variable, and the log of the time of death as an offset. The model is declared with Poisson errors and a log link as follows.

Suppose that we have two groups, each of 21 larvae of the same size and age. They are fed on leaves from two kinds of plants; the first group get leaves from plants that have been genetically engineered to express BT toxin in their foliage, while the second group get leaves from an otherwise identical, but non-transgenic, strain. The data consist of the time at death in days (when w = 1) or the time when the animal was last seen alive (w = 0). The analysis goes like this:

```
$units 42 $

$data t w $

$dinput 6 $

File name? glex34.dat
```

We begin by creating a factor to distinguish the two diets. The first 21 rows of data refer to larvae fed on transgenic leaves:

```
$calc a=%gl(2,21) $

$factor a 2 $
```

The arithmetic mean times at death for the two groups can be obtained either without weighting:

```
$tab the t mean for a $

        1      2
[ ]  8.67  17.10
```

or with the censored individuals weighted out (can you see why this is a bad idea?):

```
$tab the t mean for a with w $

        1      2
 [ ]  8.67  12.11
```

The survival time appears to be shorter for the caterpillars fed on BT-containing leaves. Before proceeding any further, we should look to see how many of the observations from each treatment were censored:

```
$tab for w $

        0.000  1.000
 [ ]   12.00  30.00
```

The whole data set contains 30 animals that died and 12 that were censored. Next:

```
$tab the w total for a $

         1       2
 [ ]  21.000  9.000
```

which shows that 21/21 of the BT animals died against only 9/21 from the non-transgenic group. All the censoring, therefore, was in the second (control) group. This explains the big difference between the arithmetic mean times at death for the weighted and non-weighted averages in the control group, above.

The survival analysis is very simple. The response variable is the censoring indicator with Poisson errors and log(t) as an offset:

```
$yvar w $

$calc lt=%log(t) $

$error p $

$link l $

$offset lt $

$fit :+a$disp e $
```

The introduction of the offset of log(time at death) introduces the variation to be explained by the model. The output is:

```
scaled deviance = 54.503 at cycle 5
           d.f. = 41

scaled deviance = 38.017 (change = -16.485) at cycle 4
           d.f. = 40     (change = -1)
```

```
   estimate  s.e.    parameter
1  -2.159    0.2179  1
2  -1.526    0.3958  A(2)
scale parameter taken as 1.000
```

The change in deviance is significant at $p < 0.005$, and the parameter estimates show that the hazard is significantly lower for the group of caterpillars fed on non-transgenic leaves. The BT toxin is evidently effective and caused a reduction in mean age at death from 39.85 days to 8.66 days (you should check how these figures were obtained). Notice that while the mean for the group with no censoring is virtually identical to the arithmetic mean obtained earlier, the mean age at death for the group with heavy censoring is much higher than the arithmetic mean value (17.1 days). Make sure you understand why this is so. •

18.8 Weibull distribution

The origin of the Weibull distribution is in *weakest link analysis*. If there are r links in a chain, and the strengths of each link Z_i are independently distributed $(0, \infty)$, then the distribution of weakest links $V = \min(Z_j)$ approaches the Weibull distribution as the number of links increases.

The Weibull is a two-parameter model that has the exponential distribution as a special case. Its value in ecology is that it allows for the death rate to increase or to decrease with age, so that all three kinds of survivorship curve can be analysed. The density, hazard and survival functions with $\lambda = \mu^{-\alpha}$ are:

$$f(t) = \alpha\lambda t^{\alpha-1}e^{-\lambda t^{\alpha}}$$
$$h(t) = \alpha\lambda t^{\alpha-1}$$
$$S(t) = e^{-\lambda t^{\alpha}}$$

The mean of the Weibull distribution is $\Gamma(1+\alpha^{-1})\mu$ and the parameter α describes the shape of the hazard function (the background to determining the likelihood equations is given by Aitkin *et al.*, 1989, pp. 281–3). For $\alpha = 1$ (the exponential distribution) the hazard is constant, while for $\alpha > 1$ the hazard increases with age and for $\alpha < 1$ the hazard decreases with age (Fig. 18.3).

To fit the Weibull function in GLIM we use the library macro weibull. The joint maximum likelihood estimation of the shape parameter α and the linear predictor is carried out by the method of *successive relaxation*. An exponential distribution ($\alpha = 1$) is fitted first of all, and the model is fitted with the log of the survival times as an offset (see Section 18.7). A new estimate of α is then obtained from a likelihood equation based on the fitted values, and this improved estimate of α is used to calculate a new offset, $\alpha \log t_i$. The model is then refit, and the processes repeated until convergence.

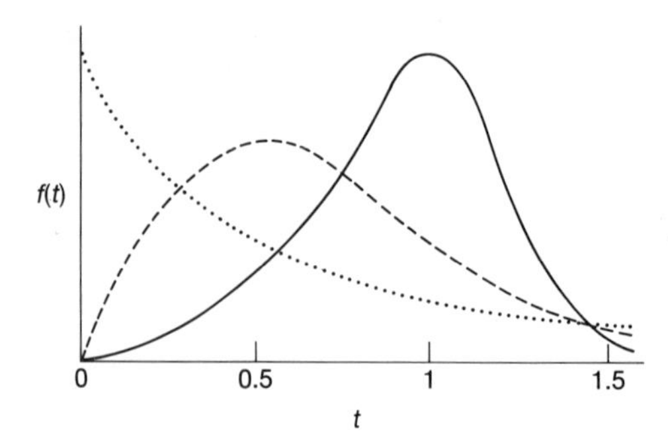

Fig. 18.3 Weibull distributions for different values of the shape parameter, α. $(\cdots\cdot)$ $\alpha=1$; $(---)$ $\alpha=2$; (——) $\alpha=3$. In all cases $\lambda=1$. For the exponential distribution $(\alpha=1)$ the mean $\mu=1/\lambda$.

The macro has three arguments: t, w and %s. The first variable contains the times to death. The second is the censoring indicator (1 if the organism was dead at the time, 0 if it was alive). The third is a scalar indicating whether a Weibull (1) or an exponential (0) fit is required. Use of the library macro must be preceded by a macro called model, which defines the model currently in force.

Because the Weibull, log-normal and log-logistic all have positive skewness, it will be difficult to discriminate between them with small samples. This is an important problem, because each distribution has differently shaped hazard functions, and it will be hard, therefore, to discriminate between different ecological assumptions about the age-specificity of death rates.

Parsimony requires that we fit the exponential rather than the Weibull unless the shape parameter α is significantly different from 1.

18.9 Age-specific death rate with censored observations

We end with an example from a study of mortality in 150 adult cockroaches. There were three experimental groups (G) and the animals were followed for 50 days. The groups were treated with three different insecticidal BT toxins added to their diet. The initial body mass of each insect (x) was recorded as a covariate. The day on which each animal died (t) was recorded, and animals that survived up to the 50th day were recorded as being censored (for them, the *censoring indicator* w = 0).

```
$units 150 $

$data t w x $

$dinput 6 $

File name? glex8.dat

$calc g=%gl(3,50) $

$factor g 3 $
```

We begin by plotting the age at death against initial body size, with plotting restricted (%re) to the non-censored animals:

```
$calc %re=w $

$plot t x '*o+' g $
```

```
    50.00 |
    47.50 |
    45.00 | 2    o                   +
    42.50 |
    40.00 | o                          o
    37.50 o   *
    35.00 |   ** *
    32.50 |
    30.00 |  +
    27.50 |
    25.00 +     2
    22.50 |      *    +
    20.00 o++  * *       o+2
    17.50 2    *           o     o
    15.00 |                     +
    12.50 |*     2    o      +
    10.00 |+2++***  +        o +
     7.50 |     3+  2                + o
     5.00 | + 2++ o +2   o   2     +        o   2
     2.50 oo+ o232o+2 o    o      +               o
     0.00 2650 72222   o o++ o+     o        +      2
----------:---------:---------:---------:---------:---------:---------:
         0.0      10.0      20.0      30.0      40.0      50.0      60.0
```

This is an extremely odd-looking graph, but it does suggest that animals with higher initial body mass were somewhat more likely to die at an early age (there are no points in the top right-hand corner of the graph). We shall test the significance of this by fitting the full model G*X using the library macro weibull. First, the preliminaries:

```
$calc %s=1 $

$input %plc weibull $

$macro model g*x $endmac $
```

We specify the model to be fitted in the macro called model, having read the library macro called weibull from channel %plc (the primary library channel). We then run the macro, specifying its three arguments in the correct sequence: the time to death (t), the censoring indicator (w, which is 1 for dead animals and 0 for live ones), and the scalar %s (we set this to 1 if we want to fit the Weibull rather than the exponential).

```
$use weibull t w %s $
```

GLIM then works (slowly on older machines) through the iterative calculations, refining the value of the shape parameter at each step. The output from the macro is as follows:

```
-- Model is g*x
```

Exponential fit

Deviance	shape parameter	df
961.18	1.0000	144

Weibull fit

941.51	0.81535	143
939.38	0.75527	143
939.23	0.73950	143
939.23	0.73577	143

Note that fitting a shape parameter using the Weibull has reduced the deviance by 21.95 compared with the exponential. The algorithm has converged on an estimate for α of 0.736, which means that the risk of death from the BT toxin declines with age. This could mean that the cockroaches were a heterogeneous bunch and the susceptible ones all died early on in the experiment, or it could mean that the potency of the toxin, or the susceptibility of the insects, declines with time for some reason or another. It is clear that the insects are not just dying from old age.

When α has been estimated, the macro prints the parameter estimates and their standard errors with the warning that the latter are underestimated. This is because GLIM estimates the parameters of the linear predictor as if the value of α were known; of course α is not known, and this inflates the standard errors.

-- Standard errors of estimates given below are underestimated

	estimate	s.e.	parameter
1	−2.906	0.3893	1
2	0.8332	0.4537	G(2)
3	1.459	0.4419	G(3)
4	0.07155	0.06669	X
5	−0.06071	0.06816	G(2).X
6	−0.06846	0.06822	G(3).X

scale parameter taken as 1.000

It is clear from the print-out that, contrary to our initial impression from the graph we plotted, initial body weight has no effect on time to death for any of the three BT toxins (remember that the interaction terms here are differences between slopes; see Section 9.3). We repeat the fitting process, therefore, using the simplified model containing only the three toxin treatments.

```
$macro model g $endmac $

$use weibull t w %s $
```

Again, after reassessing its estimate of α, GLIM will print:

```
-- Model is g
Weibull fit

Deviance  shape      df
          parameter
941.02    0.73577    146
941.01    0.73144    146

-- Standard errors of estimates given below are underestimated

     estimate   s.e.     parameter
1   -2.530      0.1666   1
2    0.6008     0.2247   G(2)
3    1.126      0.2185   G(3)
scale parameter taken as 1.000
```

Evidently the mean time to death is significantly different on each of the three treatments. It would be wrong to estimate the mean time to death by simply writing:

```
$tab the t mean for g $

       1        2        3
[ ]  23.080   14.420   8.020
```

because several of the animals did not die during the experiment. The mean age at death of the animals that did die can be obtained using the censoring indicator, w, as a weight:

```
$tab the t mean for g with w $

       1        2       3
[ ]  12.611   9.568   8.020
```

Note that the estimated mean time of death is substantially reduced by weighting in group 1 where many of the animals survived more than 50 days, but is unaffected in group 3 where all the animals died during the experiment.

To obtain a better estimate of the mean age at death, we use the parameters estimated from the Weibull model along with the estimate of the shape parameter. There are three steps:

1 work out the antilogs for each parameter;
2 raise the antilogs to the power $1/\alpha$ to get the hazard;
3 work out the reciprocal of the hazard to get mean time to death, μ.
For the intercept (G(1)), we have a parameter estimate of -2.53. This gives an antilog of 0.07966 and a hazard of $0.07966{**}(1/0.73144) = 0.03146$. The mean age of death is the reciprocal of the hazard, $1/0.03146 = 31.78$. For the second group, we have a linear predictor of $(-2.53+$

0.6008 = −1.9292) and an antilog of 0.14526. Thus the hazard is 0.07153 and the mean age at death is 13.98. The prediction for the third group is (−2.53 + 1.126 = −1.404), giving an antilog of 0.2456, so the hazard is 0.14668 and the mean age at death is 6.82 days.

Notice how the heavy censoring has caused the estimated mean age at death in the control group to be increased from 23.08 to 31.78 days (type **$hist** t to get a visual impression of the extent to which censoring has distorted the density function). Also, note how the curvature introduced by $\alpha < 1$ has moved the estimated mean in the third group (where there was no censoring) away from its true mean value (6.82 rather than 8.02 days). This probably means that a model with different values of α for each treatment might be preferable. This example highlights the folly of attempting to estimate the mean age at death from arithmetic means when there is substantial censoring. It also illustrates the magnitude of the mistake involved in ignoring the censored individuals and estimating mean age of death from the ages of those individuals that did die (i.e. using the weighted means gave an estimate of 12.61 days for the control group compared with 31.79 when the predicted ages of death of the censored individuals were taken into account).

18.10 Analysing the censoring indicator with a known shape parameter

In the unlikely event that we knew the value of α in advance, then we could fit the log-linear hazard model with w as the response variable and with α log t as an offset. Let us say we knew that α was 0.731 (as determined above), then instead of using log(t) as the offset, we would use αlog(t). The deviance is different (because the offset has been changed), but the parameter estimates and their standard errors are exactly the same as were obtained with the Weibull (check the **disp e** table below with the earlier result):

```
$yvar w $

$error p $

$link l $

$calc off=0.736*%log(t) $

$offset off $

$fit $

scaled deviance = 229.08 at cycle 5
           d.f. = 149
```

```
$fit +g*x$disp e $

scaled deviance = 199.99 (change = -29.09) at cycle 5
            d.f. = 144    (change = -5)

    estimate    s.e.      parameter
1  -2.907     0.3893     1
2   0.8333    0.4537     G(2)
3   1.459     0.4419     G(3)
4   0.07156   0.06669    X
5  -0.06071   0.06816    G(2).X
6  -0.06847   0.06822    G(3).X
scale parameter taken as 1.000
```

The problem is that changing α changes the deviance, so the models cannot be directly compared. Look what happens if we guess a value of $\alpha = 0.5$:

```
$calc off=0.5*%log(t) $

[w] -- change to data affects model

$fit :+g*x$disp e $

scaled deviance = 142.37 at cycle 4
            d.f. = 149

scaled deviance = 124.51 (change = -17.86) at cycle 4
            d.f. = 144    (change = -5)

    estimate    s.e.      parameter
1  -2.072     0.3910     1
2   0.7402    0.4543     G(2)
3   1.155     0.4425     G(3)
4   0.06074   0.06718    X
5  -0.05448   0.06861    G(2).X
6  -0.05730   0.06864    G(3).X
scale parameter taken as 1.000
```

The parameter estimates differ by more than 20% in their back-transformed means for G(2), so it is clear that guessing the value of α is not a sensible option. On the other hand, once α has been estimated using the Weibull distribution macro, the structure described here, employing Poisson errors with an offset of $\alpha \ln(t)$, makes model simplification and error checking much more straightforward and eliminates the need for repeated use of the macro. You can also try different values for α until the plots produced by kaplan.mac are optimized (see Exercise 18.5).

CHAPTER 19

Macros and GLIM programming

GLIM is a simple yet powerful programming language. As you gain in experience with the GLIM directives, you are likely to make more and more use of GLIM programs. The best way to get started is to look at macros that other people have written, and the present chapter provides a range of examples. These have been chosen to give a feel for the variety of things that can be done with macros and to provide some useful programming tips. They are not supposed to be polished or professional software; they are not idiot-proof and they have no supporting documentation. Nevertheless, they will enable you to see how macros are constructed, and how some of the fundamental computing principles such as loops and conditional branching are handled by GLIM. The code for the macros is stored on the disk supplied with this book in files whose names end with .MAC.

19.1 GLIM programs

The way to become familiar with the process of programming in GLIM is to start writing your GLIM code in a word-processor, instead of typing the directives straight into GLIM at the '?' prompt. This has the major advantages that you can use the word-processor's full-screen editing facilities, and can run the same job again and again with minor changes, without having to type in all the preliminary material every time (variable declaration, data input, factor level generation, etc.). The ASCII file containing both the code and the data is read into GLIM using **input** (instead of writing the code inside GLIM and using **dinput** to read the data from an external file). Reading a program file with the **input** directive causes any executable statements in the file to be executed at once, so you just sit back and wait for the answer to appear on the screen.

19.2 Rules for writing macros

1 Macros begin with a **macro** directive and end with **endmac**.
2 There must be a space between the end of the macro name and the $ sign at the end of the directive.
3 Macros are executed by the **use** directive (amongst other means).
4 Parameters passed into the macro by listing their names after the macro name in the **use** directive are called *arguments*; macros can have any number of arguments between 0 and 9.
5 The vector names %1, %2, ..., %9 can be used inside the macro,

and will take the values allocated during the **use** directive to the 1st, 2nd, . . . , 9th argument.

6 Any (or none) of the arguments need be provided in the list of arguments in each **use** directive; missing arguments are denoted by *.

7 Employ **switch** for conditional branching.

8 Employ **while** for loops and repeated execution of code.

In writing your own macros, you have access to a number of vectors and scalars that contain important information on the response variable, the error structure, the current model and its fitted values. These structures are listed and explained in Table 19.1.

Table 19.1 System vectors and system scalars that contain useful information for writing general post-model-fitting macros. They allow a macro to recognize the current error distribution and link function, and to recognize how many (and which) arguments of a macro have been allocated values

Identifier	Meaning
1 Scalars	
%a1 ⋮ %a9	Set to 1 (true) if the first argument of a macro has been set, and to 0 (false) if the first argument has not been set. And so on, up to %a9 for the ninth argument of a macro
%df	Current error degrees of freedom
%dv	Current (scaled) deviance
%err	Is set to a value reflecting the current error structure: 1 for normal, 2 for Poisson, 3 for binomial, 4 for gamma and 9 for own model
%link	Is set to a value reflecting the current link statement: 1 for identity, 2 for log, 3 for logit, 4 for reciprocal, 5 for square root, 6 for probit, 7 for complementary log-log, 8 for exponent and 9 for own model
%ml	The number of elements in the lower triangle of the variance/covariance matrix under the current model
%nu	Contains the number of rows in the response variable as defined in the current **units** directive
%pl	Is the length of the parameter list for the current model. After a linear regression it would take the value 2; parameter 1 is the intercept and parameter 2 is the slope. After a one-way ANOVA with five levels of one factor, it would be equal to 5. Aliased parameters are not included in %pl
%pwf	Takes the value 1 when prior weights have been set by the **weight** directive
%s1 to %s3	Contain the values of the three seeds for the random number generator
%sc	Is the scale parameter; this can be changed by the **scale** directive
%x2	Pearson's χ^2 statistic
%zl to %z9	Temporary scalars for use in macros; do not use these names elsewhere

continued on p. 334

Table 19.1 *Continued*

Identifier	Meaning
2 Vectors	
%bd	Binomial denominator. When binomial errors are declared, this contains the value of the vector listed in the **error b** directive
%di	The deviance increment contributed by each observation
%dr	The derivative of the linear predictor with respect to its mean value. A statement will need to be provided to calculate this quantity in any user-defined model
%fv	Fitted values. This is one of the most important system scalars and is used routinely following a model-fit to investigate the behaviour of the residuals (%yv−%fv)
%lp	Linear predictor. In a non-iterative model this is the same as %fv, but in an iterative model it takes the values assigned by the fit directive
%pe	Parameter estimates from the current model, stored in a vector of length %pl. Thus, in a simple linear regression %pe(1) would contain the value of the intercept and %pe(2) the slope
%re	Restricts plotting to those elements for which %re > 0. Also used with the disp W option to restrict printing of residuals to those for which %re > 0
%va	Values of the variance function
%vc	Variance/covariance matrix: thus, for a simple two-parameter regression, %vc(1) is the variance of the intercept, vc(2) is the covariance of the slope and intercept, and vc(3) is the variance of the slope; the length of the vector is %ml
%vl	Variance of the linear predictor
%wt	Iterative weights assigned by the **fit** directive
%wv	Working vector. The same as the y values %yv in non-iterative models, but current working vector as defined by the **fit** directive in an iterative model

19.3 Programming tips

1 End each line of the macro with ! (this saves space inside GLIM).

2 Use the fully abbreviated forms of commands (this also saves space).

3 Pack as many statements on a line as possible (ditto).

4 Make maximum use of the colon ':', the 'repeat directive' command.

5 Write lots of small macros rather than a few big ones.

6 Debug each macro separately to make sure it works in the way you want it to, before embedding it in a more complex program file.

7 Always end a macro by **deleting** all the vectors you have created inside the macro for temporary storage (if you do not do this, then errors may occur next time the macro is used because of inconsistent vector lengths).

8 Try to use variable names inside the macro that you would not use in normal GLIM code (e.g. end names with the underline character: x_, res_, etc.); make full use of the reserved macro scalar names %z1 to %z9.
9 If you do not put a $ at the end of each line, put $$ in front of the **endmac** directive.
10 Do not try to embed the definition of one macro inside another macro; you can invoke macros inside macros, but you cannot define them.

19.4 Sorting and shuffling

The first example is a macro whose purpose is to count the number of ties in a vector of data. The logic is to sort the vector into order, then produce a copy of the sorted vector, which is lagged by one cell, like this:

Sorted	Lagged	Equal ?
3	7	0
4	3	0
4	4	1
4	4	1
7	4	0

The number of ties is then given simply by counting the number of times the two vectors have the same value in a given position (two in this case), using the %cu function on the logical variable containing the tied = true/false information.

```
$macro ties $!                          %1 is the vector to be analysed
$sort y_ %1 $!                          y_ is the sorted version of %1
$sort y2_ y_ 2 $                        y2_ is y_ lagged by one cell
$calc %z1=%cu(y2_==y_) $                %z1 is the number of ties
$print 'Number of ties = ' *i %z1 $!    *i means integer
$del y_ y2_ $!                          delete the local variables
$endmac $!                              end the macro
$return $!                              return from the file to GLIM
```

The macro is stored in a file called ties.mac and is used as follows:

```
$units 10 $
$assign y=6,3,5,4,9,2,6,5,3,6 $
```

```
$input 6 $

File name? ties.mac

$use ties y $

Number of ties = 4
```

•

Do not forget to reset all the input channels between each exercise; the simplest way is to $stop then restart GLIM. Alternatively, you could use a different input channel number each time; note, however, that channel 6 is used in all the following exercises (remember to start at channel 7 in GLIM 4).

As a second example we look at a macro written for use in experimental design. The aim is to take the number of experimental units and to randomize their sequence (i.e. to shuffle the numbers). The shuffled numbers could then be used to allocate experimental units to locations in a field or positions on a greenhouse bench. The logic is to generate a vector containing the correct number of numbers in sequence using %cu(1), then use a vector of random numbers to sort the numbers into random sequence. This scheme has the twin advantages that no random numbers are repeated and all the numbers appear once. The shuffled vector is printed.

$macro shuffle $!	the argument defines the numbers for shuffling
$calc %z1=%1 $!	%z1 is the scalar value of the argument
$var %z1 x_ rn_ $!	two vectors of length %z1 are set up
$calc x_=%cu(1) $!	x_ contains 1, 2, ..., %z1 in sequence
$calc rn_=%sr(0) $!	rn_ is a vector of random real numbers
$sort x_ x_ rn_ $!	replace x_ by its randomized version
$print x_ $!	print the shuffled numbers
$del x_ rn_ $!	delete internal vectors
$endmac $!	end the macro
$return $!	return from file to GLIM

The macro is used to shuffle 12 numbers as follows:

```
$calc %n=12 $

$input 6 $

File name? shuffle.mac
```

```
$use shuffle %n $
```

and produces the output:

```
3.000  9.000  4.000  6.000  5.000  12.00  1.000  11.00  8.00
2.000  10.00  7.000
```

How would you modify the macro to print the numbers as integers?

19.5 Random numbers from specified distributions

In simulation models we sometimes want to generate random numbers from a specified probability distribution. The first step is to generate a cumulative probability function $P(x)$, showing the probability of getting x *or less* in an independent random trial. Next we generate a random real number between 0 and 1 (say 0.73) and ask 'what is the smallest value of x that has a cumulative probability greater than the random number?' This is the chosen random number (3 in this case).

$P(x)$	0.1	0.35	0.68	0.95	0.99	1.0
x	0	1	2	3	4	5
< or >?	<	<	<	0.73	>	>

The logic of the macro is to count the number of less-than symbols in the bottom row (three in this case, but zero if the random number had been smaller than 0.1). Suppose we want to generate a series of %w Poisson-distributed random numbers with a mean of %u. We restrict the data to lie in the range 0 to 10 (this is reasonable for most ecological examples of the Poisson, where the mean is generally less than 2).

```
$macro poisson $!           %1 is the mean, %2 the number
                            of numbers needed
$output $!                  switch off warning messages
$calc %c=0 $!               initialize the counter
$calc %u=%1 $!              set the mean to %1
$calc %w=%2 $!              set the total count to %2
$var 11 x p $!              set up two vectors of length 11
$calc x=-1+%cu(1) $!        x gets the numbers 0 to 10
$calc lx=%log(x) $!         lx is log x
$calc clx=%cu(lx) $!        clx is cumulative log x
$calc f=%exp(clx) $!        f is the antilog (i.e. x factorial)
```

```
$calc p=%exp(-%u)*%u**x/f $!
                                          p is the probability of x
$calc cp=%cu(p) $!                        cp is the cumulative probability
$calc %f=1 $!                             %f is true
$while %f prn $!                          generate numbers until %f is false
$endmac $!                                end macro
$macro prn $!                             new macro to generate random nos.
$calc %x=%sr(0) $!                        %x is random real 0<x<1
$calc t=(%x>cp) $!                        t is true if %x>cp
$calc %q=%cu(t) $!                        %q is our random Poisson integer
$output %poc $!                           output back on again
$calc %c=%c+1 $!                          increase the counter by 1
$print 'Poisson random number' *i %c ' = ' *i %q $!
                                          print the rn value
$calc %f=(%c<%w) $!                       check the counter against the total
$endmac $!                                end macro
$return $!                                return to GLIM
```

To obtain five Poisson numbers from a distribution with a mean of 1.24, the macro is invoked as follows:

```
$input 6 $
File name? poisrn.mac
$calc %u=1.24: %w=5 $
$use poisson %u %w $
Poisson random number 1 = 1
Poisson random number 2 = 2
Poisson random number 3 = 0
Poisson random number 4 = 0
Poisson random number 5 = 0
```

On this particular run, the numbers turned out to be 1, 2 and three 0's.

19.6 Estimating *k* of the negative binomial distribution

It is common in ecological survey work to obtain frequency distributions of count data for which the variance is greater than the mean. Suppose that we want to obtain the maximum likelihood estimate of the parameter k of the negative binomial for such a data set. This is obtained when the left- and right-hand sides of the maximum likelihood equation are equal:

$$n \ln\left(1 + \frac{\mu}{k}\right) = \sum_{x=0}^{\max}\left(\frac{A_{(x)}}{k + x}\right) \quad (19.1)$$

where $A(x)$ is a vector containing the frequency of values greater than x.

The logic of the macro is to take an initial estimate of k from the sample mean and variance (see Section 14.10):

$$k = \frac{\mu^2}{s^2 - \mu}$$

then to try different values of k until the sign of the difference between the left- and the right-hand side of equation (19.1) changes. At this point, the step-size is reduced by a factor of 10 and the process repeated until the sign changes again. And so on until the desired accuracy is achieved. The default accuracy in the present macro is three decimal places.

```
$input 6 $

File name? negbin.mac
```

Use **assign** to set up the vector of frequencies from 0 to 12 for which we wish to estimate k of the negative binomial (131 0's and one 12 in this case):

```
$assi s=131,55,21,14,6,6,2,0,0,0,0,2,1 $

$use kfit s $
```

The macro produces the following output:

```
Maximum likelihood estimate of k = 0.5778

mean = 1.004; variance = 3.076
variance/mean ratio = 3.063

G-test for goodness of fit. G = 2.098

with 2 d.f. from 5 comparisons
```

which shows that the distribution is highly aggregated ($k < 1$) and that the negative binomial is a good fit to the data ($G < \chi^2$ value in tables).

You should check the code of the macros called by kfit. The macro to compute the G-test is quite general and could be adapted for use with simple contingency tables. Why were there only five comparisons?

19.7 Model-checking macros

The package of macros in the file called mcheck.mac was kindly provided by J. A. Nelder. It calculates most of the model-checking vectors discussed in Chapters 5 and 13.

The default option is to plot two graphs: the standardized residuals against fitted values and the normal probability plot. You can then use other macros to do added-variable plots, partial residual plots for covariates, etc. The vectors calculated in the macros can be used to plot Cook's statistic and other leverage measures. You can explore the potential of the package by working carefully through the code of the various macros, which is in the file called mcheck.mac. Examples of its use are found in Sections 15.7, 16.3 and 17.4. If you want to **use** the macro several times in one section, you should modify the code by removing the **use** directive from the macro.

Another useful plotting macro for survival analysis is kaplan.mac (modified from Aitkin *et al.*, 1989). This gives three different error-checking plots for censored exponential and Weibull data on age at death (see Section 18.9).

19.8 Finding the minimum of a function

Suppose we have a function whose minimum we need to determine, like this double exponential:

```
$units 30 $

$calc x=%cu(1) $

$calc y=300*%exp(-0.1*x)+%exp(0.2*x) $

$plot y x '+' $
```

The minimum value of y is about 80 and it occurs close to $x = 17$. To use the routine we must define a macro called FN which contains the equation of the function in the form: %f=f(%x). You must use the names %f for the y-variable and %x for the x-variable. Do not forget to put a blank space after the end of the macro name, before the $ sign:

```
$macro fn $

$MAC? $calc %f=300*%exp(-0.1*%x)+%exp(0.2*%x)$endmac $
```

Some starting values must now be assigned for the initial guess at where on the x-axis the minimum occurs (%x), the initial step length (%l) and the precision (%e):

```
$calc %x=5: %l=1 : %e=0.0001 $
```

Then we read the file from disk and invoke the macro by typing:

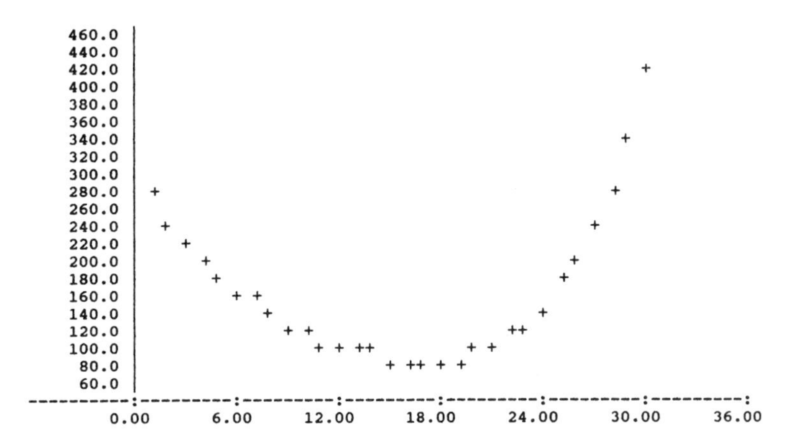

$input 6 $

File name? min.mac

$use minimum $

The minimum of the function is at 16.70: value = 84.69

You can check the result by calculus; the minimum should be at $x = [\ln(30) - \ln(0.2)]/0.3 = 16.7$. Go through the code of the macro and see if you can work out how the quadratic interpolation works. •

19.9 Non-linear models

GLIM can only deal with models that are linear in their parameters (see Section 10.1). Many important ecological functions are inherently non-linear and cannot be linearized by transformation; two examples are the hyperbolic function and the asymptotic exponential. The exponential curve is dealt with here, and the hyperbolic curve in Exercise 19.4. The rationale of the macro is that the non-linear parameter is varied step-wise and the model is fitted repeatedly to the data. A plot of residual deviance against the parameter value shows the value that gives the best fit. The macro also illustrates the way that graphs of deviance can be built up, one element at a time, using subscripted vectors.

The object is to obtain estimates of the three parameters in the function:

$$y = a(1 - be^{-cx}) \tag{19.2}$$

which is intrinsically non-linear in the parameter c. The macro works by taking successive numerical values for c, then calculating $z = \exp(-c.x)$ and fitting the linear model:

$$y = a(1 - b.z) = a - \beta z$$

where GLIM estimates the two parameters a and β by maximum likelihood. The value of c is selected that gives the minimal deviance when this model is fitted to the data. The example concerns the numbers of prey eaten per 5-hour period by individual fish in tanks supplied with water-fleas at different densities:

```
$units 50 $

$data x y $

$dinput 6 $

File name? glex36.dat

$plot y x '*' $
```

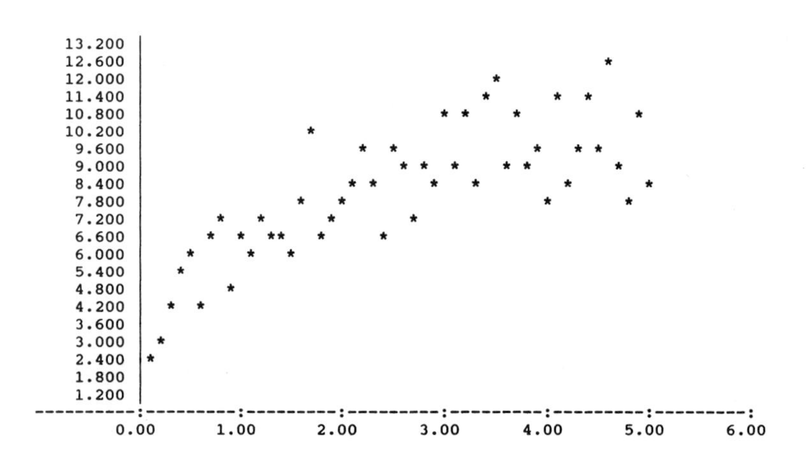

Despite the scatter there is a clear asymptotic trend with an asymptote of roughly 10 prey per hour and an intercept of about 2. This gives us guesstimates of $a = 10$ and $b = -0.8$ (by rearranging equation 19.2).

```
$input 7 $

File name? expfit.mac
```

Begin by assigning scalar values to the two arguments required by the macro called expo: (i) the maximum value of c (%m=2); and (ii) the step length for calculating and plotting deviance (%s=0.1). The macro is then invoked:

```
$yvar y $

$calc %m=2 $

$calc %s=0.1 $

$use expo %m %s $
```

After fitting the model 20 times (%m/%s), the macro plots residual deviance against the estimated value of parameter c:

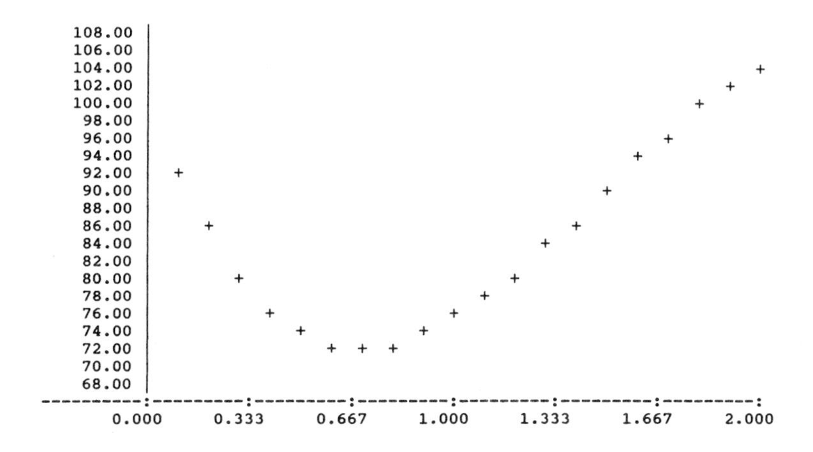

The minimum deviance of 72 occurs at about c = 0.7, so we recalculate z and then refit the model with this vector of z and inspect the maximum likelihood parameter estimates:

```
$calc z=%exp(-0.7*x) $

[w] -- change to data affects model

$fit:+z $


deviance = 259.05
    d.f. = 49

deviance = 71.580 (change = -187.5)
    d.f. = 48     (change = -1)

$disp e $
```

```
     estimate   s.e.     parameter
1     10.24    0.2541       1
2    -7.814    0.6969       Z
scale parameter taken as 1.491
```

The intercept of 10.24 is the estimate of the asymptotic y-value, $a = 10.24$. The slope parameter β is the product $a.b$ so $b = -7.814/10.24 = -0.7631$. Our best estimate for the equation is therefore:

$$y = 10.24(1 - 0.7631e^{-0.7x})$$

The data and the fitted values can be inspected by plotting both on the same axes:

```
$plot (y=0,15 x=0,5) y %fv x '*+' $
```

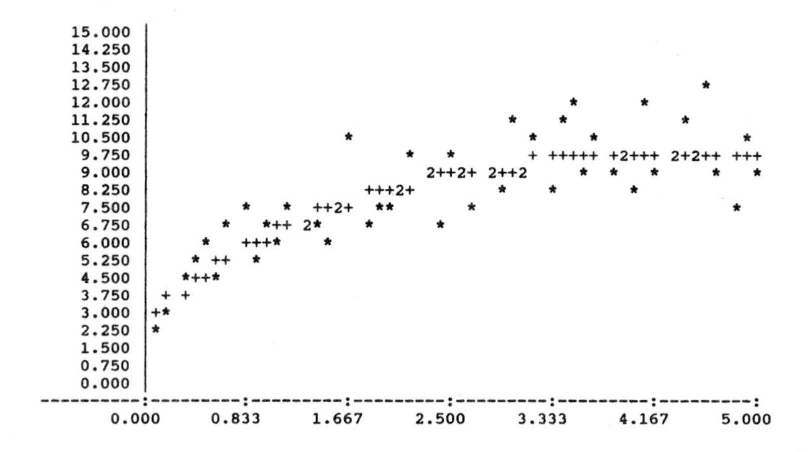

The main drawback of this method is that it does not give us an estimate of the standard error of parameter c. If it is important to have a standard error (e.g. for comparing the initial slope of this functional response with the slope observed in another experiment, or with a theoretical expectation) then a non-linear modelling package like Genstat or MLP would have to be used.

To see how the model-fitting works, you should go through the macros in the file called expfit.mac, and suggest some improvements that would make the macros more general.

19.10 User-defined models (own)

In addition to normal, Poisson, binomial, gamma and inverse Gaussian, you can write your own model for the error distribution and link function in GLIM. To do this you need to calculate four quantities:

1 the fitted values;

2 the derivative of the linear predictor;
3 the variance function;
4 the deviance increment.

In GLIM 3.77, each term is written in its own macro, then the **own** directive is invoked by naming the four macros in sequence. In GLIM 4 there is no **own** directive but the **link** and **error** functions can be modified appropriately. Here is the code for an own model with negative binomial errors that might be useful in analysing count data from aggregated populations where the variance is substantially greater than the mean; the code is in the file called ownnb.mac:

```
$print 'Own model with Negative binomial errors' $!
```

The negative binomial parameter (%k) needs to be set in advance (see Section 19.6):

```
$calc %a = 1/%k $!
```

Macro 1 gives the fitted value as a function of the linear predictor. The fitted value is the antilog of %lp:

```
$macro m1 $!
$calc %fv = %exp(%lp) $!
$endmac $!
```

Macro 2 works out the derivative of the linear predictor. The derivative of log x is $1/x$:

```
$macro m2 $!
$calc %dr = 1/%fv $!
$endmac $!
```

Macro 3 calculates the variance function:

```
$macro m3 $!
$calc %va = %fv*(1+%a*%fv) $!
$endmac $!
```

Macro 4 works out the deviance increment (see Table 19.2):

```
$macro m4 $!
$calc %di = 2*(%yv*%log(%yv/%fv)-%if(%a/=0,(1+%a*%yv)*
%log((1+%a*%yv)/(1+%a*%fv))/%a, %yv-%fv)) $!
$endmac $!
```

The negative binomial model is defined by the **own** directive:

```
$own m1 m2 m3 m4 $
```

The linear predictor %lp needs an initial value before invoking **fit**:

Table 19.2 Variance function, deviance increment and fitted-value transformation for the standard distributions as used in the model-checking macros, where standardized residuals are plotted against the appropriately transformed fitted values

Distribution	Variance function	Deviance increment	Fitted-value transformation
Normal	1	(%yv−%fv)**2	%fv
Poisson	%fv	2*(%yv*%log(%yv/%fv)−(%yv−%fv))	%sqr(%fv)
Binomial	%fv*(1−%fv/%bd)	2*(%yv*%log(%yv/%fv)+(%bd−%yv) *%log((%bd−%yv)/(%bd−%fv)))	%ang(%fv/%bd)
Gamma	%fv**2	2*(%log(%fv/%yv)+(%yv−%fv)/%fv)	%log(%fv)
Negative binomial	%fv*(1+%a*%fv)	2*(%yv*%log(%yv/%fv)−(1+%a*%yv)* %log((1+%a*%yv)/(1+%a*%fv))/%a)	%sqr(%fv)
Inverse Gaussian	%fv**3	((%yv−%fv)**2)/(%yv*%fv**2)	−1/%sqrt(%fv)

The system vectors are listed in Table 19.1. Notice that the negative binomial is the only distribution to have an external constant (%a = 1/k) in the variance function or the deviance increment. The other distributions are all standard in GLIM (although the inverse Gaussian is not in Version 3.77).

```
$calc %lp = %log(%yv + 0.5) $!

$return
```

To show how this works, we shall repeat the analysis of the over-dispersed slug data from Section 14.9.1, using an **own** model for negative binomial errors. The preamble is the same as before:

```
$units 80 $

$data c $

$dinput 6 $

File name? glex5.dat

$calc a=%gl(2,40) $

$factor a 2 $

$tab the c mean for a $

       1.000  2.000
  [ ]  1.275  2.275
```

We begin by inspecting the Poisson error fit; note the parameter estimates and their standard errors:

```
$yvar c $

$error p $

$fit:+a$disp e $

scaled deviance = 224.86 at cycle 4
            d.f. = 79

scaled deviance = 213.44 (change = −11.42) at cycle 4
            d.f. = 78      (change = −1)

    estimate   s.e.      parameter
1   0.2430     0.1391    1
2   0.5789     0.1742    A(2)
scale parameter taken as 1.000
```

The large residual deviance indicates substantial overdispersion. Now we repeat the procedure using the home-made negative binomial error model (above) which is stored in a file called ownnb.mac. Before loading this file, you need to supply a value for the negative binomial parameter %k (a good starting guesstimate and a reasonable compromise is $k = 1$; if you use the macro kfit in file negbin.mac and estimate k for slug distribution in each field separately, you will get $k = 0.2896$ for field A and $k = 1.934$ for field B).

```
$calc %k=1 $

$input 7 $

File name? ownnb.mac

$fit $

deviance = 97.67 at cycle 3
      d.f. = 79

$fit +a$disp e $

deviance = 93.493 (change = −4.175) at cycle 2
      d.f. = 78       (change = −1)

   estimate  s.e.     parameter
1  0.2441    0.2288   1
2  0.5784    0.3083   A(2)
scale parameter taken as 1.199
```

●

The parameter estimates are very close to their Poisson equivalents (0.243 and 0.5789) but the standard errors are much larger (cf. 0.1391 and 0.1742). The deletion test, however, has an F-ratio of 3.48 on d.f. = 1,78 which is significant at better than 0.05, suggesting that the fields do have significantly different slug densities. Recall that with Poisson errors and an adjusted scale parameter, we concluded that the difference fell just short of significance. For comparison, the standard errors using the adjusted scale parameter were 0.2467 and 0.3088 respectively (see Section 14.9.1).

19.11 Statistics

As an example of the kind of time-saving macro you might write for frequently used statistical tests, we take Bartlett's test for equality of variances. The macro compares the variances of k samples and computes a test statistic which is compared with the χ^2 value in tables with $k-1$ d.f. We shall use the macro to compare the five variances in the regrowth experiment described in Section 8.7.

```
$units 30 $

$data y $

$dinput 6 $

File name? glex39.dat

$calc t=%gl(5,6) $

$factor t 5 $
```

A look at the data suggests that the variance in treatment 1 is relatively high, and that this might invalidate an ANOVA with normal errors:

```
$tab the y mean for t $

        1       2       3       4       5
[ ]   553.3   569.3   610.5   465.2   610.7
```

```
$tab the y variance for t $

        1       2       3       4       5
[ ]   12134.  2303.   3593.   3319.   3455.
```

We check the significance of this five-fold difference between the largest and smallest variances with Bartlett's test, which is stored in bart.mac:

```
$input 7 $

File name? bart.mac
```

The macro has two arguments — the sample measurements (y) and the classifying factor (t):

```
$use bart y t $

Bartlett's statistic = 4.441 with 4 degrees of freedom.

Probability that variances are the same = 0.3497
```

So, despite the big difference between the largest and smallest variances, this is well within the range that could be expected by chance alone (the χ^2 value in tables with 4 d.f. is 9.49). You should check the code of the macro to see how it works. The most interesting programming tips are contained in the first two lines which use abbreviated **tab** directives to: (i) store each separate variance in a new vector var; and (ii) count the sample sizes in each treatment and store these in a new vector n. Line 3 is necessary because f_ is a factor (it was created by the **by** directive in line 1), and we need a variate fv_ from which to find the largest value in line 4. Line 4 finds the largest factor level and stores this in a_.

```
$t t %1 v f %2 i var by f_ $

$t w * f %2 u n_ by f_ $

$ca fv_=f_ $

$t t fv_ l i a_ $
```

Note the use of the additional macro that worked out and printed the associated χ^2 probability (see the macro in the file called chip.mac).

19.12 Confidence intervals for LD_{50}

The macro uses information on the slope and intercept of the regression contained in the system vector %pe to estimate the LD_{50} (the intercept is %pe(1) and the slope is %pe(2); so LD_{50} = −%pe(1)/%pe(2)). It also uses information from the variance/covariance matrix to estimate the standard error of the LD_{50} (the first element %vc(1) contains the variance of the intercept, the second %vc(2) has the covariance of the two parameters, and the third %vc(3) contains the variance of the slope). In order to get access to the vectors %pe and %vc, we need to use the **extract** directive in the second line of the macro:

```
$macro fieller $
!
! A macro to compute 95% confidence limits for LD50
! using Fieller's Theorem.
! See D. Collett (1991) Modelling Binary Data. Chapman & Hall,
! London. pp 97–101.
!
$extract %pe %vc $
$calc %g=1.96**2*%vc(3)/%pe(2)**2 $
$calc %p=%pe(1)/%pe(2) $
$calc %a=%p−%g*%vc(2)/%vc(3) $
$calc %i=(%vc(1)−2*%p*%vc(2)+%p**2*%vc(3)−%g*(%vc(1)−
%vc(2)**2/%vc(3))) $
$calc %e=(−%a+1.96/%pe(2)*%sqrt(%i))/(1−%g) $
$calc %f=(−%a−1.96/%pe(2)*%sqrt(%i))/(1−%g) $
$print 'Estimated LD50 =' −%p $
$print '95% Limits' %f 'and' %e $
$endmac $
$return $
```

The macro is invoked simply by typing the macro name; there are no arguments required, so, for the bioassay example of Section 15.5, the output is:

```
$use fieller $
```

```
Estimated LD50 = 2.923

95% Limits 2.648 and 3.220
```

The output shows the LD_{50} (2.923 on the log dose scale) and the 95% confidence intervals. These should be antilogged to obtain the doses at the extremes of the 95% interval.

19.13 Overdispersion with binomial errors

This macro (from Collett, 1991) uses Williams' procedure for adjusting for overdispersion in data with binomial errors. The example involves the germination of seeds from two genetic strains of a parasitic plant species that were treated with extracts from the roots of two potential host plants (see Collett, 1991, for details). The number of seeds germinating (y) out of each batch (n) was recorded from five (or six in one case) replicates for each of the four combinations of species and extract:

```
$units 21 $

$data y n species extract $

$dinput 6 $

File name? glex52.dat

$factor species 2 extract 2 $
```

As usual, the response variable is the number of seeds germinating, and the sample size is given as the binomial denominator:

```
$yvar y $

$error b n $

$fit:+species*extract:-species.extract $
```

Fitting the maximal model, then attempting to simplify it by removing the interaction term (species.extract), demonstrates that the interaction is significant and must be retained. We add it back, then **disp e** for the maximal model:

```
$fit +species.extract $

scaled deviance = 33.278 (change = -6.408) at cycle 3
             d.f. = 17       (change = -1)

$disp e $

    estimate     s.e.     parameter
1   -0.5582    0.1260    1
2    0.1459    0.2232    SPEC(2)
```

```
3    1.318    0.1774  EXTR(2)
4  −0.7781   0.3064  SPEC(2).EXTR(2)
scale parameter taken as 1.000
```

This shows the significant interaction between species and extract (i.e. lower germination exhibited by species 2 on extract 2 than on extract 1, compared with the higher germination of species 1 on extract 2). The effect of extract is much greater than the effect of species, but the significance of the main effects is not an issue here, since we have already demonstrated a significant interaction term. The problem is that the data are overdispersed after the maximal model has been fitted (ratio of scaled deviance to d.f. = 33.278/17 = 1.96. Williams' procedure is to determine an iterative estimate for ϕ in the equation:

$$\sigma^2_{jkl} = 1 + (n_{jkl} - 1)\phi$$

by equating the value of Pearson's χ^2 statistic for the model to its appropriate expected value, then using $1/\sigma^2_{jkl}$ as a weight for the individual binomial observations. The **weight** directive is adjusted and then the fit of the model is reassessed.

```
$input 8 $

File name? wills.mac
```

The macro is invoked by typing:

```
$use williams $
```

and produces the following output:

```
Williams' estimate of phi = 0.0249 after 3 iterations.

Full model using heterogeneity factor:

scaled deviance = 18.446 at cycle 3
            d.f. = 17

     estimate   s.e.     parameter
1  −0.5354    0.1937   1
2    0.07011  0.3114   SPEC(2)
3    1.330    0.2781   EXTR(2)
4  −0.8195    0.4351   SPEC(2).EXTR(2)
scale parameter taken as 1.000
```

The effect of the weighting on the parameter estimates is slight, but the standard errors are increased to take account of the overdispersion. The interaction term is no longer significant at 5%. In this example, the 95% confidence interval for germination of species 1 ranges from 0.28 to 0.47 when Williams' procedure has been used to allow for overdispersion, compared with a range of 0.31 to 0.42 in the unadjusted example. You should check how these confidence intervals were obtained. The effect of the adjustment has been to increase the width of the confidence interval by 73% in this case.

In terms of programming, the most interesting aspects of the macro are the iterative comparison of the scaled deviance (using Pearson's χ^2, %x2) and the use of the **weight** directive followed by the **fit +** directive, to repeat the last model fit. The estimation macro is repeated until the absolute value of the difference between the old and new estimates (%z2=%z5−%z3) is less than 0.0001.

To summarize: using the scale parameter to adjust for overdispersion in binomial data is the simplest and best procedure when the binomial denominators are equal. Williams' procedure is preferred when the sample sizes are unequal (see Collett, 1991, for details). ●

19.14 Library macros

In order to save space inside GLIM, a number of valuable, but less frequently used, procedures are stored in GLIM's own *macro library*. You can get access to these programs by reading the relevant files into GLIM through what is called the *primary library channel* whose channel number is a system scalar known as %plc. Table 19.3 gives an index of the library macros available in GLIM 3.77.

As an example of how to use library macros, let us generate a vector of random normal data, x, with mean 15 and standard deviation 2. Then we can use the library macro called summ to get a summary of the data.

```
$units 10 $

$calc x=15+2*%nd(%sr(0)) $

$look x $

      X
 1  15.45
 2  15.97
 3  12.84
 4  13.26
 5  13.83
 6  13.72
 7  18.23
 8  15.70
 9  12.95
10  16.86
```

To import and run the library macro, we just type:

```
$input %plc summ $

$use summ x $
```

Table 19.3 GLIM 3.77 macro library

Sub-file name	Macro name	Description
1 Data description, exploration and display		
summ	summ	Summary statistics of a variate
stem	stem	Stem and leaf plots
smoo	smoo	Tukey smoothing of a variate
2 Statistical utilities		
chip	chip	χ^2 probability
3 Normal models		
qplot	qplot	Normal probability plotting
qplot	stan	Standardized residuals
qplot	jack	Jack-knife residuals
tnor	tnor	Test for normality of variate by χ^2 goodness-of-fit test
tnor	wdash	Shapiro–Francia W' test for normality
normac	rsq	R^2 statistic
normac	tval	t-values of parameter estimates
lev	lev	Leverage values
boxcox	boxcox	Box–Cox transformation family on y-variate
boxcox	boxfit	Box–Cox transformation for fixed λ
press	press	Prediction error sum of squares
4 Survival models and censored data		
weib	weib	Fitting the exponential and Weibull distributions to censored data
weib	resp	Residual plotting after use of macro weib
5 Other statistical models and techniques		
tuni	tuni	Test for distribution of variate against uniform distribution by χ^2 goodness-of-fit test

and the macro produces the following output:

```
Summary statistics for X

Length         10.

Maximum        18.23     Minimum       12.84     Range              5.391

Mean           14.88     Total        148.8

Variance        3.366    Std. Dev.      1.835

1st Quartile   13.26     Median        14.64     3rd Quartile      15.97
```

The use of other library macros is demonstrated in Section 13.6. You can obtain a listing of the code by importing the GLIM file called maclib.glm into your word-processor. Alternatively, you can browse through the code using 'view mode' in your file manager.

Appendix

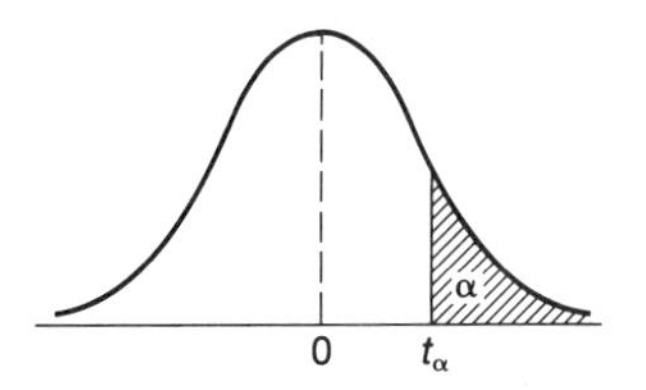

Table 1 Student's t tables: one-tailed critical values of the t distribution; to find two-tailed values (e.g. for confidence intervals or *LSD*s), look up the column headed by $\alpha/2$. For example, two-tailed 5% t-values are found in the column headed by 0.025

	α				
v	0.10	0.05	0.025	0.01	0.005
1	3.078	6.314	12.706	31.821	63.657
2	1.886	2.920	4.303	6.965	9.925
3	1.638	2.353	3.182	4.541	5.841
4	1.533	2.132	2.776	3.747	4.604
5	1.476	2.015	2.571	3.365	4.032
6	1.440	1.943	2.447	3.143	3.707
7	1.415	1.895	2.365	2.998	3.499
8	1.397	1.860	2.306	2.896	3.355
9	1.383	1.833	2.262	2.821	3.250
10	1.372	1.812	2.228	2.764	3.169
11	1.363	1.796	2.201	2.718	3.106
12	1.356	1.782	2.179	2.681	3.055
13	1.350	1.771	2.160	2.650	3.012
14	1.345	1.761	2.145	2.624	2.977
15	1.341	1.753	2.131	2.602	2.947
16	1.337	1.746	2.120	2.583	2.921
17	1.333	1.740	2.110	2.567	2.898
18	1.330	1.734	2.101	2.552	2.878
19	1.328	1.729	2.093	2.539	2.861
20	1.325	1.725	2.086	2.528	2.845
21	1.323	1.721	2.080	2.518	2.831
22	1.321	1.717	2.074	2.508	2.819
23	1.319	1.714	2.069	2.500	2.807
24	1.318	1.711	2.064	2.492	2.797
25	1.316	1.708	2.060	2.485	2.787
26	1.315	1.706	2.056	2.479	2.779
27	1.314	1.703	2.052	2.473	2.771
28	1.313	1.701	2.048	2.467	2.763
29	1.311	1.699	2.045	2.462	2.756
∞	1.282	1.645	1.960	2.326	2.576

Table 2 χ^2 tables: critical values of the χ^2 distribution

	α					
v	0.1	0.05	0.025	0.01	0.005	0.001
1	2.706	3.841	5.024	6.635	7.879	10.828
2	4.605	5.991	7.378	9.210	10.597	13.816
3	6.251	7.815	9.348	11.345	12.838	16.266
4	7.779	9.488	11.143	13.277	14.860	18.467
5	9.236	11.070	12.832	15.086	16.750	20.515
6	10.645	12.592	14.449	16.812	18.548	22.458
7	12.017	14.067	16.013	18.475	20.278	24.322
8	13.362	15.507	17.535	20.090	21.955	26.124
9	14.684	16.919	19.023	21.666	23.589	27.877
10	15.987	18.307	20.483	23.209	25.188	29.588
11	17.275	19.675	21.920	24.725	26.757	31.264
12	18.549	21.026	23.337	26.217	28.300	32.910
13	19.812	22.362	24.736	27.688	29.819	34.528
14	21.064	23.685	26.119	29.141	31.319	36.123
15	22.307	24.996	27.488	30.578	32.801	37.697
16	23.542	26.296	28.845	32.000	34.267	39.252
17	24.769	27.587	30.191	33.409	35.718	40.790
18	25.989	28.869	31.526	34.805	37.156	42.312
19	27.204	30.144	32.852	36.191	38.582	43.820
20	28.412	31.410	34.170	37.566	39.997	45.315
21	29.615	32.670	35.479	38.932	41.401	46.797
22	30.813	33.924	36.781	40.289	42.796	48.268
23	32.007	35.172	38.076	41.638	44.181	49.728
24	33.196	36.415	39.364	42.980	45.558	51.179
25	34.382	37.652	40.646	44.314	46.928	52.620
26	35.563	38.885	41.923	45.642	48.290	54.052
27	36.741	40.113	43.194	46.963	49.645	55.476
28	37.916	41.337	44.461	48.278	50.993	56.892
29	39.088	42.557	45.722	49.588	52.336	58.301
30	40.256	43.773	46.979	50.892	53.672	59.703
31	41.422	44.985	48.232	52.191	55.003	61.098
32	42.585	46.194	49.480	53.486	56.329	62.487
33	43.745	47.400	50.725	54.776	57.649	63.870
34	44.903	48.602	51.966	56.061	58.964	65.247
35	46.059	49.802	53.203	57.342	60.275	66.619
36	47.212	50.998	54.437	58.619	61.582	67.985
37	48.363	52.192	55.668	59.892	62.884	69.346
38	49.513	53.384	56.896	61.162	64.182	70.703
39	50.660	54.572	58.120	62.428	65.476	72.055
40	51.805	55.758	59.342	63.691	66.766	73.402
41	52.949	56.942	60.561	64.950	68.053	74.745
42	54.090	58.124	61.777	66.206	69.336	76.084
43	55.230	59.304	62.990	67.459	70.616	77.419
44	56.369	60.481	64.202	68.710	71.893	78.750
45	57.505	61.656	65.410	69.957	73.166	80.077

Table 2 *Continued*

v	α					
	0.1	0.05	0.025	0.01	0.005	0.001
46	58.641	62.830	66.617	71.201	74.437	81.400
47	59.774	64.001	67.821	72.443	75.704	82.720
48	60.907	65.171	69.023	73.683	76.969	84.037
49	62.038	66.339	70.222	74.919	78.231	85.351
50	63.167	67.505	71.420	76.154	79.490	86.661
52	65.422	69.832	73.810	78.616	82.001	89.272
54	67.673	72.153	76.192	81.069	84.502	91.872
56	69.918	74.468	78.567	83.513	86.994	94.460
58	72.160	76.778	80.936	85.950	89.477	97.039
60	74.397	79.082	83.298	88.379	91.952	99.607
62	76.630	81.381	85.654	90.802	94.419	102.166
64	78.860	83.675	88.004	93.217	96.878	104.716
66	81.085	85.965	90.349	95.626	99.331	107.258
68	83.308	88.250	92.689	98.028	101.78	109.791
70	85.527	90.531	95.023	100.43	104.21	112.317
71	86.635	91.670	96.189	101.62	105.43	113.577
72	87.743	92.808	97.353	102.82	106.65	114.835
73	88.850	93.945	98.516	104.01	107.86	116.092
74	89.956	95.081	99.678	105.20	109.07	117.346
75	91.061	96.217	100.84	106.39	110.29	118.599
76	92.166	97.351	102.00	107.58	111.50	119.850
77	93.270	98.484	103.16	108.77	112.70	121.100
78	94.373	99.617	104.32	109.96	113.91	122.348
79	95.476	100.75	105.47	111.14	115.12	123.594
80	96.578	101.88	106.63	112.33	116.32	124.839
81	97.680	103.01	107.78	113.51	117.52	126.082
82	98.780	104.14	108.94	114.69	118.73	127.324
83	99.880	105.27	110.09	115.88	119.93	128.565
84	100.98	106.39	111.24	117.06	121.13	129.804
85	102.08	107.52	112.39	118.24	122.32	131.041
86	103.18	108.65	113.54	119.41	123.52	132.277
87	104.28	109.77	114.69	120.59	124.72	133.512
88	105.37	110.90	115.84	121.77	125.91	134.745
89	106.47	112.02	116.99	122.94	127.11	135.978
90	107.56	113.15	118.14	124.12	128.30	137.208
91	108.66	114.27	119.28	125.29	129.49	138.438
92	109.76	115.39	120.43	126.46	130.68	139.666
93	110.85	116.51	121.57	127.63	131.87	140.893
94	111.94	117.63	122.72	128.80	133.06	142.119
95	113.04	118.75	123.86	129.97	134.25	143.344
96	114.13	119.87	125.00	131.14	135.43	144.567
97	115.22	120.99	126.14	132.31	136.62	145.789
98	116.32	122.11	127.28	133.48	137.80	147.010
99	117.41	123.23	128.42	134.64	138.99	148.230
100	118.50	124.34	129.56	135.81	140.17	149.449

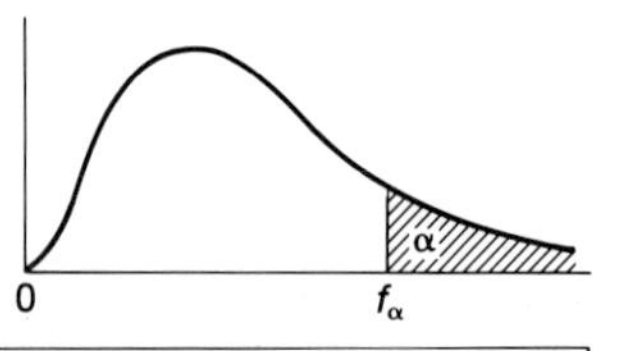

Table 3 *F* tables (5%): critical values of the *F* distribution, where ν_1 is the d.f. in the numerator and ν_2 is the d.f. in the denominator

	ν_1								
ν_2	1	2	3	4	5	6	7	8	9
1	161.4	199.5	215.7	224.6	230.2	234.0	236.8	238.9	240.5
2	18.51	19.00	19.16	19.25	19.30	19.33	19.35	19.37	19.38
3	10.13	9.55	9.28	9.12	9.01	8.94	8.89	8.85	8.81
4	7.71	6.94	6.59	6.39	6.26	6.16	6.09	6.04	6.00
5	6.61	5.79	5.41	5.19	5.05	4.95	4.88	4.82	4.77
6	5.99	5.14	4.76	4.53	4.39	4.28	4.21	4.15	4.10
7	5.59	4.74	4.35	4.12	3.97	3.87	3.79	3.73	3.68
8	5.32	4.46	4.07	3.84	3.69	3.58	3.50	3.44	3.39
9	5.12	4.26	3.86	3.63	3.48	3.37	3.29	3.23	3.18
10	4.96	4.10	3.71	3.48	3.33	3.22	3.14	3.07	3.02
11	4.84	3.98	3.59	3.36	3.20	3.09	3.01	2.95	2.90
12	4.75	3.89	3.49	3.26	3.11	3.00	2.91	2.85	2.80
13	4.67	3.81	3.41	3.18	3.03	2.92	2.83	2.77	2.71
14	4.60	3.74	3.34	3.11	2.96	2.85	2.76	2.70	2.65
15	4.54	3.68	3.29	3.06	2.90	2.79	2.71	2.64	2.59
16	4.49	3.63	3.24	3.01	2.85	2.74	2.66	2.59	2.54
17	4.45	3.59	3.20	2.96	2.81	2.70	2.61	2.55	2.49
18	4.41	3.55	3.16	2.93	2.77	2.66	2.58	2.51	2.46
19	4.38	3.52	3.13	2.90	2.74	2.63	2.54	2.48	2.42
20	4.35	3.49	3.10	2.87	2.71	2.60	2.51	2.45	2.39
21	4.32	3.47	3.07	2.84	2.68	2.57	2.49	2.42	2.37
22	4.30	3.44	3.05	2.82	2.66	2.55	2.46	2.40	2.34
23	4.28	3.42	3.03	2.80	2.64	2.53	2.44	2.37	2.32
24	4.26	3.40	3.01	2.78	2.62	2.51	2.42	2.36	2.30
25	4.24	3.39	2.99	2.76	2.60	2.49	2.40	2.34	2.28
26	4.23	3.37	2.98	2.74	2.59	2.47	2.39	2.32	2.27
27	4.21	3.35	2.96	2.73	2.57	2.46	2.37	2.31	2.25
28	4.20	3.34	2.95	2.71	2.56	2.45	2.36	2.29	2.24
29	4.18	3.33	2.93	2.70	2.55	2.43	2.35	2.28	2.22
30	4.17	3.32	2.92	2.69	2.53	2.42	2.33	2.27	2.21
40	4.08	3.23	2.84	2.61	2.45	2.34	2.25	2.18	2.12
60	4.00	3.15	2.76	2.53	2.37	2.25	2.17	2.10	2.04
120	3.92	3.07	2.68	2.45	2.29	2.17	2.09	2.02	1.96
∞	3.84	3.00	2.60	2.37	2.21	2.10	2.01	1.94	1.88

Table 3 *Continued*

ν_2	ν_1 10	12	15	20	24	30	40	60	120	∞
1	241.9	243.9	245.9	248.0	249.1	250.1	251.1	252.2	253.3	254.3
2	19.40	19.41	19.43	19.45	19.45	19.46	19.47	19.48	19.49	19.50
3	8.79	8.74	8.70	8.66	8.64	8.62	8.59	8.57	8.55	8.53
4	5.96	5.91	5.86	5.80	5.77	5.75	5.72	5.69	5.66	5.63
5	4.74	4.68	4.62	4.56	4.53	4.50	4.46	4.43	4.40	4.36
6	4.06	4.00	3.94	3.87	3.84	3.81	3.77	3.74	3.70	3.67
7	3.64	3.57	3.51	3.44	3.41	3.38	3.34	3.30	3.27	3.23
8	3.35	3.28	3.22	3.15	3.12	3.08	3.04	3.01	2.97	2.93
9	3.14	3.07	3.01	2.94	2.90	2.86	2.83	2.79	2.75	2.71
10	2.98	2.91	2.85	2.77	2.74	2.70	2.66	2.62	2.58	2.54
11	2.85	2.79	2.72	2.65	2.61	2.57	2.53	2.49	2.45	2.40
12	2.75	2.69	2.62	2.54	2.51	2.47	2.43	2.38	2.34	2.30
13	2.67	2.60	2.53	2.46	2.42	2.38	2.34	2.30	2.25	2.21
14	2.60	2.53	2.46	2.39	2.35	2.31	2.27	2.22	2.18	2.13
15	2.54	2.48	2.40	2.33	2.29	2.25	2.20	2.16	2.11	2.07
16	2.49	2.42	2.35	2.28	2.24	2.19	2.15	2.11	2.06	2.01
17	2.45	2.38	2.31	2.23	2.19	2.15	2.10	2.06	2.01	1.96
18	2.41	2.34	2.27	2.19	2.15	2.11	2.06	2.02	1.97	1.92
19	2.38	2.31	2.23	2.16	2.11	2.07	2.03	1.98	1.93	1.88
20	2.35	2.28	2.20	2.12	2.08	2.04	1.99	1.95	1.90	1.84
21	2.32	2.25	2.18	2.10	2.05	2.01	1.96	1.92	1.87	1.81
22	2.30	2.23	2.15	2.07	2.03	1.98	1.94	1.89	1.84	1.78
23	2.27	2.20	2.13	2.05	2.01	1.96	1.91	1.86	1.81	1.76
24	2.25	2.18	2.11	2.03	1.98	1.94	1.89	1.84	1.79	1.73
25	2.24	2.16	2.09	2.01	1.96	1.92	1.87	1.82	1.77	1.71
26	2.22	2.15	2.07	1.99	1.95	1.90	1.85	1.80	1.75	1.69
27	2.20	2.13	2.06	1.97	1.93	1.88	1.84	1.79	1.73	1.67
28	2.19	2.12	2.04	1.96	1.91	1.87	1.82	1.77	1.71	1.65
29	2.18	2.10	2.03	1.94	1.90	1.85	1.81	1.75	1.70	1.64
30	2.16	2.09	2.01	1.93	1.89	1.84	1.79	1.74	1.68	1.62
40	2.08	2.00	1.92	1.84	1.79	1.74	1.69	1.64	1.58	1.51
60	1.99	1.92	1.84	1.75	1.70	1.65	1.59	1.53	1.47	1.39
120	1.91	1.83	1.75	1.66	1.61	1.55	1.50	1.43	1.35	1.25
∞	1.83	1.75	1.67	1.57	1.52	1.46	1.39	1.32	1.22	1.00

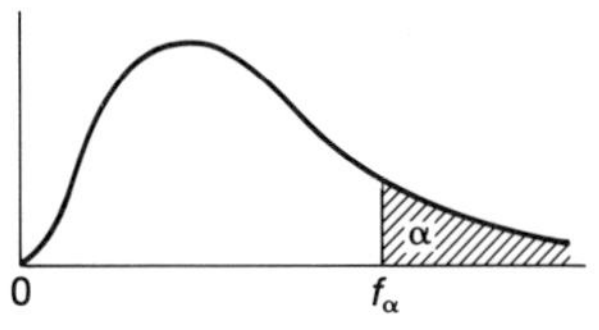

Table 4 F tables (1%): critical values of the F distribution, where ν_1 is the d.f. in the numerator and ν_2 is the d.f. in the denominator

ν_2	ν_1 = 1	2	3	4	5	6	7	8	9
1	4052	4999.5	5403	5625	5764	5859	5928	5981	6022
2	98.50	99.00	99.17	99.25	99.30	99.33	99.36	99.37	99.39
3	34.12	30.82	29.46	28.71	28.24	27.91	27.67	27.49	27.35
4	21.20	18.00	16.69	15.98	15.52	15.21	14.98	14.80	14.66
5	16.26	13.27	12.06	11.39	10.97	10.67	10.46	10.29	10.16
6	13.75	10.92	9.78	9.15	8.75	8.47	8.26	8.10	7.98
7	12.25	9.55	8.45	7.85	7.46	7.19	6.99	6.84	6.72
8	11.26	8.65	7.59	7.01	6.63	6.37	6.18	6.03	5.91
9	10.56	8.02	6.99	6.42	6.06	5.80	5.61	5.47	5.35
10	10.04	7.56	6.55	5.99	5.64	5.39	5.20	5.06	4.94
11	9.65	7.21	6.22	5.67	5.32	5.07	4.89	4.74	4.63
12	9.33	6.93	5.95	5.41	5.06	4.82	4.64	4.50	4.39
13	9.07	6.70	5.74	5.21	4.86	4.62	4.44	4.30	4.19
14	8.86	6.51	5.56	5.04	4.69	4.46	4.28	4.14	4.03
15	8.68	6.36	5.42	4.89	4.56	4.32	4.14	4.00	3.89
16	8.53	6.23	5.29	4.77	4.44	4.20	4.03	3.89	3.78
17	8.40	6.11	5.18	4.67	4.34	4.10	3.93	3.79	3.68
18	8.29	6.01	5.09	4.58	4.25	4.01	3.84	3.71	3.60
19	8.18	5.93	5.01	4.50	4.17	3.94	3.77	3.63	3.52
20	8.10	5.85	4.94	4.43	4.10	3.87	3.70	3.56	3.46
21	8.02	5.78	4.87	4.37	4.04	3.81	3.64	3.51	3.40
22	7.95	5.72	4.82	4.31	3.99	3.76	3.59	3.45	3.35
23	7.88	5.66	4.76	4.26	3.94	3.71	3.54	3.41	3.30
24	7.82	5.61	4.72	4.22	3.90	3.67	3.50	3.36	3.26
25	7.77	5.57	4.68	4.18	3.85	3.63	3.46	3.32	3.22
26	7.72	5.53	4.64	4.14	3.82	3.59	3.42	3.29	3.18
27	7.68	5.49	4.60	4.11	3.78	3.56	3.39	3.26	3.15
28	7.64	5.45	4.57	4.07	3.75	3.53	3.36	3.23	3.12
29	7.60	5.42	4.54	4.04	3.73	3.50	3.33	3.20	3.09
30	7.56	5.39	4.51	4.02	3.70	3.47	3.30	3.17	3.07
40	7.31	5.18	4.31	3.83	3.51	3.29	3.12	2.99	2.89
60	7.08	4.98	4.13	3.65	3.34	3.12	2.95	2.82	2.72
120	6.85	4.79	3.95	3.48	3.17	2.96	2.79	2.66	2.56
∞	6.63	4.61	3.78	3.32	3.02	2.80	2.64	2.51	2.41

Table 4 *Continued*

ν_2	ν_1 10	12	15	20	24	30	40	60	120	∞
1	6056	6106	6157	6209	6235	6261	6287	6313	6339	6366
2	99.40	99.42	99.43	99.45	99.46	99.47	99.47	99.48	99.49	99.50
3	27.23	27.05	26.87	26.69	26.60	26.50	26.41	26.32	26.22	26.13
4	14.55	14.37	14.20	14.02	13.93	13.84	13.75	13.65	13.56	13.46
5	10.05	9.89	9.72	9.55	9.47	9.38	9.29	9.20	9.11	9.02
6	7.87	7.72	7.56	7.40	7.31	7.23	7.14	7.06	6.97	6.88
7	6.62	6.47	6.31	6.16	6.07	5.99	5.91	5.82	5.74	5.65
8	5.81	5.67	5.52	5.36	5.28	5.20	5.12	5.03	4.95	4.86
9	5.26	5.11	4.96	4.81	4.73	4.65	4.57	4.48	4.40	4.31
10	4.85	4.71	4.56	4.41	4.33	4.25	4.17	4.08	4.00	3.91
11	4.54	4.40	4.25	4.10	4.02	3.94	3.86	3.78	3.69	3.60
12	4.30	4.16	4.01	3.86	3.78	3.70	3.62	3.54	3.45	3.36
13	4.10	3.96	3.82	3.66	3.59	3.51	3.43	3.34	3.25	3.17
14	3.94	3.80	3.66	3.51	3.43	3.35	3.27	3.18	3.09	3.00
15	3.80	3.67	3.52	3.37	3.29	3.21	3.13	3.05	2.96	2.87
16	3.69	3.55	3.41	3.26	3.18	3.10	3.02	2.93	2.84	2.75
17	3.59	3.46	3.31	3.16	3.08	3.00	2.92	2.83	2.75	2.65
18	3.51	3.37	3.23	3.08	3.00	2.92	2.84	2.75	2.66	2.57
19	3.43	3.30	3.15	3.00	2.92	2.84	2.76	2.67	2.58	2.49
20	3.37	3.23	3.09	2.94	2.86	2.78	2.69	2.61	2.52	2.42
21	3.31	3.17	3.03	2.88	2.80	2.72	2.64	2.55	2.46	2.36
22	3.26	3.12	2.98	2.83	2.75	2.67	2.58	2.50	2.40	2.31
23	3.21	3.07	2.93	2.78	2.70	2.62	2.54	2.45	2.35	2.26
24	3.17	3.03	2.89	2.74	2.66	2.58	2.49	2.40	2.31	2.21
25	3.13	2.99	2.85	2.70	2.62	2.54	2.45	2.36	2.27	2.17
26	3.09	2.96	2.81	2.66	2.58	2.50	2.42	2.33	2.23	2.13
27	3.06	2.93	2.78	2.63	2.55	2.47	2.38	2.29	2.20	2.10
28	3.03	2.90	2.75	2.60	2.52	2.44	2.35	2.26	2.17	2.06
29	3.00	2.87	2.73	2.57	2.49	2.41	2.33	2.23	2.14	2.03
30	2.98	2.84	2.70	2.55	2.47	2.39	2.30	2.21	2.11	2.01
40	2.80	2.66	2.52	2.37	2.29	2.20	2.11	2.02	1.92	1.80
60	2.63	2.50	2.35	2.20	2.12	2.03	1.94	1.84	1.73	1.60
120	2.47	2.34	2.19	2.03	1.95	1.86	1.76	1.66	1.53	1.38
∞	2.32	2.18	2.04	1.88	1.79	1.70	1.59	1.47	1.32	1.00

References

Agresti, A. (1990) *Categorical Data Analysis*. Wiley, New York.

Aitkin, M., Anderson, D., Francis, B. & Hinde, J. (1989) *Statistical Modelling in GLIM*. Clarendon Press, Oxford.

Atkinson, A.C. (1985) *Plots, Transformations, and Regression*. Clarendon Press, Oxford.

Baker, R.J. (1987) *GLIM 3.77 Reference Manual*, 2nd edn. Numerical Algorithms Group, Oxford.

Becker, N.G. (1989) *Analysis of Infectious Disease Data*. Chapman & Hall, London.

Belsley, D.A., Kuh, E. & Welsch, R.E. (1980) *Regression Diagnostics: Identifying Influential Data and Sources of Collinearity*. John Wiley, New York.

Bender, E.A., Case, T.J. & Gilpin, M.E. (1984) Perturbation experiments in community ecology: theory and practice. *Ecology*, **65**, 1–13.

Box, G.E.P. & Cox, D.R. (1964) An analysis of transformations. *Journal of the Royal Statistical Society B*, **26**, 211–52.

Box, G.E.P. & Cox, D.R. (1982) An analysis of transformations revisited. *Journal of the American Statistical Association*, **77**, 209–10.

Carroll, R.J. & Ruppert, D. (1988) *Transformation and Weighting in Regression*. Chapman & Hall, London.

Cochran, W.G. & Cox, G.M. (1957) *Experimental Designs*, 2nd edn. John Wiley, New York.

Collett, D. (1991) *Modelling Binary Data*. Chapman & Hall, London.

Connolly, J. (1987) On the use of response models in mixture experiments. *Oecologia*, **72**, 95–103.

Cook, R.D. & Weisberg, S. (1982) *Residuals and Influence in Regression*. Chapman & Hall, London.

Cox, D.R. (1958) *Planning of Experiments*. John Wiley, New York.

Cox, D.R. & Oakes, D. (1984) *Analysis of Survival Data*. Chapman & Hall, London.

Cox, D.R. & Snell, E.J. (1989) *Analysis of Binary Data*. Chapman & Hall, London.

Crowder, M.J. & Hand, D.J. (1990) *Analysis of Repeated Measures*. Chapman & Hall, London.

Day, R.W. & Quinn, G.P. (1989) Comparisons of treatments after an analysis of variance in ecology. *Ecological Monographs*, **59**, 433–63.

Draper, N.R. & Smith, H. (1981) *Applied Regression Analysis*. John Wiley, New York.

Edwards, A.W.F. (1972) *Likelihood*. Cambridge University Press, Cambridge.

Fisher, R.A. (1954) *Design of Experiments*, 7th edn. Oliver & Boyd, Edinburgh.

Francis, B. *et al.* (1993) *The GLIM System: Release 4 Manual*. Clarendon Press, Oxford.

Gurevitch, J. & Chester, S.T. (1986) Analysis of repeated measures experiments. *Ecology*, **67**, 251–5.

Hairston, N.G. (1989) *Ecological Experiments: Purpose, Design and Execution*. Cambridge University Press, Cambridge.

Hosmer, D.W. & Lemeshow, S. (1989) *Applied Logistic Regression*. Wiley Interscience, Chichester.

Hurlbert, S.H. (1984) Pseudoreplication and the design of ecological field experiments. *Ecological Monographs*, **54**, 187–211.

Manly, B.F.J. (1985) *The Statistics of Natural Selection on Animal Populations*. Chapman & Hall, London.

Manly, B.F.J. (1990) *Stage-Structured Populations: Sampling, Analysis and Simulation*. Chapman & Hall, London.

McCullagh, P. & Nelder, J.A. (1983) *Generalized Linear Models*, 1st edn. Chapman & Hall, London.

McCullagh, P. & Nelder, J.A. (1989) *Generalized Linear Models*, 2nd edn. Chapman & Hall, London.

Mead, R. (1989) *The Design of Experiments*. Cambridge University Press, Cambridge.

Nelder, J.A. & Wedderburn, R.W.M. (1972) Generalized linear models. *Journal of the Royal Statistical Society A*, **135**, 370–84.

Platt, J.R. (1964) Strong inference. *Science*, **146**, 347–53.

Rice, W.R. (1989) Analysing tables of statistical tests. *Evolution*, **43**, 223–5.

Ryan, T.A., Joiner, B.L. & Ryan, B.F. (1976) *Minitab Student Handbook*. Duxbury Press, North Scituate, MA.

Seber, G.A.F. (1982) *Estimation of Animal Abundance and Related Parameters*, 2nd edn. Griffin, London.

Sokal, R.R. & Rohlf, F.J. (1981) *Biometry. The Principles and Practice of Statistics in Biological Research*, 2nd edn. W.H. Freeman, New York.

Taylor, L.R. (1961) Aggregation, variance and the mean. *Nature*, **189**, 732–5.

Tilman, D. (1988) *Plant Strategies and the Dynamics and Structure of Plant Communities*. Princeton University Press, Princeton, NJ.

Index and glossary

A $ prefix means a GLIM directive. GLIM directives can be abbreviated to the letters shown in upper case (e.g. $PRint can be reduced to $pr). A % prefix means a GLIM function or system vector. G4 means a facility is available in GLIM Version 4 but not in Version 3.77; 'not G4' means included in Version 3.77 but not supported in GLIM 4. 'Ex.' numbers refer to exercises on the disk accompanying the book.

!, end of line marker, useful for comments within lines
&, logical 'and'; 1 if both are non-zero, 0 otherwise
/, logical 'not'; 1 if the first is true and the second false, 0 otherwise
?, logical 'or'; 1 if either is non-zero, 0 otherwise
1
  the intercept in ANOVA 130
  the intercept in output from GLIM models 108
a
  definition of the best-fit regression intercept 94
  GLIM regression output from $disp e 107
  intercepts in ANCOVA 158
%a1 ... %a9, indicates whether the *i*th argument of a macro is set (1/0)
%abs(x), absolute value of x (G4)
%acc
  current accuracy setting
  number of significant figures in current accuracy
accumulation *see* %cu function 18, Ex. 3.1
$ACcuracy, number of digits in output
accuracy (significant figures) *see* %acc, %np
additive errors 22
additive treatment effects, assumptions of ANOVA 119
additivity, in ANOVA 122
age at death, survival analysis 315
age-specific death rates, example using Weibull analysis 326
aggregation and spatial heterogeneity *see* negative binomial
aggregation of factor levels, in model simplification 190
%al, extrinsic aliasing structure of parameter list (G4)
$ALias, switch to include aliased parameters
aliasing
  definition 62
  in GLIM 66
  in model simplification 202
  specifying a ZERO vector 67, 208
aliasing structure after fit? *see* %al
analysis of covariance *see* ANCOVA
analysis of deviance table, overdispersion 224
analysis of deviance tables, multiple regression 197
analysis of variance *see* ANOVA
ANCOVA
  with binomial errors 279
  comparison with ANOVA Ex. 9.2, Ex. 12.2
  $disp s for *SE* of differences 186
  example of use 10
  in GLIM, worked example 159, Ex. 9.1
  graphical output 154
  graphical output in GLIM 161
  interpreting the $disp e table 162
  model simplification 199
  worked example 154
ANCOVA table, worked example 158
and *use* & for logical 'and' 31
%ang(x), function for arcsine (sqrt(x)) transformation
ANOVA
  calculations of *SST*, *SSE* and *SSA* 117
  $disp s for *SE* of differences 186
  factorial designs introduced 133
  introduction and statistical background 113
  list of assumptions 119
  model simplification 197
  model used by GLIM 126
  split-plot Ex. 8.5
  two-way Ex. 8.2
  worked example 123, Ex. 8.1
  worked example in GLIM 127
ANOVA table
  comparison with ANCOVA Ex. 9.2, Ex. 12.2
  with contrasts 148

$LInk R, reciprocal link (default with gamma errors)
$LInk S, square root link
$LIST, defines a name to represent a list of names (G4)
lists *see* p. 679 of G4 manual
  of current variables, $env d 21
  position of a name? *see* %in
lizard data Ex. 14.4, Ex. 15.4
$LOAd, allows alteration of WSSP matrix (G4)
load vector set? *see* %lof (G4)
%lof, 1 if load vector specified, 0 otherwise (G4)
log gamma function? *see* %lga
log-linear hazard model, for censored survival data 322
log-linear models of counts vii
  see also Poisson errors
log link
  definition 169
  ensures positive fitted values with count data 226
log odds ratio, introduction to the logit link function 171, 268
logarithmic transformation
  introduction 21
  plant reproduction data 204
logical functions, %eq, %ge, %gt, %if, %le, %lt, %ne
logical operators, examples and use 30, Ex. 3.1
logical variables, examples of use in macros 336
logistic curve, graph of 266
logistic regression *see* binomial errors
logit link, definition 170
logit link function, introduction to binomial errors 265
logit models of proportions vii
logits, calculating the binomial parameter (p) 273
logs *see* transcripts
logs to base e *see* %log
%log(x), function for natural logarithm of x (base e logarithms)
$Look, list data in columns (s = 1 in a box, s = −1 gives data only)
loops, $while for repeated use of macros 41
%lp, vector containing values of the linear predictor
%lr(0), local random real number between 0 and 1
%lr(n), local random integers between 0 and n
$LSeed, local random number seed
%lv, leverage values from the current fit (G4)
lynx data Ex. 19.1h

$Macro, define a macro name 334
macro library channel *see* %plc
macros
  introduction 4, 38, 332, 335
  text substitution 39
  text substitution in model simplification 193
  unconditional use 41
main effects, significance in the presence of interaction effect 137
$MANual, on-line help manual; type directive name (G4)
$MAP, makes a continuous variate (compare $group)
marginal totals, in multinomial contingency tables 236, 297
mark-recapture data, example of analysis of bird ringing recoveries 244
%match, matches contents of source and target macro (G4)
mathematical functions *see* functions
maximal model, includes all potential explanatory variables 188
maximum, $tab the y largest 84
maximum likelihood
  estimator of k of negative binomial 339
  introduction 173
  parameter estimation 2
mcheck
  introduction 340
  use of model-checking macro 288, 300, 313
mcheck.mac, file containing model-checking macros 14
mean age at death
  with age-specific death rates 329
  exponential distribution 317
means, comparison with unequal variances 120
median, $tab the y fifty 84
%met, scalar showing the fitting method (1 = Givens) (G4)
metapopulation models *see* binary response variables
$MEthod, Givens (default) or Gauss–Jordan for fitting (G4)
method of model fitting? *see* %met
Michaelis–Menten equation 166
minimal adequate model, simplified model whose terms are all significant 188, 210
minimum, $tab the y smallest 84
minimum of a function, macro using quadratic interpolation 340
missing values 25

%pl, number of non-aliased parameters (%pe)
platykurtosis, definition 121
%plc, primary library channel
$PLot, plot *y* against *x*, examples of use 70
plot
  example of use in data exploration 8, Ex. 5.1
  multiple factor levels 71
  multiple *y*-variables to check model fit 106
plotting selected values *see* restricted plots, %re
plotting symbol, choice of 70
%plp, after $predict has predicted linear predictor (G4)
%poc, primary output channel number
%poh, primary output channel height
Poisson deviate *see* %pd
Poisson distribution
  generating random Poisson numbers 337
  introduction 227
  maximum likelihood estimate of the mean 174
Poisson errors
  choice of scale of measurement as a compromise 211
  example of analysis of bird ringing recoveries 244
  introduction 226
  for non-negative (e.g. count) data 167
  overdispersion 223
  standardized residuals 212
Poisson probability *see* %pp
%pol, primary output channel length
polynomials
  as an example of linear models 165
  use of inverse polynomials with gamma errors 303
position in a list *see* %in
%pp (two arguments), Poisson distribution cumulative probability (G4)
$PREdict, to use the current model to predict *y* given *x* (G4)
predicted fitted values? *see* %pfv (G4)
predicted linear predictor? *see* %plp (G4)
predicted variance of linear predictor? *see* %pvl (G4)
predictor macro $method *see* %eta (G4)
$PRint, print data in a row across the page
printing fixed precision *use* ⋆r, e.g. $print ⋆r %a,8,2 $
printing integers *use* ⋆i, e.g. $print ⋆i %a $
printing variable names *use* ⋆n, e.g. $print ⋆n %yv $
printing text *use* single quotes, e.g. $print 'This is a text label' $
prior weights, use in standardized residuals 211
prior weights? *see* %pw
prior weights set? *see* %pwf
probability distribution, current model is described by $disp m 184
probit link, definition 171
program macros, introduction 40
programming in GLIM, tips on writing macros 334
proportion data
  analysis in GLIM 268
  introduction 55, 265
  and non-normal errors 167
%prt, printing frequency in iterative fitting
pseudoreplication
  detection of 56
  handling it in GLIM 151, Ex. 8.6
  in nested designs 52, 151
  non-independence of errors 121
pulse and press experiments 45
%pvl, after $predict contains variance of the linear predictor
%pw, vector of prior weights (set by $weight)
%pwf, 1 if prior weight specified, 0 otherwise

q, the probability of failure $(1-p)$ 267
quadratic terms, use as a test of non-linear responses 193

r^2
  calculation of example 110
  definition and use 101
random effects, in experimental design 61
random number generation, real and integer numbers with %sr 17
random numbers
  generating random numbers from given distributions 337
  generation of 24, Ex. 3.1
  in shuffling for randomizing designs 336
random sampling 119
random variables, linear components of GLM's 165
randomization 45
  methods 58
  to reduce bias 121
rank correlation Ex. 19.1d
ranking, examples and use 35
ranks, replacing variates by their ranks 36
%re, for restricted plots with $plot and residual lists with $disp w 74, Ex. 5.3

$Read, read data from keyboard (follows $data directive) Ex. 4.1
reciprocal link 169
reciprocal transformation, inverse polynomial functions 166
recoding *see* $group and $map
rectangular error distribution 77
$RECycle, use %fv as starting values for model-fitting
%ref, the reference category of an identifier (G4)
reference category of variable *see* %ref
regression
  background and statistics 92
  efficient experimental designs 53
  in GLIM 103, Ex. 7.1
  model simplification 192
  significance testing 97
regression sum of squares *see* *SSR* 97
$REInput, rewinds a channel number then re-inputs GLIM code
repeated measures 56
repetition symbol, use colon ':' to repeat the last directive, e.g. $f: + x $
replication 45, 56, 121
  how many samples? 58
reproductive effort, trade-off with mortality (binary response) 292
residuals
  current values *see* %rs (G4)
  effects of leverage 79
  error checking in ANCOVA 202
  introduction and examples 211, Ex. 20.14
  obtained by $disp r 185
  plots against explanatory variables 212
  plots against fitted values 13
  plots with cumulative normal distribution 13
  print selected *see* %re and $disp w
  standardized 81
  use of $disp w with %re to restrict printing 187
resinosis, example of gamma errors 311
response variable
  binary data (yes/no data) 291
  with binomial errors 265
  current model is described by $disp m 184
  declaration with $yvar 105
  definition 54
response variable? *see* %yv
$REStore, activates a dumped program
restricted plotting *see* %re 74
$RETurn, go back from a file to main program
$return directive, end of file macros 40
$REWind, go to start of specified file
rounding up and down, use %tr(x + 0.5)
%rs, Pearson residuals of current fit as viewed in $disp r (G4)

%s1, %s2 and %s3, current random number seeds
sample size, degrees of freedom 48
saving a GLIM session, use $dump then $restore to restart later on
%sb, with model terms shows the position in %pe (G4)
%sc, current value of the scale parameter
scalars
  defined by the $number directive (G4)
  reserved macro scalars, %z1 to %z9
  single letters preceded by '%', e.g. $ca %t = %cu(x) $
  system scalars 333
$SCale, set the scale parameter
scale
  in ANOVA for hypothesis testing with deletion test 132
  example of use with normal errors 110
  factorial ANOVA 143
  overdispersion correction with binomial errors 278
scale 1, gamma errors for the exponential distribution 318
scale directive, introduction 110
scale of measurement, as a compromise 211
scale parameter
  in ANOVA 130
  current model is described by $disp m 184
  in factorial ANOVA 140
  introduction 105
  use in standardized residuals 211
scale parameter set? *see* %sc
scaled deviance, introduction with Poisson errors 229
scatter, in regression 101
%scf, 1 if scale directive invoked, 0 otherwise
scrolling, control using $page
%se, standard errors of parameter estimates (G4)
seed
  for random number generator 24
  use $env r to view current random number seeds 24
seeds for random numbers *see* %s1 and $sseed
selection of observations *use* $weight with a logical vector
  use %re for plotting
  use %re with $disp w for residual listing
$SEt, batch or interactive mode
sex ratios, use of binary response variables

subscripted variables
  creating shorter vectors Ex. 3.1
  in data editing, $calc x(18) = 0 21
summary statistics, examples of use of library macro called summ 353
survival analysis, introduction 314
survival analysis in GLIM, exponential distribution 318, Ex. 18.1
survivor function, exponential distribution 317
survivorship, survival analysis 314
$SUSpend, revert to primary output
$SWitch, branching and conditional use of macros 41
system scalars, table of scalars 333
system vectors
  information available after model-fitting 181
  table of vectors 334

t distribution deviate *see* %td
t distribution probability *see* %tp
t tables 355
t-tests
  factorial ANOVA 142
  in GLIM 182
  in hypothesis testing in GLIM 131
$Tab By, store the dimensions of the output classification
$Tab For, counting data into contingency tables 87, 229, 253
$Tab Into, save a summary table for subsequent analysis 88
$Tab The, one of Mean, Total, Variance, Smallest, Largest, Percentile
$Tab Using, saving frequency distributions or output weights 90
$Tab With, weighted average and other statistics 87
$Tab With For, counting frequencies in complex data sets 256
tables of χ^2 356
tables of Student's *t* 355
$Tabulate, make tables of means, etc. 83, Ex. 6.1
tabulate directive, example of use 9
tabulation, multiple entries per cell *see* $tprint 89
tabulation of vectors *see* $tprint 30
tannin 99
Taylor's power law 61, 120
%td (two arguments), Student's *t* distribution deviate (G4)
temporal pseudoreplication, repeated measures 56
termination *see* $stop, $exit, $skip (all versions), $newjob (G4)
$TERms, used to specify the model formula (G4)
text substitution macros 39
  model simplification 193
$TIDy, deletes all the identifiers in a LIST (G4)
timber data, example of link functions and transformation 214
time series analysis Ex. 19.1h
time-to-death data, survival analysis 314
%tol, aliasing tolerance in $cycle
toxicological assay, analysis of proportion data 265
%tp (two arguments), Student's *t* probability integral (G4)
$TPrint, print a vector in tabular form 253
tprint, definition and use 88
%tra, current amount of transcription
trade-off, reproduction and mortality 292
transcript file, glim.log 4
$TRanscript W H O F I V, specifies what to write to GLIM.LOG 4
transcription information *see* %tra
transformation
  arcsine on proportion data 287
  effects on error structure 214
  example from timber data 215
  examples 21, 205
  of explanatory variables to reduce deviance 275
  influence 208
  by the link function in GLIM 166
  modulus and folded power 220
  plant reproduction data, log transformation 204
  of *x*-axis to get *SE* of predicted values 276
  square root for count data 263
transitional dynamics, short-term experiments 45
%trg, trigamma function (G4)
trigamma function *see* %trg
truncation *see* %tr
%tr(x), truncation function; take the integer part of x
two-by-two contingency tables, introduction 232
%typ, the type of an identifier (G4)
Type I survivorship curve, survival analysis 315
type of an identifier? *see* %typ

unfriendliness of GLIM vi, 16
$UNits, define length of vectors (rows of data)
units directive
  example of use 103
  introduction 15
$Use, run a macro

GLIM is supplied by NAG:
NAG Ltd, UK Tel: +44 (0)865 511245 Fax: +44 (0)865 310139
NAG Inc, USA Tel: +1 708 971 2337 Fax: +1 708 971 2706
NAG GmbH, Germany Tel: +49 (0)89 3207395 Fax: +49 (0)89 3207396

Request for Information

If you are in the USA, Canada, or Mexico and wish to receive further information about GLIM (including prices, licensing and ordering procedures) then please call NAG Inc or complete the details below and return to NAG Inc.

Please send me further information about GLIM - The Generalised Linear Modelling System.

Name: ______________________________ Dept: ________________

Organisation: ____

Address: ____

Tel: _______________ Fax: ________________ e-mail: ____________

NAG Ltd, Wilkinson House, Jordan Hill Road, OXFORD, OX2 8DR, UK Tel: +44 (0)865 511245 Fax: +44 (0)865 310139
NAG Inc, 1400 Opus Place, Suite 200, Downers Grove, IL 60515-5702, USA Tel: +1 708 971 2337 Fax: +1 708 971 2706
NAG GmbH, Schleißheimerstr. 5, 8046 Garching bei München, Deutschland Tel: +49 (0)89 3207395 Fax: +49 (0)89 3207396

GLIM is supplied by NAG:
NAG Ltd, UK Tel: +44 (0)865 511245 Fax: +44 (0)865 310139
NAG Inc, USA Tel: +1 708 971 2337 Fax: +1 708 971 2706
NAG GmbH, Germany Tel: +49 (0)89 3207395 Fax: +49 (0)89 3207396

Request for Information

If you are **NOT** in the USA, Canada, or Mexico and wish to receive further information about GLIM (including prices, licensing and ordering procedures) then please call NAG Ltd or complete the details below and return to NAG Ltd.

Please send me further information about GLIM - The Generalised Linear Modelling System.

Name: ______________________________ Dept: ________________

Organisation: ____

Address: ____

Tel: _______________ Fax: ________________ e-mail: ____________

NAG Ltd, Wilkinson House, Jordan Hill Road, OXFORD, OX2 8DR, UK Tel: +44 (0)865 511245 Fax: +44 (0)865 310139
NAG Inc, 1400 Opus Place, Suite 200, Downers Grove, IL 60515-5702, USA Tel: +1 708 971 2337 Fax: +1 708 971 2706
NAG GmbH, Schleißheimerstr. 5, 8046 Garching bei München, Deutschland Tel: +49 (0)89 3207395 Fax: +49 (0)89 3207396

Please affix postage

NAG Inc
1400 Opus Place, Suite 200
Downers Grove, IL 60515-5702
USA

Please affix postage

NAG Ltd
Wilkinson House
Jordan Hill Road
OXFORD
UK OX2 8DR